FRANCE
SINCE
THE POPULAR FRONT

FRANCE
SINCE
THE POPULAR FRONT

Government and People 1936–1986

Maurice Larkin

Clarendon Press · Oxford
1988

Oxford University Press, Walton Street, Oxford OX2 6DP
Oxford New York Toronto
Delhi Bombay Calcutta Madras Karachi
Petaling Jaya Singapore Hong Kong Tokyo
Nairobi Dar es Salaam Cape Town
Melbourne Auckland
and associated companies in
Berlin Ibadan

Oxford is a trade mark of Oxford University Press

Published in the United States
by Oxford University Press, New York

British Library Cataloguing in Publication Data
Larkin, Maurice
France since the Popular Front:
government and people 1936–1986.
1. France, 1936–1986
I. Title
944.08
ISBN 0–19–873034–9
ISBN 0–19–873035–7 Pbk

Library of Congress Cataloging in Publication Data
Larkin, Maurice.
France since the Popular Front: government and people, 1936–1986
Bibliography: p   Includes index.
1. France—Politics and government—1914–1940. 2. France—
Politics and government—1940–1945. 3. France—Politics and
government—1945– I. Title.
DC369.L27   1988   944.08—dc19   88–2742
ISBN 0–19–873034–9
ISBN 0–19–873035–7 (pbk.)

Set by Joshua Associates Ltd., Oxford
Printed in Great Britain
at the University Printing House, Oxford
by David Stanford
Printer to the University

PREFACE

HISTORY is commonly called the mythology of advanced societies. Its form and content are shaped by the questions we ask of it, while its direction and force stem from the assurances we seek in it. This book is meant for the reader who is looking for a history of the interaction of people, government, and material change in France during the last fifty years, and who is less immediately concerned with the rich diversity of life that lies beyond the direct influence of government and economic forces. The cultural and leisure activities of the French have been well described in several recent books in English, while the broad structural changes of post-war France have been thematically examined in a number of useful publications that dissect the country's development since the Liberation and look individually at its basic elements.[1] Yet, paradoxically, there are few English histories of France that span the great divide of the Second World War—the point at which the chronological accounts of the historians give way to the theme-by-theme analyses of social scientists and area-studies writers. There are admittedly more histories of this kind in French. But they are inclined to treat France in isolation, and it is rare for them to consider how far the developments they are describing are peculiar to France and how far they reflect wider European currents. In seeking to fill some of these gaps, I have attempted to set the changing fortunes of France in a comparative, international context, juxtaposing the French performance with that of her neighbours, in particular Germany, Britain, and Italy. But, since the reader's prime concern is likely to be with France, I have tried to do this as concisely as possible, allowing the tables of comparative data to speak for themselves, rather than intrude continuously into the discussion of French affairs in the main text.

Given the book's main focus on the interrelation of people, government, and economy, other themes receive sparser treatment. If man is spirit as well as consumer, then spirit gets somewhat short shrift in the pages that follow. I have included very little on France's many distinguished writers and creative artists, except when they explicitly enter the political arena; and if Michael Foucault was right in claiming in 1970 that 'our entire epoch struggles to disengage itself from Hegel', the reader may be forgiven for failing to pick up the point from what is offered here. Nor will he find an exhaustive analysis of the private and collective neuroses of the population in thought-provoking chapters with

---

[1] Among the first category, see T. Zeldin, *The French* (London, 1983), and J. Ardagh, *The New France: A Society in Transition, 1945–1977* (London, 1977), subsequently reissued as *France in the 1980s* (London, 1982); and among the second category, see D. L. Hanley, A. P. Kerr, and N. H. Waites, *Contemporary France: Politics and Society since 1945*, 2nd edn. (London, 1984).

appetizing titles such as, 'From anomie to bonhomie', 'Menu peuple; menu touristique', etc.

The complexity of the French experience during the last fifty years makes it difficult to summarize the book's principal themes in a brief introduction. The modernization of the French economy is inevitably a central issue; but by putting it in a broad comparative setting, rather than in the customary two-horse race with Britain, its limitations as well as its strengths should become clearer. France's social record is likewise examined in an international context—with its wages, working conditions, and welfare benefits placed in comparison with those of her neighbours. Changes in education, demographic growth, and the disabilities of women are also outlined, together with patterns of religious observance and the role of the Church. While the book's main concern is with the broad forces producing change and consolidation, the major political events of the period are treated in chronological sequence, notably the fortunes of the Popular Front and the crises of 1940, 1958, and 1968. The social programmes of the Popular Front, the Liberation, and the early Mitterrand years are also considered—together with the sad irony that all three ventures coincided with periods of economic difficulty, in which their aims were frustrated by the exigencies of solvency and survival. Specific chapters are devoted to foreign and colonial issues, where French policy is assessed in the light of the opportunities afforded by the international circumstances of the time. Other chapters survey the impact of war and liberation on living standards and on social and political behaviour.

In the field of politics, the evolution of France from 'la République des députés' to what Maurice Duverger has called 'la République des citoyens' occupies a prominent place among the book's principal themes. The alignment of political forces has markedly changed from a multipolar to a bipolar pattern, with the result that since the 1960s the French voter has increasingly been offered clearly defined alternatives, with the choice of the majority essentially determining who is in charge of government. Under previous regimes, the multiplicity of parties and the fragility of political alliances had made the choice of government largely dependent on back-stairs negotiations by politicians, rather than on the decision of the electorate. Political scientists, such as Duverger, have tended to concentrate their attention on the importance of constitutional change in bringing about this development. Undoubtedly, the adoption of a semi-presidential system of government has been a major factor, since it encourages the disparate political forces in the country to band together in two opposing camps behind the surviving finalists in the presidential election. Yet, as this book argues, other factors have also played a substantial role: notably, the waning of the old divisive issues that had recurrently frustrated the formation of coherent parliamentary majorities in the past. The later chapters, therefore, trace the decline of the clerical, constitutional, and colonial issues—which, together with 'class', might be termed 'the four Cs' of

French politics. 'Class' has remained as the dominant question in political debate, embracing not only the distribution of national income between employer and employee, but also the adjacent problems of state welfare and national productivity. In this respect, French politics have come closer to those of Britain and Scandinavia. The net result has been a stability of government that France has not known since the Second Empire. And, unlike the Second Empire, this stability has been won without impairing fundamental democratic safeguards—even if parliament's prerogatives have been over-rigorously pruned.

Some readers may think that 1936 is a curious year with which to start a history of contemporary France; the First or Second World War has more usually been the point of departure. Yet the period 1936–86 has a logic and symmetry that transcends the mere aesthetic attraction of it being a clear-cut half-century. It began with the country's first taste of Socialist government, and ended with its last—both demonstrating the problems of achieving a national programme of social reform within an unfavourable international context, beset by economic problems. Between these uneasy experiments, there occurred the modernization of the French economy, partly arising from the change in attitudes engendered by war and reconstruction, and further stimulated by EEC membership. And, interwoven with these developments, there ran the chequered histories of decolonization and the search for a more stable political system. None of this can properly be understood without reference to pre-war France and the changes brought about by war. To go back further, however, to the First World War, would inevitably pose problems of length and necessitate a much briefer coverage of the book's principal concerns. Even so, the opening chapters on the 1930s attempt to give some indication of what France inherited from earlier decades.

Since socio-economic and political changes rarely follow the same rhythms and patterns, there is much that does not fit comfortably into the familiar chronological divisions of French history. For this reason, the socio-economic expansion of 1947–73 is covered in a separate chapter, while the subsequent recession is dealt with in the general chapters on the Giscard and Mitterrand presidencies. Convenience of discussion has also resulted in some issues being treated in a thematic rather than a chronological fashion—particularly the opening years of the Fourth and Fifth Republics, when new institutions, attitudes, and forms of political behaviour were coming into being. To avoid repetition, the general characteristics of these regimes are examined in the chapters on their formative years, making for a certain disparity in length between these early sections and those that follow.

The following have very kindly given me permission to use copyright material from their publications in the maps, figures, and tables listed next to their names. (Precise references will be found at the foot of each item.) I am indebted to Éditions Ouvrières for Map 1.1 from Fernand Boulard and Jean Remy, *Pratique religieuse urbaine et régions culturelles* (Paris, 1968); to Messrs Cassell for Map 4.1 from Brian Liddell Hart, *A History of the Second World War* (London, 1970); to Messrs Hodder and Stoughton for Map 4.2 and material in Table 6.1 from Colin Dyer, *Population and Society in Twentieth-century France* (London, 1978); to Penguin Books Ltd for Figure 9.1 from Philip Williams, *Crisis and Compromise: Politics in the Fourth Republic*, 4th edn. (London, 1972); to the Economist Newspaper Ltd for Figures 10.1 and 10.2 from *Europe's Economies* (London, 1978); to Professor Jean-Marcel Jeanneney and the Presses de la Fondation Nationale des Sciences Politiques for material used in Appendix Figures 1 to 3 from J.-M. Jeanneney and Élizabeth Barbier-Jeanneney, *Les Économies occidentales du xix siècle à nos jours*, i (Paris, 1985); to Professor Alfred Sauvy for Table 1.2 from his *Histoire économique de la France entre les deux guerres*, ii: *1931–1939* (Paris, 1967); to Messrs Macmillan for Table 1.5 from *The Statesman's Year-Book, 1939* (London, 1939); to Publications de la Sorbonne for material used in Table 4.1 from R. Frankenstein, *Le Prix du réarmement français, 1935–1939* (Paris, 1982); to Messrs Shenkman for Tables 10.2, 10.4 to 10.6, and 10.9 from Sima Lieberman, *The Growth of European Mixed Economies, 1945–1970* (New York, 1977); to Oxford University Press for Table 10.8 and Appendix Tables 4 and 6 from Andrea Boltho (ed.), *The European Economy: Growth and Crisis* (Oxford, 1982); to Éditions Bordas for material used in Table 10.14 from Yves Trotignan, *La France au xx siècle*, i (Paris, 1976); to Messrs Routledge and Kegan Paul for Table 10.15 from D. L. Hanley, A. P. Kerr, and N. H. Waites, *Contemporary France: Politics and Society since 1945*, 2nd edn. (London, 1984); to Presses Universitaires de France for Table 11.1 from Fernand Braudel and Ernest Labrousse (eds.), *Histoire économique et sociale de la France*, iv, pt. 3 (Paris, 1982); to Messrs Allen and Unwin for Table 15.3 from Vincent Wright (ed.), *Continuity and Change in France* (London, 1984); to the Board of Trustees of the Leland Stanford Junior University for Appendix Table 5 from J.-J. Carré, P. Dubois, and E. Malinvaud, *French Economic Growth* (London, 1976); to Columbia University Press for Appendix Table 7 from François Caron, *An Economic History of Modern France* (London, 1979); and to Messrs Secker and Warburg for several extracts from Herbert Lüthy, *The State of France*, trans. Eric Mosbacher (London, 1955).

I am indebted to many people for their help, advice, and encouragement while writing this book, but I wish particularly to express my gratitude to Malcolm Anderson, Robert Anderson, Bill and Sheila Bell, Roger Bullen, Eric Cahm, Brian Darling, Neil Fraser, John Frears, Ronald Irving, Jean-Marcel Jeanneney, Douglas Johnson, Roderick Kedward, Malcolm Maclennan, Richard McAllister, James McMillan, Jean-Marie and Françoise Mayeur, Peter Morris, Pascal Petit, Sian Reynolds, Jean-Pierre Rioux, Vaughan

Rogers, Donald Rutherford, Ted Taylor, Peter Vandome, Neville Waites, and Vincent Wright. I must also pay tribute to the organizers and editorial committee of the Association for the Study of Modern and Contemporary France, whose labours have so greatly enhanced the opportunities in Britain for keeping in touch with French affairs. Ivon Asquith of Oxford University Press has exercised a remarkable blend of patience and realism, and I am also grateful to the Press's readers and production staff for helpful comments. Like the hand-loom weavers of old, a variety of typists have coped with the unpredictable nocturnal demands of my putting-out system; and I should particularly like to thank Veronique Magennis, Alison Munro, May Norquay, and my wife, Enid Larkin. Indeed, my tributes to the help and forbearance of my wife and family have now acquired the weary droop of a habitual drunkard's peace-offering of flowers. They are none the less deeply felt.

June 1987   Maurice Larkin
Edinburgh

CONTENTS

LIST OF FIGURES

LIST OF MAPS

LIST OF TABLES

Appendix Tables

ABBREVIATIONS

ACJF	Association Catholique de la Jeunesse Française
APPCA	Assemblée Permanente des Présidents de Chambre d'Agriculture
CAP	Common Agricultural Policy
CEG	Collège d'Enseignement Général
CERES	Centre d'Études, de Recherches, et d'Éducation Socialistes
CES	Collège d'Enseignement Secondaire
CFDT	Confédération Française Démocratique du Travail
CFLN	Comité Français de Libération Nationale
CFTC	Confédération Française des Travailleurs Chrétiens
CGP	Commissariat Général du Plan
CGPF	Confédération Générale de la Production Française
CGPME	Confédération Générale des Petites et Moyennes Entreprises
CGT	Confédération Générale du Travail
CGTU	Confédération Générale du Travail Unitaire
CNI(P)	Centre National des Indépendants (et Paysans)
CNJA	Centre National des Jeunes Agriculteurs
CNPF	Conseil National du Patronat Français
CNR	Conseil National de la Résistance
CRS	Compagnies Républicaines de Sécurité
CODER	Commission de Développement Économique Régional
DATAR	Délégation à l'Aménagement du Territoire et à l'Action Régionale
DOM-TOM	Départements et Territoires d'Outre-Mer
EDC	European Defence Community
EDF	Électricité de France
EEC	European Economic Community
ENA	École Nationale d'Administration
FFI	Forces Françaises de l'Intérieur
FGDS	Fédération de la Gauche Démocrate et Socialiste
FIDES	Fonds d'Investissement et de Développement Économique et Social des Territoires d'Outre-Mer
FLN	Front de Libération Nationale
FNSEA	Fédération Nationale des Syndicats d'Exploitants Agricoles
FO	Force Ouvrière
GDP	Gross Domestic Product
GNP	Gross National Product

GPRA	Gouvernement Provisoire de la République Algérienne
HLM	Habitation à Loyer Modéré
IFOP	Institut Français d'Opinion Publique
INSEE	Institut National de la Statistique et des Études Économiques
JAC	Jeunesse Agricole Chrétienne
JEC	Jeunesse Étudiante Chrétienne
JOC	Jeunesse Ouvrière Chrétienne
MLN	Mouvement de Libération Nationale
MRG	Mouvement des Radicaux de Gauche
MRP	Mouvement Républicain Populaire
MTLD	Mouvement pour le Triomphe des Libertés Démocratiques
NATO	North Atlantic Treaty Organization
OAS	Organisation de l'Armée Secrète
OECD	Organization for Economic Co-operation and Development
OEEC	Organization for European Economic Co-operation
ORTF	Office de la Radiodiffusion—Télévision Française
PCF	Parti Communiste Français
PDM	Progrès et Démocratie Moderne
PPF	Parti Populaire Français
PS	Parti Socialiste
PSF	Parti Social Français
PSU	Parti Socialiste Unifié
PTT	Postes, Télégraphes, et Téléphones (subsequently Télécommunications)
RDA	Rassemblement Démocratique Africain
RGR	Rassemblement des Gauches Républicaines
RI	Républicains Indépendants
RPF	Rassemblement du Peuple Français
RPR	Rassemblement pour la République
SAFER	Sociétés d'Aménagement Foncier et d'Établissement Rural
SFIO	Section Française de l'Internationale Ouvrière
SMAG	Salaire Minimum Agricole Garanti
SMIC	Salaire Minimum Interprofessionnel de Croissance
SMIG	Salaire Minimum Interprofessionnel Garanti
SNCF	Société Nationale des Chemins de Fer
SOE	Secret Operations Executive
SOFRES	Société Française d'Enquêtes par Sondages
STO	Service de Travail Obligatoire
UDCA	Union de Défense des Commerçants et des Artisans
UDF	Union pour la Démocratie Française
UDMA	Union Démocratique du Manifeste Algérien
UDSR	Union Démocratique et Socialiste de la Résistance
UEC	Union des Étudiants Communistes

UER	Unité d'Enseignement et de Recherche
UGTT	Union Générale des Travailleurs Tunisiens
UNEF	Union Nationale des Étudiants de France
UNR	Union pour la Nouvelle République

1
France in the 1930s

FRANCE in the 1930s was geographically the most varied country in Western Europe, stretching from the wind-swept agricultural plains of the north to the sun-baked scrubland of the Mediterranean. Its lateral sweep was no less chequered, ranging from the vineyards and sandy forests of the Atlantic seaboard to Europe's highest mountain range in the east. If its mineral deposits were poor, its diversity of crops was prodigious, extending from potatoes and beet in the north to rice and olives in the south. Its wild animals included the seals of the Somme Estuary and the bears of the Pyrenees; there were wild boar within thirty miles of Paris, and scorpions in the gardens of Nice. And all this within a country that lay between narrower latitudes than the British Isles. Its variety of life and spectacle was such that even half a century later the great majority of its people chose to spend their holidays in France.

Yet if geography still flattered France in this period, history had been less kind. In the time of Louis XIV, one European in five was French, and a third of the wealth of the continent lay in France. By the 1930s, less than one European in twelve was French, and France held no more than an eighth of the wealth of Europe. Militarily, Louis XIV's France and the brief Napoleonic Empire had been the foremost power in the world, but the Third Republic of the 1930s was to end with France being invaded and occupied by Germany for the third time in seventy years.

Inter-war France has frequently been described as a 'stalemate society' or a 'société bloquée'.[1] The temptation is to assume that an apt description is also an explanation. 'Une société bloquée' suggests that a potentially healthy growth was being thwarted by the interlocking of French social forces in a vicious circle; it implies that what was essentially needed was the intervention of exterior factors to release these forces into normal productive activity. 'A stalemate society' is less charged with implication, yet still suggests a strong element of abnormality in the French situation. No historian would deny the galvanizing impact of the events of the 1940s on French economic attitudes and activity—as later chapters will seek to demonstrate. Yet it is easy to forget that a large section of French society was reasonably content with things as they were. If there undoubtedly existed frustrated forces seeking to burst out of the vicious circle, they represented a minority; and if the industrial working class was prominent among the frustrated, it too was a minority. Writing as late as 1953, the Swiss journalist, Herbert Lüthy, commented:

France produces fewer goods and less horse-power per man-shift and works fewer hours per unit of population than most industrial countries, but she is full of men who preserve their own mind, their own individuality, their own fortunes and misfortunes, and are not organisable units of population, but individual men. They revolt with all the instinctive clarity of an ancient civilisation against being turned into modern mass-men. They refuse to take part in the Darwinian struggle for life, and at heart they look more with sorrow than with anger at the success of the nations which have so efficiently, progressively, busily, restlessly, organisedly, and competently undermined the former supremacy of the French nation; for they regard them as hordes of semi-animal, Neanderthal men making onslaughts on the painfully acquired human values the home of which is France, *la patrie de l'homme* ... A Frenchman who is not in love with his country is hard to find. He may curse beyond redemption the French state and its institutions and the morals and manners of his fellow-countrymen, he may dismiss the French Government as a gang of crooks and the French administrative system as a racket, the most commonplace article of foreign manufacture may cause him to break out into a tirade and declare that only foreign-made goods are worth having, but at the end of it all he will passionately announce that France is the only country in which it is possible to live and breathe freely. What do technical achievements and social progress, efficient plumbing and lifts that work, amount to in comparison with the pleasure of being an unhampered individualist?[2]

This was the attitude celebrated in the evergreen film comedies of Jean Renoir and René Clair.

Attitudes in inter-war France were still heavily influenced by the legacy of the previous century. The French social structure had mainly rested on a sullen consensus of land-holding peasants and self-employed members of the middle classes. Both were direct and indirect beneficiaries of the Great Revolution of the 1790s; and both were determined that their hard-won prizes should not be put at risk by over-generous attempts to better the lot of other classes. This defensive mentality was prepared to tolerate upward mobility within the existing structure of society, but it was strongly opposed to reshaping the structure itself.

AGRICULTURE

In popular mythology, the tight-fisted peasant, tenaciously holding on to his independence, was the most representative figure of France—even in the 1930s. A third of the population still worked on the land (Table 1.1), and agriculture provided nearly a quarter of the nation's wealth. Memoirs and biographies of public figures made much of their peasant antecedents—and, as Richard Cobb reminds us, they followed the fashion of the Ancestral School of popular literature, boasting 'plusieurs générations formées dans le dur labour

Table 1.1.
Population, Employment, and Nationalities in 1936

	Both sexes	Proportion women (%)
Total population in France	41.18 million	51.9
Actively employed	19.35 million	33.1
Employment[a]	*Proportion of active population: both sexes (%)*	*Women as proportion of each employment (%)*
Agriculture, forestry, fishing	32.5	32.0
Industry and transport	38.3	26.1
Commerce	14.6	42.1
Domestic servants	3.9	87.3
Liberal professions and ancillary staff	3.8	50.3
State employees	4.2	29.4
Armed services	2.6	
Non-French population resident in 1936 (both sexes)[b]		
Italians	720,900	
Poles	422,700	
Spaniards	253,600	
Belgians	195,400	
Swiss	78,900	
Russians	64,000	
Germans	58,100	
British	30,200	
TOTAL	1,823,800	

[a] These categories do not correspond to the post-war INSEE classifications. The percentages are based on the revised statistics contained in Alfred Sauvy, *Histoire économique de la France entre les deux guerres*, 3 vols. (Paris, 1984). Before Sauvy's revision, a broadly held division of the active population was: agriculture, etc., 37%; industry, 30%; services, 33%.

[b] *Source*: Colin Dyer, *Population and Society in Twentieth-Century France* (London, 1978).

de la terre', and proudly if fancifully claiming their sturdy characteristics from the 'rudes contours du Lubéron' or the 'douces vallées de la haute Seine'.[3] Yet the stagnation and impoverishment of the rural population continued to be a major factor in restricting French economic growth between the wars.

There is a sense in which European farmers never recovered from the flood of cheap imports of transatlantic grain and meat that started in the 1870s. The situation worsened during the First World War, when overseas farmers increased their production to feed a belligerent Europe, where so many peasants were in uniform and their land a theatre of war. But when peacetime farming was resumed, there was no corresponding reduction in overseas production; and the result was a glut of food and a fall in farm prices which left agriculture as the poor relation of the world economy. A further dimension to the problem in France was that the Napoleonic inheritance laws insisted on the division of land between heirs, in accordance with Revolutionary principles of equality. Holdings became smaller and poorer as land passed from father to children during the course of the nineteenth century. Fortunately, the growth of towns helped to stave off complete disaster, in that the attraction of urban wages induced some sons and daughters to relinquish their share of the land and try their luck in the factories or behind a counter. But the slow development of the French urban economy made it an inadequate safety-valve for a potentially desperate situation. Even in the 1930s, when the urban population belatedly reached parity with its rural counterpart, over half of French farms were under five hectares (Appendix II, Table 2). Indeed, there were now nearly four million holdings, of which at least 60 per cent were owner-occupied. Of the rest, 30 per cent were rented, and the remaining 10 per cent were let on a share-cropping basis.

The perpetuation of such a damaging system of inheritance reflected a widespread respect in Republican France for its evident fairness—a respect that was even stronger among younger brothers and sisters, who stood to lose by a system of primogeniture (and, being a majority of the electorate, their views inevitably triumphed). It was argued in any case that it was always open to younger children to transfer their share to their elders, should the farm become unviable. Many peasants, however, still preferred being 'cock of their own dung-heap' to working for someone else, and there was a strong inclination to accept the land, however small the resulting portion. A decree-law of 17 June 1938 belatedly attempted to strengthen the peasant farm as a stable concern by allowing a greater share of the inheritance to go to the heir who principally worked the land; but its effects were immediately lost to view in the smoke and upheaval of war and occupation.

The peasant was obviously under strong pressure to keep his family small, if his children's holdings were to be economically viable. The second half of the nineteenth century had therefore seen a dramatic decline in the peasant birth-rate—especially in regions where church influence was weak. Clergy, economists, and military leaders gloomily compared this situation with the large peasant families of Germany, where the inheritance laws were less egalitarian and the influence of the Churches stronger. Available evidence indicates that peasant birth-control in France mainly took the form of onanism or premature withdrawal—what the Dutch peasantry called 'leaving church before the singing'. The use of contraceptive devices, on the other hand, was mainly confined to a small section of the middle and upper classes; it was not until the First World War that mass-produced sheaths became available at a working-class price. This far-reaching development was largely the result of armies issuing them in order to reduce the incidence of venereal disease among their troops.

Despite these marital stratagems, fragmentation continued. In the 1930s, most farms were so small as to require no hired labour; and they were run with such a minimum of modern machinery that agricultural productivity in proportion to manpower was only half of what it was in Britain. On the eve of war, there were only 35,000 tractors in the whole of France, compared with perhaps 100,000 in Britain. Not only were the farms small, but their amenability to modern methods was further reduced by the fact that the peasant's holding often consisted of a widely scattered collection of strips and plots, to compensate for the varying fertility of the soil. Moreover, in a period when self-sufficiency played a greater role in rural life, this holding often included a diversity of types of land—pasture and orchard, as well as arable for grain or grape. All this involved the peasant in time-consuming walks from one part of the holding to another. Attempts were made in the inter-war period to consolidate holdings by the mutual exchange of parcels of land; but since the relevant law of 27 November 1918 required the agreement of two-thirds of the farmers affected, this *remembrement* was only proceeding at a mere 15,000 hectares a year in the 1930s.

As long as the drift to the towns remained modest and farms stayed relatively small, the inexpensive if time-consuming methods of the past were bound to stay in favour. Pre-war experience of the misery of debt had discouraged borrowing for new equipment; and the attitude remained ingrained, despite the fact that the inflation of the post-war years had wiped out the real value of much of this debt. Table 1.2 indicates how poor the resulting French yields per hectare were when compared with their northern and eastern neighbours; only

Table 1.2.
Average Agricultural Yields in Selected European Countries, 1933–1937,
in order of Overall Performance (Quintals per Hectare)

	Wheat	Oats	Barley	Potatoes
Holland	29.5	23.0	28.5	178.0
Denmark	29.2	26.0	28.3	170.0
Belgium	26.5	26.4	26.0	202.0
Sweden	23.6	18.0	20.0	143.0
UK	22.8	20.0	20.0	168.0
Switzerland	22.3	20.0	16.6	157.0
Germany	22.1	19.9	20.0	164.0
Norway	19.1	20.0	19.4	180.0
Czechoslovakia	17.5	16.0	17.0	129.0
France	15.3	14.0	14.3	108.0

Source: Alfred Sauvy, *Histoire économique de la France entre les deux guerres*, ii: *1931–1939* (Paris, 1967), 541.

the desiccated Mediterranean countries did worse. Contrary to popular assumption, France spent more than three times as much on imported food and wine than she gained from selling her own produce and vintages abroad—despite all her advantages of climate and soil. And if this partly reflected French respect for the pleasures of the table, inefficient production was the main cause.

INDUSTRY AND POPULATION

French industry enjoyed far fewer natural advantages than its agriculture did. The country's shortage of minerals was a constant lament, threading its way through speeches, textbooks, and military thinking—while its effects were sadly reflected in the industrial statistics of the period (Appendix II, Table 3). With only 14 per cent of Germany's coal resources, France was obliged to import nearly a third of the coal she needed. And although the gradual development of hydroelectric power in the inter-war years helped to give industry an alternative source of energy, the total French output of electricity in the 1930s was still ony two fifths of Germany's. Nor could France rival Germany's iron-ore and chemical deposits. Even after the French recovery of Alsace-Lorraine in 1919 with its prosperous iron industry, German pig-iron production in the late 1930s was well over twice that of the French.

Even so, France had been among the early pioneers of the Industrial Revolution in the nineteenth century, despite these disadvantages; the rapid expansion of her manufactures during the Second Empire (1852–70) had thrived on a relatively prosperous rural market at home and a paucity of major competitors abroad. The late nineteenth century, however, saw foreign competition increase substantially, as Germany and America became major industrial exporters; and it also saw an impoverishment of the French domestic market, as falling farm prices and a sagging birth-rate weakened the capacity of rural France to absorb the produce of French factories. The combination of all these factors resulted in the French share of world industrial production dropping from 10.3 per cent to 6.4 per cent between 1870 and 1914.

The First World War had ambivalent results for French industry. If it brought back Alsace-Lorraine with its two million inhabitants, it entailed an appalling loss of life, which diminished both the labour force and the domestic market for French goods. The military losses of almost 1.4 million men represented 10.5 per cent of the working male population—a higher proportion than in any other belligerent country except Romania. (Those for Germany and Britain were 9.8 per cent and 5.1 per cent.) And of the 4.3 million French wounded, more than a million were permanently disabled. Moreover, the separation of couples during the war resulted in a drop in births of perhaps over 1.7 million, while the death or disfigurement of so many potential fathers resulted in a much greater loss over the inter-war period, despite a brief baby-boom following the reunion of couples after the armistice. It also made for a population with a high percentage of elderly members (Appendix II, Table 1).

Conscious that the German population was increasing twice as fast as its own, the French government tried hard to shorten the odds by encouraging a substantial influx of foreign labour and by instituting a series of penalties and inducements to further the birth-rate. A law of 31 July 1920 prescribed up to six months' imprisonment for even the mere offer of contraceptive information, and up to three years in prison for attempted abortion—a sentence that was increased to ten years in July 1939. Yet, despite these draconian prescriptions, the law on contraception was very loosely applied. It is true that many of the coyly discreet advertisements for contraceptives disappeared from the newspapers, or became so arcane as to be intelligible only to the initiated; and they likewise disappeared from the safety curtains of theatres, where they had always been something of an embarrassment to parents, parrying the questions of puzzled children during the entr'acte. Yet the sale of sheaths continued largely unmolested, provided it was discreet; they were tacitly recognized as keeping down illegitimacy and helping to prevent the spread of venereal diseases. Indeed, sales increased as sheaths became cheaper and more reliable, notably after 1929 with the surreptitious import of the mass-market American 'Dreadnought'. Even so, all the available evidence indicates that premature withdrawal and various forms of onanism continued to be the most practised form of birth-control—and here the State, like the Church, was powerless.

Not surprisingly, the areas with the highest birth-rates were those where Catholic religious observance was most marked. However, the findings of Ogino and Knaus on the 'safe period' in 1930–1 provided a form of birth-control which was acceptable to the Church—and which was cautiously welcomed by the Vatican after the fiasco of earlier ecclesiastical patronage of safe-period methods, in which well-meaning parish clergy had been bitterly blamed for unwanted pregnancies. The State made no attempt to interfere with publications propagating the Ogino–Knaus system—perhaps surmising that it might not prove much more successful than its ill-starred predecessors.

Nor did the fearsome penalties for abortion act as an effective deterrent: a third of French pregnancies were terminated in this fashion during the inter-war period. And with abortions averaging 400,000 per annum, the *faiseuses d'anges* eliminated more French lives in a year than the German armies did in the worst year of the war. In addition to these factors, the birth-rate was also restricted by the economic depression of the 1930s which discouraged early marriage; indeed, the birth-rate was to fall below the level of the death-rate during the Popular Front era (Appendix I, Fig. 1). What prevented this overall demographic situation from being a disaster for the economy was the arrival of nearly three million foreign immigrants in the 1920s, mostly Italians, Poles, Spaniards, and Belgians. Settling mainly in the frontier regions and in Paris, the Italians and Poles mostly gravitated towards industry, while the Spaniards and

Belgians were principally employed on the land. Not only did they mitigate the labour shortage, but they increased the French marriage-rate by 10 per cent in the late 1920s and tended on balance to have larger families. The Depression and the international tensions of the 1930s, however, saw a diminution in their numbers (Table 1.1).

If the government was loud in its plaudits and threats on demographic issues, it was remarkably slow to ease the financial burden of raising a family. A law of 11 March 1932 had obliged private firms to offer their workers a contributory scheme of family allowances; but most workers preferred not to part with their money—their main concern being survival until next pay-day. And it was only with the Family Code of 29 July 1939 that there emerged a comprehensive state system of substantial allowances for families with two or more children. A second child brought with it a sum equivalent to 10 per cent of the notional average wage, while each subsequent child resulted in a further 20 per cent.

Although French industry was undeniably handicapped by demographic problems and the country's lack of mineral resources, it was also the victim of French entrepreneurial attitudes. In the 1930s, over 70 per cent of the country's manufacturing firms had fewer than half a dozen workers on their payrolls; and the industrial sector as a whole employed little more than a third of the population (Table 1.1). Although manpower was cheaper in France than in Britain, modern production methods were not easily applicable to the small-scale structure of the typical French business, with its vesting of all effective power in the hands of the family who had founded and still owned it. This was the industrial counterpart of the main problem in agriculture—that many French farms were too small for modern techniques. It can be argued that much of this, like many of France's problems in politics, sprang from a strong attachment to personal independence. A family used to managing a business was very loath to expand it at the cost of sharing power with shareholders' representatives. The outcome was that in the 1930s the proportion of mechanical horsepower to manpower was only a third of that in Britain.

Critics have been eloquent in their denunciation of the French family firms; so much so that there has arisen a school of cautious apologists, meeting rhetoric with polite requests for supportive evidence. The essence of the charge is that these firms were more concerned with the defence of the family's social position than with seeking to improve productivity. Charles Kindleberger summarized the argument thus:

They minimize risks rather than maximize profits, and hence save in liquid form as insurance against adversity rather than invest in product or process innovation. They produce to fill orders rather than for stock. They are characterized by secrecy and mistrust; they fear banks, government, and even the consuming public. They hold prices high. Turnover is permitted to languish as the larger firms restrain from expanding output and sales in ways which would embarrass the small-scale inefficient producers at the margin.[4]

In other words, the large firms were tempted to continue enjoying the high prices established by their smaller brethren, rather than seek to compete by producing at the lower prices which their size and relative efficiency would certainly have permitted. Not only did this make for a comfortable, easygoing life for the larger firm, but it helped to preserve the fabric of the social order. The situation was to be summed up twenty years later in an official report which claimed that:

Malthusianism is the principal cause of the lag of the French economy. Industrialists and agriculturalists have always been haunted by the spectre of overproduction and have feared a collapse of prices. To protect their interests they are organized into coalitions. These have as their purpose to maintain production at a relatively low level and to assure high prices for sales. They thus assure survival of the least profitable units . . . and occasionally even require the state to finance activities which have no interest for the national community . . . Mechanization and rationalization are held back; investment is limited . . . Prices are no longer competitive with foreign prices . . . Since the national market is limited, the forecasts of overproduction become justified along with the Malthusian measures which the industrialists and the agriculturalists demand.[5]

It has also been argued that the absence of primogeniture in France left firms vulnerable to internal family rivalries, which usually ended in a weary agreement to observe the familiar routine and methods of the past. At the same time, the loss of life in the First World War was an important factor in accounting for the elderly composition and ethos of much of French business management.

The 1920s were unquestionably a period of missed opportunity for French industry. The fact that France had been a theatre of war meant that there was much reconstruction to be done, and indeed the resultant head of steam and the recovery of Alsace-Lorraine pushed industrial production beyond the pre-war national level as early as 1924. The price, of course, was paid by those who had been the victims of devastation—and by the public exchequer in so far as it helped them. But although France eventually received a total sum of £1,600 million from Germany, her debts to other countries reduced her net gain to about £600 million, a mere fraction of the overall cost of the war to France. The installation of new equipment, however, gave many French factories an acceleration that soon carried them beyond many of their undamaged competitors. Yet the tragedy of reconstruction was that it left the pre-war structures of industry largely intact. Electricity replaced steam in many rebuilt textile mills, and cutting machinery replaced men in many renovated mines; but the mergers and rationalization of structures that were to take place in the post-Second World War era had little counterpart in the 1920s. Government help, such as it was, had few if any conditions attached to it, unlike the allocation of Marshall Aid in the 1940s. Replacement rather than restructure was the hallmark of economic renovation, and the old family firms re-emerged as before, better equipped but with much the same attitudes.

The origins of the world Depression of the 1930s lie outside the scope of this book. The European economy in the 1920s had expanded rapidly on borrowed American money, while the gradual rise in real wages increased consumer demand and encouraged greater industrial productivity. In America itself, production was clearly outstripping demand by 1929, and there followed a crisis of confidence among financiers and speculators. With credit subsequently hard to come by, it was difficult for industry on either side of the Atlantic to tide itself over the period of slackened demand without dismissing large numbers of workers; and with every worker out of a job, there was one less buyer in the market. At the same time, this international downward spiral of industry was exacerbated by the general impoverishment of overseas farmers, following the collapse of the short-lived demand for increased food production during the First World War (pp. 3–4). Overseas farmers were in no position to buy the industrial goods that could no longer find a market in the depressed economies of Europe and America.

France did not feel the worst effects of the world Depression until 1932. To some extent, her economy was still invigorated by the tail-end breezes of the reconstruction era. But, more significantly, she was less heavily dependent on industrial exports than Germany, Britain, and America, whose very success as manufacturing countries made them the first to suffer when the bottom fell out of the world market in 1929–30. On the other hand, their latent strength enabled them to recover more quickly, despite the unparalleled severity of the Depression within these industrial giants. France, by contrast, suffered less acutely; but the effects of the Depression on her economy were to last well into the late 1930s, whereas in most other countries recovery was well under way by the middle of the decade. In 1935, French industrial production was a quarter less than it had been in 1928, while industrial exports were down by nearly half, reducing the French share of total world exports from 6 per cent to well under 4 per cent.

Once again, opportunities were missed. Under the conservatives and Radicals, the challenge of the Depression was met merely with deflationary policies—notably a firm commitment to balanced budgets and a determined search for wage reductions. Few people expected them to pursue the more vigorous alternatives of devaluation and increased state intervention, since these might threaten the equilibrium and ethos of the prevailing order. The propertied classes would certainly oppose any depreciation of the franc, with its threat to their savings, while the peasantry would assume that a devalued franc would reduce French domestic purchasing power and thereby depress sales of farm produce. They were deaf to long-term argument that increased industrial sales abroad would enrich the home market for foodstuffs.

Another wry irony of these years was that the mergers and rationalization that the Depression might usefully have brought about were held at bay by the simple fact that many proprietors of declining firms had enough savings to tide them through the difficult times. And, to give a further twist to the irony, they had made these savings in many cases precisely through starving their factories of the modern equipment they needed. The average age of industrial machinery in the late 1930s was twenty-five years—or twelve in the case of the newer electrical and chemical industries—while in Britain it was only seven.

When it came to the Popular Front governments of 1936–8, they initially made the mistake of subordinating economic recovery to a dramatic shortening of hours and a reapportioning of existing income—with the uneasy consequences described in later chapters (notably pp. 55–60). Their miscalculations undermined the economic gains made in 1935–6, with the result that industrial production in 1938 was still 17 per cent lower than in 1928, whereas in Britain and Germany it was higher by 22.5 per cent and 28 per cent. Paul Reynaud—as Daladier's Minister for Finance (November 1938–March 1940)—endeavoured to salvage the economy by tempering a number of the Popular Front's less well-judged measures; but, as in the past, government intervention remained minimal.

Like the other old-timers of nineteenth-century industrialization, France's position in the international league table of manufacturing countries was progressively eroded by the development of the newly industrializing nations of Eastern Europe and overseas. Her share of world industrial production fell from 6.4 per cent in 1913 to 4.5 per cent in the late 1930s; and her only consolation was that Germany's share had fallen from 15.7 per cent to 10.7 per cent in the same period, and Britain's from 14 per cent to 9.2 per cent. More disturbingly, her gross domestic product was growing at an average annual rate of only 1.8 per cent between the wars, as against the European average of 2.5 per cent (Appendix II, Table 4).

As in the pre-1914 years, France counted on her invisible exports, such as investments abroad, to cover an adverse balance of trade. A dangerously large proportion of these, however, came from the tourist trade—three-quarters in the later 1920s—and inevitably, when the Depression came, the flow of tourists fell to a trickle, bringing in less than a tenth of the spendings of former years. Similarly, German reparations had been an additional source of income that disappeared with the Depression. The 1930s saw exports fall to less than two-thirds of imports; and, deprived of tourists and reparations, France had no alternative counterweight. The result was that the balance of payments changed from a surplus, averaging £44 million between 1926 and 1930 (excluding reparations), to a deficit, averaging £29 million between 1931 and 1937.

Eventually 1938 brought hopeful signs of recovery, especially after the third devaluation of the franc in May (p. 61). This helped to increase exports, and the overall payments deficit was reduced to £1.5 million. Belgium, Britain,

Switzerland, Germany, and the United States continued to be France's best foreign customers, until war once again broke the network of international trade.

Despite this litany of decline and disappointment, the French industrial record for these years was not without positive achievement. The chemical industry developed rapidly—especially in the field of synthetic textiles, artificial fertilizers, and electrochemistry—until it was providing nearly 10 per cent of France's total exports in the late 1930s. The traditional textile industries, on the other hand, were suffering precisely from the competition of these synthetic fibres, as well as from the development of modern textile factories in former foreign and overseas markets—factors that admittedly also affected France's long-established competitors. With 9.8 million spindles in the 1930s, France remained in fourth place on the European ladder of cotton manufacturers, with Britain (37.3 million spindles), Germany (10.3 million), and Russia (10 million) above her. These same countries were also ahead of her in steel production (Appendix II, Table 3), which at home was increasingly dominated by the firms of de Wendel and Schneider. Concentration was likewise a feature of the vehicle industry, where half the French output was made by Renault and Citroën. Production remained static between the wars, however, averaging about 200,000 vehicles a year, as against Britain's and Germany's rapid rise in the late 1930s to 450,000 and 304,000 respectively.

THE SOCIAL CONSEQUENCES

The main victims of economic stagnation were inevitably the work-force and the French consumer. Despite the impetus of reconstruction in the 1920s, it was not until the end of the decade that real wages in many industries rose above their pre-war level (Table 1.3); and throughout the 1930s they remained a quarter below those in Britain. As for unemployment, the official figures for France presented a relatively reassuring picture when compared with those of her neighbours (Table 1.4); and in so far as the Depression hit France in a less violent but more prolonged fashion than in Germany or Britain, the official figures reflected the broad pattern of reality—but not its depth. Precisely because there was no central state scheme of unemployment benefit, many of the unemployed slipped through the mesh of official figures. Thus, although the official average for 1936 was 431,000, it was estimated to be 756,000 in reality. Only one worker in twenty had the foresight or the money to be voluntarily insured against unemployment, which meant that the rest had to resort to whatever poor-relief the *bureaux de bienfaisance* could give.

Despite French claims to being an advanced democracy, her social security system compared badly with those of other countries with fewer humanitarian pretensions. State old-age pensions (1910) and accident compensation (1898)

Table 1.3.
The Growth of Real Weekly Earnings, 1913–1936: France and Selected Countries (Index Numbers)

	1913	1924	1928	1932	1936
France	100	100	99	109	133
UK	100	111	117	129	134
Germany	100	70	108	94	106
Switzerland	100	116	123	142	140
Sweden	100	148	161	183	178
Denmark	100	118	129	151	135
Czechoslovakia	100	115	114	131	124
USA	100	131	138	122	144

Note: This table merely indicates the degree of change, since the pre-war baseline of 100 corresponds to a different level of wages in each country.

were its main pre-war features; and those post-war improvements that had been made were of a piecemeal, perfunctory nature. A series of bills of 1928–30 attempted to schematize matters by bringing a third of the work-force into a unified system of insurance for sickness and old age. Its variable benefits were equivalent to about half of the current notional wage for each category; and contributions were calculated at 8 per cent of this wage—of which the worker paid one half and his employer the other. For agricultural workers, the combined contribution was only 2 per cent—and the benefits correspondingly modest. The scheme was also intended to cover about 80 per cent of the worker's medical expenses, but this too was based on notional costs that were well below the harsh realities of surgery and pharmacy expenses.

Table 1.4.
Unemployment, 1926–1938: France, UK, Germany, Italy

	1926	1928	1930	1932	1934	1936	1938
France	11,000	15,000	13,000	308,000	345,000	431,000	375,000
UK	1,062,000	980,000	1,467,000	2,272,000	1,801,000	1,497,000	1,423,000
Germany	2,011,000	1,353,000	3,139,000	5,579,000	2,718,000	1,592,000	160,000
Italy	113,000	324,000	425,000	1,006,000	880,000		

Source: League of Nations, *Statistical Yearbook*.

Relationships

The attitudes of mind that weakened economic enterprise in France had their counterparts in other spheres of commercial and political life. The role of the trade unions—and their failure to achieve better conditions for French workers— is discussed in Chapter 3; but part of the problem of labour relations in France stemmed from deep-seated patterns of behaviour that extended far beyond the realm of the shop-floor.

Conflicts between superiors and subordinates were rarely resolved in France by joint discussions seeking compromise. It was much more usual for each side to stick firmly to what it believed to be its rights and principles; and then, after much vehement posturing, for an eventual solution to be sought by appeal to higher authority or expert opinion. Thus each side was able to uphold its position until the last moment—when either victory, or an ostentatiously dignified acceptance of a disappointing verdict, concluded the matter. Each side felt that it had retained its integrity—and morally had the better part. This ritual of confrontation had been a major factor in endowing French institutional life with its meticulously detailed regulations and saddling it with its dauntingly formal structures of command. Both of these acted as obstacles to evolutionary change, with the result that France notoriously lacked the grey areas of Anglo-Saxon tradition that were such a fertile ground for salutary innovation. Consequently, when change came in France, it was often as the result of violent if not revolutionary action.

The citizen's counterpart to the regulations and structures of authority were his *droits acquis*. These were the hard-won rights which his antecedents and representatives had wrested from authority, and which it was his solemn duty to uphold, irrespective of their inconvenient consequences in the current situation. These, too, tended to be obstacles to change. But beyond the formal confrontation of the *règlement* and the *droits acquis*, there existed 'le système D'—the art of bypassing the regulations through subterfuge and the discovery of loopholes. This, like the regulations themselves, was inherently time-consuming and, paradoxically, tended to strengthen the status quo, since these loopholes acted as safety-valves for the pent-up frustration that might otherwise have swept away the status quo.[6]

Even so, formalism and suspicion were the hallmarks of functional relations, especially where money or entitlement to material benefits was at stake. Negotiations between workers and management, applications by individuals for state benefits—even cashing a cheque at a bank—were all dominated by formalism and suspicion. In drawing money from a bank, the client, after long queuing, generally had to deal with at least two clerks, the first of whom scrutinized his cheque and looked up the branch code in a large catalogue, as though it might be the number of a stolen car. If the cheque survived the various tests to which it was subjected, the client eventually made his way to a cashier—or, rather, to a second queue at the cashier's kiosk. The cashier, determined not to be thought as gullible as his colleague, looked at the cheque with equal suspicion and, with an air of resignation and distaste, eventually handed over the money—his weary expression suggesting that if the client had succeeded in fooling the bank, he had certainly not fooled the cashier. Obtaining a permit or authorization from the police or any public administrative office was far worse. The suppliant would queue at a succession of *guichets*, some in different buildings or even in different districts, and lay out his sheaf of *pièces justificatives* under a succession of sceptical eyes, like an art student trying to sell sketches during the Depression. The whole situation under the Third and Fourth Republics was aggravated by the fact that the staff were usually overworked and underpaid—causing their professional suspicion to be further soured with personal discontent and anxiety.

Education

The tensions between the individual, work, and society were fundamental to French thinking on education. The mainstream of French thought, both secular and religious, saw the individual as basically impervious to the attempts of society to civilize him.[7] And, despite the advances of society in overcoming this resistance, there remained an inner core that civilization never completely tamed—as exemplified in theatre queues or on the *routes nationales*. It was a commonplace of *belles lettres* that the archetypal formal French garden was a metaphor for society's taming of the unruly impulses of nature—including man. Symptomatically, one of the high priests of French economic planning, Pierre Massé, was to describe state planning as 'un jeu contre la nature'; and although economic planning, as it was later practised under the Fourth Republic, was merely the dream of a few individuals in the 1930s, state concern for the functioning of society had been a constant feature of French life since the seventeenth century, if not earlier. Although the Radicals, and their theorists such as 'Alain', claimed to deplore state intervention, they were as *étatist* as anyone else when it came to the protection of French economic interests against foreign competition; and, indeed, they were the leading exponents of *étatism* when it came to the formation of future generations in the educational system.

The prime emphasis in French education was on the training and discipline of the intellect. Much less time was spent on participatory activities such as team-games, drama, choral and orchestral work, areas in which foreign schools attempted to develop the attitudes of mind needed for the blending of individual skills in joint, constructive enterprise. This might seem paradoxical in a nation that claimed to be so conscious of the tension between the individual and society. But a voluble awareness of one's problems does not guarantee that adequate efforts will be made to deal with them; and it was perhaps a measure of the deep-rooted nature of this tension that it was spoken about so much, but so little was done to eradicate it.

Even on its chosen terrain of training the intellect, French educational method was much criticized by Anglo-American observers. They took issue with its preference for starting with principles, and then demonstrating their application—instead of allowing the child to discover principles by personal experiment. Apologists for the French system argued that time was wasted by allowing the child to explore unproductive avenues—where, moreover, the 'wrong' answers might become imprinted on the child's consciousness. Adult experience was the child's short cut to accurate knowledge and effective methods, and was best given at the earliest opportunity. Critics replied that this created passivity, and that second-hand experience made much less of an impression on children's minds than what they acquired through their own exploration, albeit under the teacher's close supervision.

Primary education in France was free and compulsory—and had been so since the 1880s. The minimum leaving age was thirteen, which the Popular Front government raised to fourteen in August 1936; and the first six years of the programme were provided by the 70,000 state *écoles primaires*. The last two years, however, were supposedly spent in either a *cours complémentaire* or an *école primaire supérieure*—although many village children stayed on at the *école primaire*. Before leaving the *école primaire*, pupils took a public examination at about the age of twelve, the *Certificat d'Études Primaires*, which was the minimum educational qualification for the lower levels of public employment, such as a postman; indeed, the framed certificate was often to be seen happily displayed in the living-rooms of farms and cottages. About half the pupils succeeded in passing it—the less able choosing not to enter for it. The *école primaire supérieure* differed from the *cours complémentaire* in being somewhat more academic, with closer affinities to what went on in the lower classes of the secondary *lycées* and *collèges*. Until the early 1930s, they had also been fee-paying and could lead directly or indirectly to a greater variety of white-collar jobs. These opportunities included a career in primary-school teaching itself, since pupils were eligible at fifteen for entry into primary-teachers' training colleges.

As in all countries in the 1930s, only a very small proportion of French children received secondary education—about 7 per cent, which was much the same as in Germany. The fortunate few entered it directly from the *école primaire* at the age of eleven or twelve. In principle, there was little difference between the curricula of the two hundred state *lycées* and the three hundred municipal *collèges*. But the *lycées*, being larger and more prestigious, tended to offer a greater range of options and generally had staff of a better quality. The traditional fees in secondary education had been phased out between 1929 and 1934, which was expected to encourage a substantial influx of pupils from lower-paid families. To control the tide, the government put up a new barrier in the shape of an entrance examination in 1933. But traditions and attitudes of mind took time to change, and the sheer length and non-vocational character of secondary-school studies were a deterrent to many parents who had no personal experience of what they had to offer. The result was that less than 3 per cent of secondary-school children in the late 1930s came from working-class homes, and well under 2 per cent from peasant backgrounds. The predominantly middle-class character of secondary-school pupils was further strengthened by the fact that a sizeable minority of them were not products of the primary schools at all, but had come up through the *lycées'* own preparatory schools, the so-called *petites classes*, which until recently had been fee-paying and whose whole programme was directed to preparing children for the *lycées*.

Given the entrance examination and the advantages of secure middle-class backgrounds, it was not surprising that about half the secondary-school population succeeded in obtaining the fairly stiff *baccalauréat*, which among other things entitled its recipients to university entrance. But preparing for it was a hard slog, and school life far from idyllic. *Lycée* pupils who came from a distance were generally boarders, and, as Henri Dubief has written:

the chosen few, heirs to the future, paid dearly for their future privileges and might well envy the ones whose backwardness in class had set them free at the age of thirteen. In spite of reforms the drum-beat that told the time of day in boys' *lycées* up to 1940 was not the only Napoleonic relic. For those who endured boarding-school life, sometimes coming straight from a nursery school at the age of five or seven, the routine was hopelessly ill-designed to inculcate civilised behaviour. Getting up at half past five in summer, 6 o-clock in winter, frozen dormitories, vile food that was worse than barrack-room fare, the unimaginable boredom of disciplined walks in crocodile, made *lycées* unspeakable prisons, occasionally rocked by rebellion.[8]

Not all schools were so grim, and the teachers included a number of remarkable individuals who made a lasting intellectual impression on their charges. But, on the whole, the lot of a day-pupil in a city *lycée* was the best, combining the stimulus of good teachers with the cultural amenities of town and the comfort and refuge of home.

Religion

Outside the state system, there existed the parallel if smaller system of private schools, most of them Catholic. In the 1930s, perhaps a fifth or a quarter of the adult French population could be regarded as practising Catholics, in that they went to Mass regularly on Sundays and outwardly conformed to the other prescriptions of the Church, such as Eastertide Communion. Observance was highest in the more remote pastoral areas of France, such as the Breton Peninsula, the Massif Central, and the eastern uplands, where there was less sustained contact with the changing patterns of secular behaviour and attitudes, and where traditions lasted longer (Map 1.1). But it was also high in

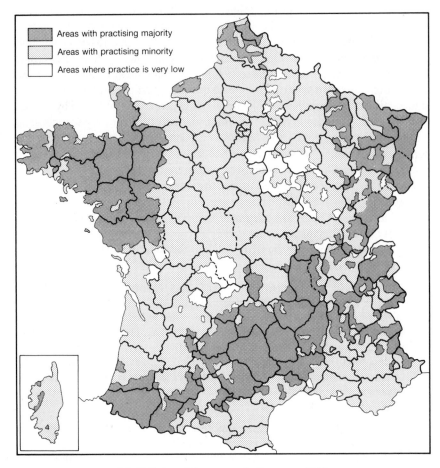

Areas with practising majority

Areas with practising minority

Areas where practice is very low

1.1 Religious Observance in France, *c*.1960

Source Fernand Boulard and Jean Remy, *Pratique religieuse urbaine et régions culturelles* (Paris, 1968), map C *hors-texte*

those areas of France that bordered on Belgium and the German Rhineland, where Catholicism was traditionally strong-rooted, and which had not been subject to anticlerical programmes on the scale of those of Republican France. Indeed, Alsace-Lorraine had escaped the worst of French anticlericalism during its brief annexation by Germany between 1871 and 1918.

In France as a whole, well over 90 per cent of the population had been baptized as Catholics; and as children a majority of them were sent to Mass and, indeed, a sizeable proportion to catechism classes, until their *communion solennelle* at about the age of nine. The *communion solennelle* was traditionally an occasion for elaborate social ritual. But, on reaching adolescence, the great majority ceased to go to Mass, and their visits to church were largely restricted to the *rites de passage* of marriage, christening, and burial. This was a situation which was also reflected in other denominations, but which was arguably more significant and disturbing for Catholics, who had been taught that deliberate non-attendance at Sunday Mass was a matter of grave sin.

Much of the conflict between Church and State under the Third Republic had been for intellectual control of the rising generation; and although the bitterness of the pre-1914 era had been partially softened by the national camaraderie of the war years, many scars remained. The importance and stridency of religious issues in French politics partly reflected the fact that the dominant spokesman of religious interests in France was the highly disciplined and doctrinally monolithic Catholic Church, with its world-wide commitments and complex diplomatic concerns. France lacked the religious pluralism of several of her neighbours, where denominational diversity tended to blur and soften the confrontations of Church and State. The limited headway made by the Reformation in sixteenth-century France had left Protestantism vulnerable to persecution by Catholic monarchs, with the result that it had only a small numerical base. Even in the 1930s, there were probably less than a million baptized Protestants, of whom the majority belonged to the Calvinist Église Réformée, and most of the others to the Lutheran Église de la Confession d'Augsbourg. The fact that Protestantism lacked the administrative and doctrinal unity of the Catholic Church allowed a greater diversity of opinion within it, which rendered it less intransigent towards the ideas and attitudes of secular Republicanism. If this allowed it to cohabit reasonably comfortably with the militant secularism of the late nineteenth century, it nevertheless left it vulnerable to the slow erosion that was the price of loosely defined doctrinal frontiers. These were the uncertain slopes of religious allegiance, where departure would seek to justify itself as 'a deeper understanding' of the human truths contained in the metaphors of traditional religious belief. The Catholic Church, on the other hand, with its clear-cut doctrine and intransigent discipline, was less able to accommodate the secular attitudes of the nineteenth century. As a consequence, it undoubtedly lost large numbers of exasperated members who left slamming the door, and a far greater number who drifted away as the social buttresses to religious observance disappeared. But it was less subject to the half-conscious seepage that afflicted Protestantism, where believers imperceptibly became unbelievers without being sure at exactly which point they had crossed the line.

The conflict between secular thought and Catholicism in the nineteenth century had been accompanied in its last decades by a mounting campaign to reduce clerical influence in politics. During the early years of the Third Republic, the Church had openly sympathized with the monarchists—seeing them as a bulwark against the secular ideals of the more militant Republicans. Many Republicans saw the prime purpose of the Republic as the formation of

future generations of Frenchmen who would think rationally and be equipped
to lead society on to a higher level of material and moral well-being. In their
view, the chief obstacle to such a programme was the fact that a large minority
of the children of France had been educated in Catholic schools, where they
were subjected to irrationalist Christian concepts such as Revelation and
imbued with hostile attitudes towards the Republican establishment. The result
was a long struggle, culminating in the forcible closure of many Catholic
schools and the separation of Church and State in 1905.

The First World War helped to soften religious issues as a divisive element in
French politics. With Catholic priests serving in the trenches, anticlericalism
seemed temporarily to be an irrelevance. Thereafter, the State in practice no
longer enforced the pre-war legislation which had prohibited members of
religious orders from teaching, and which had exiled a large number from the
country. During the First World War, the Minister of the Interior had
instructed prefects to turn a blind eye to the clandestine re-establishment of the
religious orders in private education; and this indulgent attitude lasted into the
years of peace. At the same time, the recovery of Alsace-Lorraine confronted
secular militants with provinces that had escaped the French anticlerical
legislation of 1880–1905, when they had been part of the Second Reich. The
French government was anxious to reintegrate and make welcome what in fact
was a heavily practising Christian population; and so it left intact the privileges
of both the Catholic and Lutheran Churches in these territories. This accom-
modating attitude helped to foster a greater tolerance between Church and
State during the inter-war years—despite the abortive attempt of the Radicals
to breathe life into old quarrels in the mid-1920s. Even so, elsewhere in France
the Church still took the precaution of placing the administrative and financial
affairs of its schools in the hands of laymen; and a large proportion of the nuns,
brothers, and other members of religious orders who taught there wore a
discreet if drab assortment of lay clothing.

In the late 1930s, about 17 per cent of the French primary-school population
was taught in private schools, rising to well over 30 per cent in the strongly
Catholic Massif Central and as much as 50 per cent and above in the Breton
Peninsula. Although there were no private *écoles primaires supérieures*,
Catholics were busy establishing *cours complémentaires*, especially for girls; and
by 1938 these schools accounted for just over a quarter of the total national
enrolment in *cours complémentaires*. The position of the private sector in
secondary education was even stronger, despite the fact that the state schools
were now free, an attraction that made surprisingly little difference to the ratio
of pupils in the public and private sectors. In the lower forms, the private sector
accounted for about two-fifths, but in the senior forms it dropped to little more
than a quarter. This reflected the growing tendency of many Catholic parents
to transfer their children to the state sector once the *baccalauréat* began to
loom large in their preoccupations. Only a few Catholic schools could match the

academic record of the best state *lycées*, even if their attention to character formation and the pupils' overall welfare was considered superior in a number of respects. For the celibate nun or teaching brother, the children they taught were their only wistful contact with the world they had renounced. For many of them, classes were a welcome, anarchic release from the set routines of the monastic day; and their pupils were the bright side to their lives. Parents contrasted this with the clock-watching *professeurs* at the *lycées*, whose minds perhaps were on their families, their *thèses de doctorat*, or the woman next door. Even so, the spectre of the *baccalauréat* brought about increasing transfers in the higher forms, once the child was considered to have acquired a solid religious and social foundation in the lower classes.

Women

The educational system reflected not only the divisions of wealth and belief in French society, but also the underprivileged position of women. At secondary level, boys and girls attended separate schools—girls accounting for only a third of the state enrolment—whereas at primary level their numbers straight-forwardly reflected their ratio in the population. Although girls were better represented in the private sector of secondary education, the academic stan-dards of the *bonnes sœurs* varied considerably; and the fact that their schools continued to be fee-paying put them out of the reach of the poorest families— modest though many of these fees were.

Needless to say, this situation of sexual inequality reflected the traditional distinction made between the roles of men and women—the injustices of which were even less clearly perceived in the 1930s than now. The feminine pre-dicament was well expressed by Huguette Bouchardeau, writing forty years later:

The vast majority of women accept the roles that have been assigned to them, and seek to fulfil them to their best ability, in the way that society has conceived those roles. But they also seek to make use of them—like the actor who transforms himself into the character he is playing, so as to profit from what the role affords him in developing his potential . . . Since society wants us to be wives, mothers, lovers, self-sacrificing, soothing, useful in the home, etc. we conform. In some periods all women conform; at other times only the majority of them.[9]

This was true of all countries, but was exacerbated in the case of France by various civil and legal disabilities. These disabilities were an unhappy feature of those parts of Europe in which there had traditionally been tension between Church and State, or where the Code Napoléon had left a deep impression. French women were the victims of both situations. The proportion of practising Catholics among them was almost twice that among men; and watch-dogs of Republican secularism had always been afraid that a female vote would result in the election of a conservative majority, which would then set about repealing the anticlerical legislation of the pre-war era. It was this that left women without the parliamentary vote in France until 1944, just as similar considerations left Italian and Belgian women disfranchised until the post-war period. Moreover, the granting of the vote to Spanish women in 1931 had been followed by right-wing successes in the elections of 1933, all of which was invoked by French anticlericals as incontrovertible proof of the folly of enfranchising women. At the same time, however, any politician with pretensions to democratic principle felt increasingly embarrassed by the patent injustice of the situation. They looked uncomfortably at such neighbouring countries as Germany and Britain, which had given women the vote following the First World War—admittedly without the deterrent of a clerical problem. Consequently, left-wing politicians in the Chamber of Deputies went through the motions of granting female suffrage, confident that any measure they passed on the subject would be rejected by the Senate with its old-style Radical majority. Even a Chamber vote of 488 to 1 in July 1936 cut no ice with the Senate, much to the relief of many of the deputies who had voted for it in the lower house.

French women were also victims of Napoleon's obsessive concern with restoring hierarchy and stability to all aspects of society after the Revolution. The Napoleonic Civil Code in effect made married women legal minors, subject to their husband's authorization in many matters—until the law of 18 February 1938 repealed its worst aspects. Before 1938 a wife could not run a business or even obtain a passport without her husband's consent; but the new law granted married women full legal capacity and abolished their formal duty of obedience to their husbands. Even so, the benefits of the change were offset in practice by the fact that the husband was still regarded by the courts as the 'chef de la famille', which disposed magistrates to uphold his decisions on specific issues if he claimed that his course of action was in the family's interest. Apart from these specific, national disadvantages, women suffered from customary worldwide discrimination in matters of work and wages. In the 1930s, only about a third of the French work-force were women, a proportion well below whatever was explicable by the pressures of caring for small children (Table 1.1, p. 3). Admittedly, it had been slightly higher before the Depression hit France; but the male-dominated unions saw women as potential competitors, and had paid little more than lip-service to improving their conditions and inferior rates of pay.

Leisure and Culture

As in other Western countries, the eight-hour day of the inter-war era brought the working classes greater leisure. The spread of football from Britain made Europe increasingly a continent of games-watchers; and those who could not get away to watch followed the fortunes of their teams in the press and on radio. The speed, simple rules, and size of the ball made football a game that could be followed easily, even at a distance—whereas the French pre-war passion for cycle racing, although still strong in the summer months, reduced the spectator to the role of an astronomer, waiting for the object of his interest to orbit into view.

The number of radio sets in France increased from half a million to over five million in the 1930s, and brought people into contact with a wide range of subjects which they would never have bothered to read about when skimming through newspapers. Waiting for sports results obliged listeners to hear news bulletins, which made them at least aware of events that were outside their normal sphere of concern—though it was only with the spread of television in the 1950s that this awareness was later transformed into something approaching real interest. The private sector of broadcasting was largely in the hands of newspaper companies, while the state-owned sector was initially part of the PTT (Postes, Télégraphes et Téléphones), until it acquired separate status in 1939. Although its political potential was already appreciated by Prime Ministers such as Tardieu and Doumergue in the early 1930s, the dangers of state control of news programmes did not become an issue until after the war, when state monopoly and memories of the Occupation were to make it a highly sensitive issue (pp. 290–1). Radio, like cinema, also acquainted remote rural populations with the language and accents of the capital, carrying a stage further the 'peasants into Frenchmen' process that universal military service and free and compulsory primary education had launched in the previous century.

Radio was not yet the rival of newspapers that television was to be. It was a tribute to improved literacy and rising standards of living in France that the number of newspapers sold in the 1930s, in proportion to the population, was nearly four times greater than it had been fifty years earlier—and 15 per cent higher than it was to be forty years later. Yet it was a sign of what was to come that provincial newspapers, with their local news and advertising, withstood the competition of radio much more effectively than the Paris-based national dailies, whose overall circulation was now no bigger than that of the provincial press—whereas it had been nearly 30 per cent higher in 1914. The popular evening paper, *Paris-Soir*, sold 1.75 million copies in the late 1930s, and the morning *Petit Parisien* a million; but none of the other dailies exceeded 400,000—while a quality paper such as *Le Temps* sold only 70,000 (as compared with *Le Monde*'s 500,000 in the 1970s) and its degenerate rival, *Le Figaro*, 80,000. The bulk of the large-circulation dailies with no specific party allegiance tended to lean to the Right, their aggressive, no-nonsense style attracting readers who sought excitement rather than appeals to their social conscience.

Yet among the politically committed papers, the Communist *L'Humanité* and *Ce Soir* both cleared well over a quarter of a million copies in 1939, while the Socialist *Le Populaire* (160,000) and broad Left *L'Œuvre* (235,000) held their own against the Catholic *La Croix* (140,000) and the right-wing Catholic *Le Jour—Écho de Paris* (185,000). The Radical cause, as one would expect, was most faithfully upheld in the southern provinces, notably by *La Petite Gironde* (325,000), *La Dépêche de Toulouse* (260,000), and *Le Progrès de Lyon* (220,000)—while among other large regional papers, *L'Ouest—Éclair* (350,000) was Catholic and *Le Réveil du Nord* (200,000) Socialist. The extreme Right fared better among the weeklies, the muck-raking of *Gringoire* (650,000) and *Candide* (500,000) giving them a readership that was much wider than the nucleus who sympathized with their politics.

Serialized novels survived longer as a feature of the popular press in France than they did in Britain—although, as in other countries, strip cartoons were increasingly dominating the entertainment pages. Yet France was still a long way behind Britain in the provision of public libraries and cheap editions of popular and serious literature; even thirty years later, library loans per capita were under 8 per cent of those in Britain. As in most countries, the best sellers of the 1930s were undemanding novels such as Margaret Mitchell's much-translated *Gone with the Wind* (1936)—together with utilitarian manuals such as *Le Petit Larousse* (1905) and Ginette Mathiot's *Je sais cuisiner* (1932). Antoine de Saint-Exupéry's *Vol de nuit* (1931) was the only book by a major contemporary French writer to sell over a million copies in France in the thirty years after its publication.

It is not easy to outline the main trends in serious literature in a general history of this kind, which is devoted to government and people, and in which the principal concerns are of a socio-economic and political nature. Attempts to do so run the danger of either being little more than a list of names or an embarrassingly simplistic resumé of complex matters. Readers who nevertheless prefer something on these lines to nothing at all will find a thumb-nail sketch of literary and artistic developments in Appendix IV, together with a brief discussion of French music and theatre.

If playwrights in France had traditionally catered for a privileged, urban section of society, radio and cinema were introducing drama to people who had never set foot inside a theatre—apart from the occasional visit to a music-hall. But, as in the case of news programmes, it was to be television in the 1950s that made vicarious experience a major ingredient of daily life. As elsewhere in Europe, Hollywood overshadowed the cinema of the 1930s. Apart from Hollywood's wealth and assembled talent, the French cinema's reliance on American investment was intensified by the launching of sound in 1929–30, which involved further outlay on expensive new equipment. The making of multi-language versions of American films, shortly followed by synchronized dubbing, overcame the language problem, with the result that a quarter of the French-speaking films shown in France in the early 1930s were made abroad.

A further 20 per cent of French films were made by refugee Germans fleeing from Hitler—many of whom were eventually to make their way to America. Yet it was also the golden age of indigenous French film-making—just as the 1920s had been for Germany.

An affectionate, exuberant anarchy characterized many of its best products, even when dealing with fundamentally serious themes. This is reflected in such widely differing films as René Clair's *A nous la liberté* (1931), Jean Vigo's *Zéro de conduite* (1933) and *L'Atalante* (1934), Marcel Carné's *Drôle de drame* (1937), and Jean Renoir's *Boudu sauvé des eaux* (1932), *Le Crime de Monsieur Lange* (1936), and *La Règle du jeu* (1939). They have also provided later generations with a record of the outstanding talents of stage actors such as Michel Simon, Louis Jouvet, and Jules Berry, whose performance as the unscrupulous *patron* in *Le Crime de Monsieur Lange* distils the essence of boulevard-theatre professionalism at its best. Two other great stage performers immortalized on celluloid are Jules Raimu and Fernand Charpin, whose evocation of Marseilles café life in Marcel Pagnol's trilogy exemplifies a very different style of acting, notably in *Marius* (1931) and *Fanny* (1932), directed by Alexander Korda and Marc Allegret respectively. Among younger actors who achieved fame primarily through cinema, Jean Gabin and Pierre Brasseur were closely linked with Marcel Carné, whose so-called 'réalisme poétique' in films such as *Quai des brumes* (1938) and *Le Jour se lève* (1939) was later to turn to a romanticized past in the difficult years of the Occupation, notably in *Les Enfants du paradis* (1945).

If film expanded the orbit of drama, radio and the gramophone did even more for music, especially with the improvement in electrical recording attained between the wars. In a country where nearly half the population lived in communities of less than two thousand, high standards of performance were brought to people who had rarely had the chance of hearing anything other than local or amateur players with very restricted repertoires. The effect of the gramophone on music itself was far-reaching. With the opportunity it brought for repeated listening to rarely performed works, composers were no longer obliged to make the greater part of their impact on first hearing; and, like the poet and painter, they could hope for the gradual appreciation of their work. While this may have encouraged music to be less immediately accessible to the innocent ear, the consequences in France were less marked than in Vienna or Germany. The bulk of inter-war French composers were wedded to a playful neo-classicism (p. 400); and the musical importance of Paris in the 1930s largely rested on the talents and personalities of a handful of outstanding teachers and executants. The legendary Nadia Boulanger marked several generations of performers and composers, not only pianists, while the reputation of French woodwind instructors attracted large numbers of foreign students. As for French orchestras, the Colonne, Pasdeloup, Lamoureux, and Paris Conservatoire orchestras gave underrehearsed, sometimes disastrous concerts; but with a conductor who fired their imagination, they were capable of electrifying performances—their panache owing something to the ambience of danger and uncertainty that stemmed from their lack of preparation. Opera, however, was too complex a confection to profit even occasionally from challenges of this kind. While the eyes and attention of performers were riveted on the insufficiently familiar figure of the conductor, collisions and mishaps on stage were frequent, while the music of the first act always had to compete with the chinking of money in the vestibules, as the *receveuses* counted their takings.

The spectacles that attracted most foreign visitors, however, remained the Folies Bergère, the Moulin Rouge, the Casino de Paris, the Lido, and the archetypal night-spots, where the traditional display of thighs and breasts, bedecked with feathers and spangles, made these lavish entertainments unnervingly suggestive of a poulterer's window on Christmas Eve. The comedians and singers who punctuated these expensive routines were notable for their personalities and singularity of voice rather than for the wit and humour that were to be found much more abundantly in the smaller cabarets, where quality was not judged in terms of the cost of production. Nevertheless, Paris continued to symbolize glamour and vice in the imagination of foreign visitors; and French prostitution remained the subject of much mythologizing, especially by Anglo-American journalists. The more sober estimates of social historians have suggested that there were perhaps more than 25,000 full-time prostitutes in France in the 1930s, with possibly twice that number of part-time and casual performers. As in a number of other Latin countries, notably Italy, Spain, and Belgium, *maisons de tolérance* were legally permitted, subject to a periodic but somewhat perfunctory medical examination of their inmates. Similar provision was made for registered free-lance prostitutes—the total of legalized women representing perhaps a third of the overall number of full-time prostitutes. The Popular Front Minister of Health, Henri Sellier, introduced a bill outlawing brothels, but came up against the opposition of the brothel-keepers' organizations, notably the Amicale des Maîtres d'Hôtels Meublés de France et des Colonies, and the Amicale des Tenanciers. The latter disbursed fifty million francs in its political and press campaign against the law, and Sellier was the recipient of innumerable threats of violence. The traditional guardians of freedom and the little man, *les vieux messieurs* of the Senate, remained true to their record and duly postponed discussion of the bill. No more was heard of it, despite its approval by the lower house; and when Sellier resigned, the Amicale openly boasted of its victory.

The ravages of venereal disease, which had claimed so many of the leaders of French cultural achievement in the nineteenth century, were less lethal in the inter-war period, thanks to the development of curative treatment and the wider, if technically illegal, use of condoms. Alcoholism, on the other hand, had grown since the early nineteenth century, with the French per capita consumption three times that of Italy and six times that of Holland. Indeed, this was a major factor in giving Frenchmen a death-rate in the 1930s that was over half as high again as that of women. Another factor was the joy of eating. With highly developed culinary skills and a variety and quality of wines unrivalled in the world, meals in France were a prolonged occasion of pleasure rather than an unavoidable halt for refuelling. Men were the principal addicts and victims of this heaven on earth. While their wives' appetites were often tempered by a managerial anxiety as to the success of the meal, the home-coming head of the household might already have completed a similar gastronomic marathon earlier in the day with colleagues from work. Caterers' labour costs and food prices in the 1930s were relatively low, compared with subsequent decades, and a substantial lunch in a nearby restaurant was a more regular event for the white-collar classes than was later the case.

FRANCE OVERSEAS

The hazards of over-indulgence were scarcely a problem for the bulk of the native inhabitants of the French colonial empire—even if alcoholism was a familiar danger to these French expatriates who served there. France ruled an overseas population of sixty-seven million, one and a half times the size of her own domestic population. In area, her empire was nearly twenty-two times as big as France, since it included, among much else, the larger part of the Sahara Desert (Table 1.5). But economically its value reflected the fact that so much of it consisted of desert and jungle. The oil and gas resources of the Algerian Sahara were to be discovered and tapped only on the eve of its political independence (pp. 261—2).

Before the Depression, Indo-China had traditionally been the most lucrative part of the empire, playing an equivalent, but much humbler, role to that of India in the British colonial economy. Although Algeria conducted a greater volume of business, it had a trade deficit, whereas Indo-China, even in the late 1930s, had an annual surplus of £5.6 million. With the Depression, however, French exports to Indo-China dropped below the level of her imports in exchange; and Indo-China was no longer the easy, steady market for French manufactures that it had been. This happened despite the heavy tariffs which the French authorities in Indo-China continued to put on the cheaper manufactures of other nations, keeping their share of Indo-Chinese imports down to half. French trade balances with other colonies were likewise in deficit; taking the empire as a whole, her exports in the late 1930s covered only two-thirds of her imports from the colonies, leaving her with a colonial trade deficit of over £20 million a year. This was all the more worrying now that a period of collapsing alternatives meant that France was increasingly dependent on these markets. Nearly a third of her exports were reliant on the empire, compared with under a fifth before the Depression.

Table 1.5.
French Colonies and Dependencies in 1939

	Year of acquisition	Area in square miles	Population
1. Colony under the jurisdiction of the Ministry of the Interior			
Algeria	1830–1902	847,500	7,234,684
2. Protectorates and Mandated States attached to the Ministry of Foreign Affairs			
Tunis	1881	48,313	2,608,313
Morocco	—	162,120	6,242,706
Syria and Lebanon (Mandated Territories)	1922	57,900	3,630,000
3. Territories under the Colonial Ministry			
(A) Colonies under a Governor-General and Mandated Territories			
FRENCH WEST AFRICA		1,815,768	14,944,830
Senegal	1637–1889	77,730	1,666,374
Mauritania	1893	323,310	370,764
French Sudan	1893	590,966	3,635,073
French Guinea	1843	96,866	2,065,527
Ivory Coast	1843	184,174	3,981,459
Dahomey	1893	43,232	1,289,128
Niger	1912	499,410	1,809,576
Dakar and Dependencies	—	60	126,929
Togo (Mandated Territory)	1919	21,893	737,056

FRENCH EQUATORIAL AFRICA		979,878	3,418,066
Gabun	1884	93,219	408,516
Middle Congo	1884	166,069	744,503
Ubangi-Shari	1884	238,767	833,041
Chad	1884	461,202	1,432,006
Cameroons (Mandated Territory)	1919	106,489	2,513,517
MADAGASCAR AND DEPENDENCIES	1643–1896	241,094	3,797,936
INDO-CHINA		281,174	23,853,429
Cochin-China	1862	26,476	4,615,968
Cambodia	1863	67,550	3,046,000
Annam	1884	56,973	5,989,302
Laos	1893	89,320	1,011,695
Tonking	1884	40,530	8,970,464
Kwang-Chau-Wan (Leased Territory)	1898	325	220,000
(B) *Colonies with Autonomous Government*			
Somali Coast	1864	8,492	44,240
Réunion	1643	970	208,858
French India	1679	196	295,508
St. Pierre and Miquelon	1635	93	4,175
Martinique	1635	385	246,712
Guadeloupe	1634	583	304,239
French Guiana and Inini	1626	65,041	37,005
New Caledonia and Dependencies	1854–1887	8,548	53,245
New Hebrides (Anglo-French condominium)	1887	5,790	45,000
French Establishments in Oceania	1841–1881	1,520	43,608

Source: The Statesman's Year-book, 1939 (London, 1939), 919–20.

Not only were the French colonies walled with tariffs against foreign manufacturers, but it was government policy to neglect the development of those
colonial industries that would compete with French production—a stratagem
also followed by the British in India. Colonial industry remained geared to
simple processing, building, and engineering for domestic needs; and, as far as
exports were concerned, it was confined to the preparation and packaging of
otherwise perishable goods. Yet the capacity of the colonies to absorb French
manufactures was severely limited by their own underdeveloped economies,
made worse by the world-wide fall in agricultural and raw-material prices on
which their purchasing power depended. Even the traditional jewel of the
empire, Indo-China, was greatly weakened as a market by the fall in the price of
rubber. At the same time, the purchasing power of the colonies had been
continually retarded by the low wages paid to native workers as well as the low
prices paid for peasant produce.[10]

Taxation varied from colony to colony, but in Indo-China it represented 10
to 12 per cent of the peasant's gross income. Admittedly, this was little different
from the average of 9 per cent in British India or the 13 per cent in sovereign
Japan, but for peasants living on the subsistence line, this seemingly modest
percentage could make the difference between independence and bondage.
Colonial taxation created debt and led many Indo-Chinese peasants to sell their
land, becoming either hired labourers or tenant farmers and paying up to
40 per cent of their annual crop in rent to their new landlords. By the late
1930s, the small peasant proprietor, who represented 70 per cent of the colony's
landowners, owned only 15 per cent of the arable area.

French officialdom attempted to attract private investors for colonial
development; but it favoured French financiers, many of whom charged higher
interest rates than could have been got elsewhere. In fact, French investment in
the colonies grew from 9 per cent of all her exported capital in 1914 to well over
40 per cent in the late 1930s—North Africa seeing the most significant
increase. It was a familiar jibe in Britain that the French employed as many
white officials in Indo-China as the Indian Civil Service did for a population
fifteen times the size. While the British administration was arguably under-
staffed, the abundance of French officials created discontent among educated
natives who found it difficult to find government employment; and it also
created an echelon of low-paid white bureaucrats who were tempted to engage
in questionable activities to make up for the inadequacy of their salaries.
Corruption among white officials was more common in Indo-China than it was
in India; and unscrupulous investors were quick to turn this to their advantage.

Whereas Britain had traditionally chosen to justify her imperial activities in
terms of bringing peace and justice to untutored, warring savages, France
prided herself on her 'mission civilisatrice'. Given the prestige of French
culture, this was a tempting line in propaganda. The pre-war Republic had
initially adopted the French Revolution's aim of 'assimilating' her colonial

subjects, claiming that the Black African should ultimately become as much a Frenchman as the Breton fisherman or Provençal peasant. This, however, was viewed as an ideal to be obtained in an unspecified future; and, in the meantime, the legal and electoral rights of French citizenship were restricted to Europeans and to the tiny minority of natives who fulfilled the educational and other qualifications stipulated by the particular colony. Since French citizenship required the renunciation of such native customs as polygamy, even those who had the necessary qualifications frequently chose not to become French citizens; this was particularly true of the Muslims. Even in Algeria, where there was most evidence of Westernization, there were only 1,700 Muslim requests for citizenship between 1919 and 1936, and many of these were turned down. Although Algeria was nominally regarded as part of France, this was a fiction that barely existed even on paper.

The inter-war period saw what theorists have claimed was a large-scale switch from the idea of assimilation to the idea of association in French colonial practice. The idea of association recognized that it might not be possible or even desirable to achieve identity between colonial subjects and the metropolitan Frenchman. While it accepted that France had a duty to develop the native's judgement and responsibility, it conceded that the road along which he had to travel to self-fulfilment might be a very different one from that of the Frenchman. The native was therefore encouraged to participate in the self-government of his own territory, according to the laws that most suited him. But there was to be no extension of the nineteenth-century system of limited colonial representation in the French parliament, nor was he necessarily to be subject to the same laws as Frenchmen.

Accordingly, the administrative councils of the various Black African colonies were enlarged in 1920 to include representatives of the native population. The native members were restricted to about a third of the total membership, and the native franchise was restricted to those who fulfilled a property qualification or who had been engaged in public service. French civil status was not itself a necessary qualification for the native voter—illustrating the associationist view that French citizenship on the one hand, and the right to have a voice in the territory's affairs on the other, were not the same thing. These African councils, however, were purely administrative concerns and, unlike the British African councils, they had no legislative functions.

Outside these various privileged groups, the vast majority of Black Africans were subject to the *indigénat*, with its summary administrative justice, collective fines, and, worst of all, its forced labour. The Governor-General of French Equatorial Africa had predicted in 1926 that building the Congo–Ocean railway with forced labour would 'require 10,000 deaths', an estimate that was far surpassed by the grim reality that ensued. Throughout Black Africa, thousands fled across the frontier to British colonies to avoid forced labour and other official burdens. Nor was forced labour confined to Black

Africa; the expanding rubber plantations of Indo-China made even harsher use of it, with the owners circumventing the labour code that had been introduced in 1927 to prevent its worst excesses.

Except in the mandated territories of Syria and Lebanon, the French refused to consider the question of ultimate independence. Nevertheless, the inter-war years did see a significant growth in native nationalism, which, as in other colonial empires, was most developed in Asia and least sophisticated in Black Africa. In Indo-China, the most dynamic elements of native nationalism were the right-wing Vietnam Nationalist Party on the one hand, and Ho Chi Minh's much smaller Indo-Chinese Communist Party on the other. Both had participated in the large-scale native revolt in Tonkin (1930)—which was repressed with many of the brutal features of the Algerian war of the 1950s. In Black Africa, by contrast, the only significant native political party in the inter-war period was the Senegalese Federation, which, formed in 1936 as a branch of the French Socialist Party, laid no initial claim to being pro-independence. It was in fact essentially the war, and the division between Vichy and Gaullist France, which enabled the first significant steps to be taken in the development of political self-consciousness in French Black Africa. Even in Algeria, Messali Hadj's Étoile Nord-Africaine was the only notable pro-independence group— and most of its effective supporters were migrant workers in France. The other militant Algerian movements were mainly concerned with obtaining equal rights for Muslims within the context of French rule—but without sacrifice of Islamic customs.

French Attitudes to the Empire

As for the French themselves, the empire was for most of them a source of reassurance and pride. During the period when France had been slipping down the international ladder of economic and military strength, she had acquired the second greatest empire in the world. When nationalists lamented the numerical superiority of Germany, imperialists pointed to the fact that the new 'Greater Germany' had a mere 76 million inhabitants against the 109 million of France and her empire. Anti-colonialism in France was restricted to a small segment of the population, and was to continue that way until the maintenance of the empire came to require a military and economic outlay that created serious tensions within French society in the 1950s.

If the mass of the population felt a vague warm pride in the empire, it was largely based on the quietly self-congratulatory accounts in school textbooks and the glimpses of imperial pageantry in cinema newsreels. As in Britain, most boys' brief interest in the empire was the product of adventure stories and stamp-collecting. Old Moyen Congo stamps depicting prowling leopards, later overprinted with 'Cameroun. Occupation Française', were a cheering reminder of the imperial gains of the First World War. For the minority of Frenchmen with first-hand knowledge of life in the colonies, the empire might be associated with escape from a former constricted life; for many it was a miserable exile. Richard Cobb has evocatively described the last melancholy hours of what could be any official, soldier, or company employee about to embark from Marseilles in the last decades of the empire. He might take refuge

in a large restaurant facing on to the Vieux Port; ... he might be reading the evening paper, in an effort, perhaps to cling on to scraps of local news ... like a drowning sailor who clings on to a piece of floating wood; as if some minor *revolverisation* in the Quartier du Panier or an accident involving a trolleybus ... could actually put off the imminence of departure, granting him, as it were, a tiny bonus of territoriality.

Later, perhaps in an encounter with 'the commercially offered femininity' of the Rue des Couteliers, the unburdening of 'poor little secrets ... on the sweaty, coverless bolster, or murmured while buttoning up ... a sort of last-minute clutching at reassurance and intimacy, threatened by the menacing call of ships, in a lilac-papered and red-tile-floored bedroom on the third floor'. Then, 'from the deck of the ship ... the city diminishing slowly, falling into itself: first, the two forts, then the dome of the cathedral, then the tiers of red-roofed houses, then the Virgin above Notre-Dame-de-la-Garde and, finally, even the brown and russet hills, before all disappeared in the sea sternwards'.[11]

2

'La République des Députés'
Politics in the 1930s

To foreign observers, French politics and government seemed impossibly quarrelsome, unstable, and plagued with a multiplicity of parties. Forty years later, the President of the Republic was still lamenting that 'it is as though political debate was not a competition between two points of view, but the confrontation of two rival and mutually exclusive truths . . . Its style is that of a war of religion'.[1] And in the 1930s the debate was rarely between less than six points of view. Indeed, Elliot Paul was driven to comment: 'political parties, in the interests of clarity, should not have names but numbers, like football players on the field'.[2] The impossibility of finding stable majorities on which to rest an effective government was the source of innumerable weary jokes: 'American tourists go to London to see the changing of the guard, and then on to Paris to see the changing of the government', etc., etc.

Anglo-American visitors often forgot that a multiplicity of parties characterized most European parliamentary democracies in the 1930s, and that only the Scandinavian monarchies came anywhere near to Britain's virtual two-party system. Even so, no Scandinavian party actually achieved an overall majority in parliament in the 1930s, as the British Conservatives did on no less than five occasions between the wars. But in many European countries, the consequences for the stability of administration were not as serious as they were in France. Governments were not necessarily expected to resign following a hostile vote in parliament, and their constitutions often afforded them a certain measure of protection against the whims of unstable coalitions. In other countries, a broad consensus on national priorities might be enough to hold together parties that differed on secondary issues. Unfortunately, these various saving factors were mostly absent in France.

Constitutional practice left the government more or less completely at the mercy of parliament. France's unhappy experiences in the nineteenth century had created a deep-rooted fear of dictatorship, and by extension a brooding suspicion of strong government of any kind. At the same time, her multiplicity of parties corresponded to equally significant divisions of opinion in French society at large, and were not just embodiments of the personal following of rival political leaders who could come to an agreement when the national interest demanded it. Paradoxically, there was a broader consensus on basic socio-economic issues in the French parliament than there was in many neighbouring countries. But the traditional dominance of rural and middle-class interests, which was the main arch of this consensus, was increasingly under challenge from the growing electoral strength of the urban working class. At the same time, hard-fought constitutional and religious issues played a far greater role in French politics than in most of the northern democracies, where politics tended to be bipolarized on matters of social policy and the distribution of income—much as they are today. As Philip Williams has commented:

in France three issues were fought out simultaneously: the eighteenth-century conflict between rationalism and Catholicism, the nineteenth-century struggle of democracy against authoritarian government, and the twentieth-century dispute between employer and employed. On the Continent, Right and Left defined positions in relation to the philosophical and political struggles, which turned in normal times on educational policy and in crises on the structure of the regime; the social contest over the distribution of the national income provided a new topic of division, already foreshadowed in the Revolution, which after 1848 cut the political Left in two.[3]

Whereas countries like Britain and the Scandinavian democracies theoretically required only two parties to ventilate their prime concerns in politics, France would logically have seemed to require at least six.

But in practice the situation was both simpler and yet more complex than this. Differences over religious and constitutional issues tended to cut broadly in the same direction, although not entirely; propertied conservatives in parliament were often spokesmen for the traditional camps on all three issues—religious, constitutional, and socio-economic. Likewise, the Socialists tended to be on the opposite side on all three of these questions. But within each phalanx there were important elements that might think very differently on one or more of them, thereby making cohesion and Westminster-type discipline impossible—and, indeed, inappropriate. Philip Williams observed that:

between these rivals was a great amorphous mass of peasants and small businessmen, who were social and economic conservatives yet ardent Republicans and anticlericals. As they owed their position to the Revolution it was indeed the heritage of revolution that they were determined to conserve. This was the basic source of the curious contradiction between their words and their deeds: revolutionary language in the political field, conservative actions in the social. Best represented by the Radical party, they dominated French politics between 1900 and 1940. Since Left and Right divided on issues other than those predominant in Britain, it is not surprising that they had a different electoral basis.[4]

The religious issue has already been discussed in the previous chapter. But concern for the constitution was an obvious product of the turbulent history of France in the previous century, when political change had been accomplished by a series of violent jerks, each accompanied by a new constitution. All three Republics had been challenged by dictators or would-be dictators, leaving French democrats with a deep-seated suspicion of strong personalities and a

determination to defend the constitution in all its detail, even when the requirements of modern government made major amendment desirable. On the other
side of the house, this very refusal to make necessary adjustments intensified
the conviction of the extreme Right that the constitution was a prime source of
the nation's ills.

For verbal convenience, this triumvirate of strongly felt preoccupations has
often been abbreviated to 'the three Cs'—constitution, clericalism, and class—
even if the abbreviation sacrifices accuracy to alliteration. The three Cs had
deeply split the electoral base that inter-war France had inherited from the
past, making stable political majorities hard to find, and burdening the nation
with its main political problem: the instability of cabinets. There were no less
than forty-two governments between the wars, averaging a mere six months
each—an even worse record than the ill-fated Weimar Republic or the equally
ill-fated parliamentary government of Italy. Such brevity meant that it was
extremely difficult for a government to undertake any reform programme that
would take time or was likely to meet with opposition in parliament. Ironically,
the constitution had given the President of the Republic the power to dissolve
an uncooperative Chamber of Deputies before its four-year mandate elapsed,
and to hold new elections—provided that the Senate agreed. But in practice this
power had long since been abandoned after its ill-judged use by President
MacMahon in 1877. Symptomatically, when majorities could not be found for
the electorally unpopular tough measures that were needed to deal with the
economic and financial problems of the 1920s and 1930s, parliament saw no
alternative but to hand over legislative authority in such matters to the government; in other words, it was prepared, exceptionally, to let the government
legislate by decree. But, equally symptomatically, when the Senate refused to
grant the government this recourse in 1937 and 1938, the result was to be
virtual breakdown (p. 61).

In the Third Republic, as under many Western democracies, the public's
participation in government only became meaningful at election-time. And
even the results of its choice, however firmly declared, did not guarantee that
any favoured individual would be called on to form a government. In a multi-
party state in which no party ever enjoyed an overall majority, each government was of necessity the result of negotiation and compromise between
competing groups. In consequence, the Third Republic, like the Fourth, was
essentially 'la République des députés', in that the person of the Premier and
the nature of his cabinet depended more or less entirely on wheeling and dealing in the corridors of parliament. In these circumstances, it was impossible for
the electorate to make conscious choices between alternative governments, as
was normally the case in Britain. Nor could the electorate express its opinion
through referendums or by calling for a dissolution of parliament, in the event
of parliament or government pursuing unpopular policies. Referendums were
associated with plebiscitary dictatorship, and dissolution with the misguided

attempts of a pro-royalist President to challenge the parliament in 1877. The everyday business of government was largely left to civil servants, with ministers seeking to impart a general direction to it during their brief sojourn in power. In practice, parliament was mainly concerned to safeguard the public from over-zealous policies by those in power—be it the civil servants or the governments with their mayfly terms of office.

Even the most fervent Republican had to admit that parliament was full of provincial mediocrities. France was still predominantly a rural and small-town country, and the electoral system tended to underrepresent the industrial cities. The distribution of seats in the Chamber failed to take account of the growth of these towns, while the method of election to the Senate, which gave the delegates of municipal councils a large share of the votes, failed to differentiate sufficiently between the voting power of large and small towns.[5] The result was a disproportionately large rural vote, a potentially dangerous situation in the 1930s, with its socio-economic and international tensions, since the country voter was generally more interested in local issues than in national ones.

The type of deputy elected was often a local lawyer or small-town mayor with a taste for politics, a man who would promise to get a new bridge built or bring about the extension of a branch-line to improve the commerce of the region. Once this scheme was fulfilled, he could live on his constituents' gratitude for life.

This predominance of local issues in national politics was to some extent the outcome of the excessive centralization of France. Many issues which in Britain would be dealt with by county and municipal authorities, were in France the direct responsibility of the central government. The result was that if a district wanted some local improvement, it had to raise the matter at a national level, which might mean competing for the favours of the minister. In the last resort, the best chance of success for a scheme was for a friend of the district's parliamentary representative to be either in or close to the government. This bore out the sad unspoken truth that for many people the rapid turnover of ministries in France was a desirable thing, since the faster the turnover, the greater chance there was of someone sympathetic coming to power. Admittedly, the turnover of ministers was not nearly so fast as the turnover of governments, since an incoming Premier generally kept on a sizeable number of the old cabinet in order to retain the support of their parties for his coalition. Even so, 278 men and 3 women served in government between 1918 and 1940; and although not all of these were members of parliament, deputies and senators had a one-in-ten chance of achieving cabinet rank at some point or other in their careers. Clearly, it required considerable altruism to vote for the reform of a system which offered such handsome prizes. Even if the cabinet you were in only lasted a few months, you could build up credit for life in that short time.

The peasant proprietors on the one hand, and the industrial and commercial middle classes on the other, could rely on the Radicals of the Centre Left and the conservatives to oppose any substantial increase in direct taxation. They wanted government interference kept to a minimum, except in the case of effective tariffs to protect their products against foreign competition. It was a tribute to their political muscle that in the inter-war years the average French taxpayer paid only about two-thirds as much direct tax as his British counterpart. Conversely, indirect taxes were more numerous, especially on foreign imports; and, as is so often the case with indirect taxes, it was the urban poor who suffered from them most. French governments, for their part, welcomed the fact that indirect taxes did not involve prying into citizens' personal affairs and were easier to impose and cheaper to administer—all of which had obvious attractions for governments that were unsure both of their majority in parliament and of the honesty of their taxpayers. The constant skirmishing between the self-employed and the taxmen was a permanent feature of middle-class and rural life, which encouraged these classes to practise the art of inconspicuous consumption. This turning of Veblen on his head sprang from the tax authorities' practice of collecting details of the life-style of their victims as a counter-check on the plausibility of their declarations of income. Middle-aged couples who had salted away millions during a lifetime of successful business would refer self-deprecatingly to 'nos petites économies'.

The loyalty of the peasants and middle classes to the parties of 'moderation' also reflected their fear that an authoritarian government of the Left or of the revolutionary Right would embark on expensive programmes of improvement. Although the peasant landholder might not have very much money, he did have land; and, like the middle class, he feared that if the champions of the urban proletariat came to power, they would establish an expensive programme of social betterment which would result in heavier taxes. The main benefits of such a programme would go to the industrial worker, who paid next to nothing in direct taxes, even if he was the victim of heavy indirect taxation; but the peasant who helped pay for it would gain little for himself from a programme of factory legislation and slum clearance. Even when the Socialists advocated measures of social insurance that would help the peasantry, those likely to benefit generally preferred to face possible misfortune with the solid strength of personal savings, rather than with promises of future help from a government they distrusted. 'Méfiez-vous' was the motto of most Frenchmen, but of none more so than the peasant. At the same time, he was scarcely likely to be won over by talk of the eventual collectivization of land. Admittedly, the mainstream of the Socialist party, and in practice many Communists, made a positive attempt to attract the peasant proprietor by laying emphasis on collective protection for the peasant rather than on collective ownership, which had long since ceased to appear on Socialist electoral programmes—and which few Communists publicly advocated.

The fall in world agricultural prices in the 1920s saw many poor peasants and agricultural workers turn from the Radical party to the Socialists—and some to the Communists, notably in the handful of rural *départements* in the south whose left-wing sympathies went back to the French Revolution (p. 45). Indeed, their votes were to be a significant contribution to the Popular Front victory of 1936. But the fact remained that the bulk of the peasantry in the nation as a whole still voted for the Radicals and conservatives. The poor record of the Third Republic with regard to social reform was a measure of their political success—and also that of the middle classes. Even more strikingly, the general economic record of the conservatives and the Radicals showed a similar unfailing solicitude for peasant and middle-class interests. The power of these interests in parliament was a strong unbroken force that lay beneath the succession of political parties, and continued unaffected by the shifts and changes that were taking place on the parliamentary surface. Though from the political angle the conservatives and the Radicals were sharply distinct, between them they developed a comparatively coherent economic programme. It was this stability of common economic interests that helped to preserve a political system that seemed to outsiders to be the very essence of chaotic instability. Indeed, the longevity of the Third Republic owed much to the relative modesty of its aims. It sought to stabilize rather than to transform society, and the energies of its more restless supporters were traditionally expended in anti-clerical and other activities that left the basic structure of society untouched.

The Radical party had never been very radical. It is easy for the foreigner to be misled by French political labels. On the one hand, the gradual leftward slide of events tended to leave parties stranded to the right of their original labels; while on the other it could be said that the Revolutionary tradition favoured progressive-sounding titles. All of which contrasted with the Frenchman's passionate clinging to his individuality, and his reluctance to part with his money. As the threadbare proverb said: 'If a Frenchman's heart is on the left, his wallet is on the right.' Then, as now, French politics were plagued with party initials, whose meaning many people had forgotten—largely because they signified dynamic progressive virtues which had either been belied in reality or had never been true in the first place. The radicalism of the Radicals normally consisted of a pronounced anticlericalism and opposition to strong government. The strong individualism that characterized their political doctrine ran counter to any idea of an ambitious scheme for social reform; as the Radical philosopher 'Alain' declared in an oft-quoted statement: 'Resistance to power is more important than reforming action.' By the time the separation of Church and State had come about in 1905 and income tax had been established during the First World War, most of their political doctrine was outdated, because it had already been realized in practice. Thereafter, they were essentially the party of 'the little man', protecting the farmer and corner shopkeeper against higher taxation and the attempts of larger, more efficient producers to enlist the government in modernizing the economy.

The conservatives, for their part, were much more loosely organized than the

Radicals, since they were dispersed in parliament under a variety of titles. The Catholic connections of some of them and the big-business connections of others were points of major difference with the Radicals; and the 1920s and early 1930s had usually found them in mutual opposition, the conservatives subscribing to the Union Nationale and the bulk of the Radicals to the Cartel des Gauches. The continuing Depression, however, together with a desire to protect the franc and middle-class savings against working-class demands for better wages, security, and working conditions caused the right wing of the Radicals to draw closer to the conservatives. The result was a series of governments in the mid-1930s with Radical and conservative ministers committed to deflationary policies. In many ways this was a formal expression of what had long been true in fact: that despite the alternation in office of governments waving rival banners of Left and Right, the French Republic was largely characterized by government from the Centre. On the one hand this was a familiar consequence of a multi-party system, where coalition and compromise were the prerequisites of forming governments, and where centre groups were given a pivotal role provided they acquired sufficient electoral support. On the other hand, it was also the outcome of a fundamental paradox in inter-war France. On the institutional level, acceptance of the constitution was the touchstone of political respectability, so that all parties, except the revolutionary Right, were regarded as entitled to play the parliamentary game—even to the point of being eligible for ministerial office—if they could find sufficient support. Even the Communists were eligible in principle, provided they paid lip-service to the constitution. This was a legacy of the pre-1914 era, when the principal menace to the regime was traditionally seen as coming from the revolutionary Right—from the various leagues that sought to subvert the constitution. This was the era of 'no enemies on the Left'. On the level of society, however, the situation was reversed. There the touchstone of respectability was acceptance of the social status quo. On this criterion, the revolutionary Right were seen as respected members and guardians of the existing social structure: a number of their leaders came from wealthy families, and their militant followers were secretly viewed by many conservatives as a potentially useful counter-force should the extreme Left ever seek to take power by violent means. The Communists, by contrast, were the pariahs of society because they openly professed their intention to overthrow it—in the long term, if not necessarily in the short. This reversal of roles left the centre parties as the common denominator of respectability on both levels.

It was often a matter of surprise to Anglo-American observers that the French parliament and, by extension, the ministers themselves, were more extensively drawn from the professions (especially law) and what might loosely be termed the intellectuals (journalists and teachers) than was the case in Britain. Industry, land, and labour, though exercising a strong influence on parliament and government, had far fewer practitioners sitting in the legislature. This was a feature of many Continental countries, especially those whose parliamentary traditions pre-dated extensive industrialization, or where early revolutionary activity and reform had initially been led by civil servants, lawyers, and the like. It also reflected the fact that, on the Continent, young men who wanted to make a career in politics often chose law as an initial apprenticeship; not only because it provided knowledge and skills that were useful to a politician, but also because it was a profession in which the hours and pressure of work could be regulated to fit in with the uncertainties of parliamentary life.

The professional background of so many members of parliament, combined with the distinctive characteristics of French public relations (p. 14), tended to give political discourse in France a somewhat abstract quality. Relatively minor matters of a material nature were often debated in terms of high principle, admittedly often masking straightforward self-interest. Compromise on such issues was made more difficult, often absurdly so, by the fact that the disputing parties had pitched their arguments at such an exalted height. Even so, the realities of the situation and mutual convenience generally triumphed at the end of the day; and it was a matter for wonderment among foreign observers that men who had earlier sworn readiness to defend to the death some basic liberty allegedly enshrined in a routine issue, were prepared to change their minds without fuss after a brief conferment in the corridors. Nevertheless, it was widely felt that French political life suffered from the stridency of parliamentary debate, and could do with more of what Miliukov called 'the cement of hypocrisy', which enabled British political institutions to work so well.

The erudition of French members of parliament likewise ensured that debate on any issue was constantly illustrated by reference to past parallels in history. It is perhaps a characteristic of societies that have progressed through widely spaced revolutionary leaps rather than through steady evolution to make appeals to the past. In a famous phrase, de Gaulle was to speak of 'Vieille France, accablée d'histoire', and J. E. C. Bodley described France as 'the land of political surprises, where lost causes come to life again'. The politicians of the French Revolution had constantly likened themselves to the politicians of the ancient Roman Republic. Those of 1848 and the Commune of 1871 referred back to the 1790s—and, more recently, a number of the student militants of 1968 were to resurrect the language and concerns of nineteenth-century Utopian socialism, while others took old Revolutionary slogans as titles for their broadsheets.

In contrast to the constitution, the highly centralized administrative and judicial structure of France had survived relatively unchanged and unchallenged since the time of Napoleon. While a matter of pride to most Frenchmen, democrats also regarded it as a reason for vigilance towards the government in office, given the strength of the instruments at its disposal. In a well-known dictum, 'Alain' remarked in 1906: 'In France there are a great many radical electors, a certain number of radical deputies and a very small number of radical ministers: as for the heads of the civil service, they are all reactionaries. He who properly understands this has the key to our politics.' Like many such comments, it had an element of truth in it. One might object that at the time 'Alain' was writing, it was difficult for practising Catholics and men suspected of right-wing sympathies to gain entry and promotion in the politically sensitive branches of the public service. Yet it was also a fact that caution and attachment to familiar routines were a safe ladder to advancement, once one had survived the vetting processes and the competition of other claimants, with their wads of letters from deputies and senators, testifying to their impeccable Republican sentiments.

A further factor in this conservatism of outlook was the wealthy family background of many senior civil servants. The higher levels of the *fonction publique* were more easily scaled by those who were products of the *grandes écoles*, the élite professional training institutions, such as the Polytechnique, the École Centrale, the École des Mines, and the Fondation Nationale des Sciences Politiques. While both entry and success in the *grandes écoles* required a high level of ability, preparation for their entrance examinations was a speciality of certain prestigious *lycées* and a handful of private schools—which for boys not of the locality meant the expense of boarding and (until the 1930s) fees, expenses which grew heavier once the successful candidates obtained entrance to the *grandes écoles* themselves and were living in the capital. This, and the desirability of well-connected sponsors, gave boys from bourgeois backgrounds a distinct advantage. The class links of the senior civil servants were similarly strengthened by the practice of *pantouflage*, the movement of civil servants into highly paid managerial posts in the world of private business.

The provincial executants of the policies of Paris were still primarily the Napoleonic prefects and sub-prefects, punctiliously responsive to the will of central government and no more than politely attentive to the *départemental* councils and other organs of local democracy. As a prefecture official in the 1950s remarked: 'Si le public dit du mal de moi, je l'emmerde sereinement. Cela ne fait que prouver la valeur de ma section et de mes méthodes. Plus le public est emmerdé, mieux l'État est servi.'[6] The absence of a strong tradition of democratic local government was the source of many problems in France. It allowed animosities to develop between Paris and the provinces, and encouraged the type of posturing described above. Although a large number of deputies and senators were *départemental* or municipal councillors, or both, the experience this provided was minimal compared with British local government. It did not provide them with a preliminary schooling in the difficult art of combining the role of critic with the requirements of effective government. By British standards, local councils met infrequently and with very limited powers, and were regarded by both intending and serving politicians as a means of developing a personal following in the constituency.

In the communes, it was generally the tenured town clerk who exercised authority and influence, while the mayor was often a politician whose parliamentary duties kept him in Paris for much of the time. It was the town clerk who was the recipient and executant of the government's instructions from the sub-prefect and *départemental* prefect above him. In small villages, it was often the primary schoolteacher who fulfilled these functions; and in conservative, Catholic regions, primary schoolteachers frequently came to be regarded as both missionaries and executants of Paris-based Republican orthodoxy. This could be a very lonely existence, knowing that they owed their functions to their education rather than to the trust of the villagers or even the mayor, who had little alternative but to appoint a teacher as the clerk, given the dearth of suitably qualified people.

Foreigners often contrasted the powerful, centralized administrative machinery of government with the weakness and brevity of the cabinets whose wishes this machinery supposedly implemented. In a famous passage, the Swiss journalist, Herbert Lüthy, asserted:

The real decision in all matters, from the budget of the smallest village in the Pyrenees which desires to build a bridge or lay a water-pipe to the control of the national budget and the government of a great empire and the long-term conduct of foreign affairs, is in the hands of an administrative hierarchy which is practically immune from political fluctuations and, though it may sometimes be forced to compromise with a Minister whom Parliament sets before it, is always able to wait patiently until he and his ideas have departed ... France is not ruled, but administered, and it is her apparent political instability which guarantees the stability and permanence of her administration. Thanks to this division of labour, politics remains with impunity the playground of ideology, abstraction, extremism, verbal tumult, and pure demagogy, because all these things hardly touch the life of the French state; and no counterweight to them is required, because they cancel themselves out. The republic reigns, but does not rule. Both the presupposition and the consequence of unbridled rhetoric is that it should work in a void.[7]

While, as always with Lüthy, there is a strong vein of acutely observed truth in this assessment, one ought not to infer from it that the administration and judiciary were an effective government, where the real decisions lay. While the other half of his proposition—that parliament and ministers were unable to exercise effective authority—was partly, perhaps largely, true, it did not follow that the administration took over this vacant role. Because the vehicle of State was moving, despite the evident incapacity of the driver, it did not follow that there was an invisible instructor, with a secret panel of controls, keeping it in motion. In many respects, the State was merely rolling down the incline of economic and social contingency. Many would claim that there was a vacancy of power. Yet the concept of what constituted a proper level of government activity was so much more modest than in many other countries that there were still Radicals who thought that France was overgoverned. At the same time, the broad continuity of views between one government and another on basic socio-economic matters made the rapid turnover of ministers less of a vacancy of power than was often alleged. With governments lasting a mere six months on average, it was inevitable that the administration should implement the policies of previous governments—especially since the more substantial of newly passed laws required supplementary administrative regulations before they could be put into effect. Governments of a different political colour from their predecessors would often seem to be at odds with the administration, still wrestling with their predecessor's legislation. Experienced senior civil servants would recognize when to mark time on such tasks, pending a change of direction from the new government—and, consequently, accusations of dilatoriness or of the cynical burying of legislation were levelled at them regularly.

3

The Popular Front
Legend and Legacy

France was traditionally a country of challenge and frustration for socialists. It had Europe's longest record of universal male suffrage, and had provided socialism with a large proportion of its leading thinkers in the nineteenth century. Moreover, the Paris Commune of 1871 had created the myth of Europe's first socialist government, a victory of international brotherhood destroyed by the self-interest of the possessing classes with the connivance of German militarism. Initially a figment of French conservative propaganda, this myth had been developed by the Left in a variety of conflicting forms, each suited to the polemical needs of the particular sect that propagated it. And yet France was to offer little again in the way of precept or example. It was Germany that provided the best-known theorists and political leaders, and whose growing urban masses offered most hope of a large parliamentary socialist party. Similarly, Britain's socialists, for all their pragmatism and looser links with the international movement, seemed to promise more in the way of material gains for the working classes than their French counterparts could. The obvious problem was that France was less industrialized than either of these countries, and the combined electoral strength of the rural population and the possessing classes would continue to outweigh the power of the urban work-force to demand social reform at the taxpayer's and consumer's expense. This was to remain the greatest obstacle to socialism in France until the urban workers became a larger section of the electorate in the post-Second World War period.

As the preceding pages have indicated, the misfortunes of agriculture in the inter-war years provided the French Socialist party with a growing amount of support among the disgruntled peasantry; Communism, too, picked up rural votes, notably in Cher, Allier, Lot-et-Garonne, and along the eastern Mediterranean coast. Indeed, in these *départements*, the PCF took over 20 per cent of the vote in the Popular Front elections of 1936. Rural Communism, however, was essentially heir to the traditional protest vote, picking up disciples in the milieux where anticlericalism, socialism, and syndicalism had been particularly strong in the past. It was basically a gesture of defiance against the government, and an expression of hostility towards the moderate voters of the *département*.

Had there been a party to the left of the Communists, they would have voted for that—just as some chose to vote for the ephemeral parties of the extreme Right. In addition to the traditional underdogs of agriculture, such as the share-croppers of the central provinces, the PCF also found support among the Protestant peasant landowners of the Gard, who in the distant past had tradi-tionally suffered from discrimination at the hands of the government and their Catholic neighbours. In a characteristically vigorous half-truth, Herbert Lüthy observed:

Only the slightest ideological nuances divide the harmless Radicals ... from the Com-munists ...; and in fact whole groups of electors, particularly in the word-intoxicated south, went over unitedly from the Radical to the Communist ticket, apparently without noticing the difference ... Uncompromising rhetoric and the daily compromise with reality advance on parallel lines and never meet.

   La Révolution est un bloc. That famous phrase of Clemenceau has remained a dogma of the politicians ... The foundation of Communist ideology in France is Jacobin Utopianism, not belief in the revolutionary mission of the proletariat ... For the revolu-tion which started in France a hundred and fifty years ago and was never completed is believed to have been completed in Russia.[1]

Certainly, Marxism as a system of thought had only a limited appeal among the French working class, its main impact being on a selection of the intelli-gentsia. Teachers and the lower echelons of the civil service were fertile fields for socialism between the wars—just as, to a more limited extent, they were for Communism after 1944. Pétain was to accuse the primary schoolteachers of having demoralized the nation with socialist pacifism—ignoring the fact that it was the Popular Front that had stepped up rearmament (pp. 67—8).

The secession of the French Communists from the Socialist party in December 1920—taking with them two-thirds of the Socialists' material assets—had been a major blow to the SFIO. Although it removed many of the intransigent elements that had resisted pragmatic strategies, there were still divisions of opinion within the party; and hopes of reunion with the PCF tended to keep open options and attitudes that should arguably have been abandoned at the split. Such hopes, however, invariably foundered on the question of loyalty to Moscow. By the late 1920s no one could pretend that the Comintern was anything other than a mere instrument of Soviet policy. Exasperation with the internal switches of strategy that Russia imposed on Western Communists, together with revulsion at the ruthlessness with which they were expected to carry out these policies, precluded the SFIO from subscribing to Comintern direction. In the eyes of Léon Blum and his colleagues, methods permissible in overthrowing the Tsarist autocracy and perhaps even the rootless Kerensky government were not appropriate to oust-ing Western governments mandated by democratic election; and Blum was fond of citing the comment on Soviet-style Communism by Jules Guesde, the aged standard-bearer of French Marxism: 'How can we build the new society

... if on the day of victory you have corrupted all the human material!' PCF members, for their part, remained awed by the immensity of Soviet Russia's material achievements and its steadfastness against the hostility of other governments. Although the price paid in human suffering and moral integrity was great, the results by the mid-1930s already seemed to many Western Communists to justify it.

Of more immediate import to the working classes was the corresponding split in the union movement, the Confédération Générale du Travail, when the seceding Communists formed the CGT Unitaire in January 1922—'unity' being the watchword of any self-respecting splinter group in France. The failure of an attempted general strike in May 1920 had already demoralized the labour force; and when the employers increasingly turned to non-union workers, it was the more militant CGTU that was the principal victim. Union membership fell from perhaps two million to little more than 600,000 in 1921, and the Communist share of it fell from two-thirds to a mere quarter.

The division and mutual intransigence of the Left were reflected in its poor showing in parliamentary elections. In the later 1920s and early 1930s, Socialist candidates obtained barely a fifth of the popular vote, and held little more than a hundred seats in a Chamber of just under six hundred members—rising to 131 in 1932, with the onset of the Depression and claims of financial mismanagement levelled against conservative governments. The Communists, for their part, were supported by only a tenth of the electorate and, given the difficulty and reluctance they had in forming tactical alliances with other left-wing parties, they obtained only about a dozen seats.

THE FORMATION OF THE POPULAR FRONT

What transformed this mediocre run of results into the decisive combination of 1936 was the force of external circumstances: the world Depression, and the menace of Nazi Germany, which in turn engendered an exaggerated if understandable fear of the revolutionary Right at home. There is a sense in which Russia and the PCF brought the Popular Front into being, just as, conversely, their behaviour in the past had been the greatest obstacle to working-class unity. Hitler's advent to power rekindled Russian fears of German eastward expansion, and convinced Stalin of the necessity of creating an alliance with the Western democracies to keep Germany firmly in check. It was therefore very much in Russia's interests for the PCF to adopt a more conciliatory attitude to other French parties, in the hope of encouraging such an alliance. Conciliation would logically begin with the SFIO, in the expectation of eventually extending it to embrace the Radicals, who were a respectable party of government with hands on the levers of power. Exactly when Russia first instructed the PCF under Maurice Thorez to be more co-operative towards the SFIO is a matter of dispute. Some claim that it was as early as 12 February 1934, when Communist demonstrators joined with Socialists to protest against the recent riots of the revolutionary Right. This, it is alleged, was the result of a telegram from Moscow to the PCF leadership. Others argue that this demonstration of left-wing solidarity was a grass-roots phenomenon, owing little, if anything, to the party leadership. Whatever the truth of the matter, the summer of 1934 saw consultation between the higher echelons of the two parties, and by 1935 this *entente* was viewed with growing interest by a large segment of the Radicals, who were becoming increasingly apprehensive about the activities of the revolutionary Right.

The extreme Right were numerically weak, but, seen against the background of fascist activity in other countries, it was understandable that they should seem to represent a more serious threat to the Republic than their relatively modest strength would justify. It is true that Colonel de La Rocque's ex-servicemen's movement, the Croix-de-Feu, claimed with its offshoots to have 450,000 members in the mid-1930s, while its self-styled parliamentary successor, the Parti Social Français, eventually claimed 800,000—figures that are admittedly hard to verify. Although these numbers, if correct, overshadowed membership of the SFIO (260,000) and the PCF (350,000) in the late 1930s, the strength of the Left, and indeed of most parliamentary parties, lay in their voters rather than in their formal membership. But for the movements of the extreme Right—which in the last resort claimed to see the streets rather than parliament as their road to power—their committed membership was supposedly their principal weapon. Most of their members, however, were not prepared to take the conventional revolutionary rhetoric of their leaders seriously and 'descendre dans la rue'. Nor, for that matter, were the bulk of their leaders. They inwardly acknowledged the truth of Hitler's dictum that 'the machine gun has destroyed the romance of the barricades': a government prepared to use the military forces at its disposal could always beat the canes, knuckledusters, and small arms of even experienced amateurs. It was loss of nerve by the government not armed force that had enabled Mussolini to take power; and most of the rhetoric of the revolutionary Right in France was intended to create such a loss of nerve.

The extreme Right cut little ice with the electorate. In affecting to despise parliament, and in threatening to abolish or curtail its powers, they were condemning themselves to the non-democratic road to power, which few of them had the means or the will to take. France lacked the principal ingredients of a strong fascist movement. She had not suffered the national humiliation of defeat in 1918, as Germany had, or the frustration of demoralizing reverses in war and of scant reward at the peace, as Italy had. At the same time, the economic *malaise* of the inter-war period had not hit her so hard as Germany and Italy, while her parliamentary government, for all its shortcomings, had avoided some of the pitfalls of its foreign counterparts. As a consequence, the bulk of the middle classes felt that they could continue to defend their privileges within the framework of its provisions, without having to resort to a dictator to keep the Left in check. Moreover, the interests of the small shopkeeper and peasant proprietor had been effectively represented by the Radical party, which had no equivalent of comparable strength in Germany and Italy. The result was that an important potential recruiting ground for fascism elsewhere was already pre-empted in France by a party that was committed to the existing regime. Had industrialization in France been more rapid, and had the urban working class become a much larger percentage of the electorate, then the middle and rural classes might have felt less confidence in the propensity of the regime to function in their interests. But, as the vicissitudes of the Popular Front were shortly to show, their confidence was on the whole well justified. Yet it must be recognized that the economic crisis of the 1930s persuaded an influential minority of them to look with interest towards spokesmen of the extreme Right who promised constitutional changes that would permit firm government action within the context of the existing social order. At the same time, the advances of the Left in the 1932 elections caused others to wonder whether the existing parliamentary system could always be relied on to work in the interests of social conservatism. Both these factors were undoubtedly important in the creation of new movements of the extreme Right in the early 1930s, and the numerical growth of others.

The revolutionary Right was deeply divided between nostalgia for the imagined virtues of earlier regimes, and a desire to harness modern technology and methods of state control to new expressions of national sentiment. Charles Maurras's neo-royalist Action Française had some good publicists and was more articulate than many of its numerically stronger rivals. But the avid readership of its newspaper was more a reflection of public taste for its muck-raking revelations about Republican politicians than serious public interest in its highly cerebral, archaic concepts of what a restored monarchy might do for France—so archaic that the royalist pretender refused to subscribe to them. Indeed, the pretender's son formally disowned the movement in December 1937, following the increasing resort of its younger members to violence, exemplified by the physical attack on Léon Blum by Camelots du Roi in February 1936. By 1939, the circulation of *L'Action Française* was a very modest 45,000. Nor was there much of a following for the neo-Bonapartist Solidarité Française, founded in 1933 and largely sustained by the ample funds of the perfume manufacturer, François Coty. Nostalgia of a related kind was also to be found in Pierre Teittinger's Jeunesses Patriotes, which looked back to Déroulède's Ligue des Patriotes of the Dreyfus era, and claimed in the mid-1930s to have 100,000 supporters in the country as a whole. Even so, like other backward-looking organizations, it was taken much less seriously than the Croix-de-Feu and its successor, the Parti Social Français.

Yet the PSF too, despite its strength, was to suffer from the competition of newcomers. The increasingly circumspect behaviour of its leader, Colonel de La Rocque, in the late 1930s was to cause part of its clientele to gravitate to the new star in the right-wing firmament, Jacques Doriot, whose Parti Populaire Français was France's closest equivalent to a fascist movement. A renegade from the Communists, Doriot's change of skin came too late to play a significant part in the 1936 elections; and although he was to obtain ten million francs from big business in 1937, his membership was never to reach 100,000 in the pre-war period. He had to await the Occupation before achieving major significance, and even then his power was largely illusory (p. 105).

In 1934, however, it was still the discordant voices of the Croix-de-Feu and those nostalgic for the dynasties or the ghost of Déroulède that spoke for the extreme Right. Episodes such as the Stavisky scandal could unite them in joint demonstrations of hostility to the regime, as on 6 February 1934, when the alleged connections of government members with the shady financier, Alexandre Stavisky, offered a useful rallying cry. Stavisky had taken refuge with his mistress in a chalet near Chamonix. When the police broke in, they found him dead—or so they claimed. Stavisky probably had committed suicide; but Maurras insinuated that it was the police who had murdered him—on the orders of the government, which, it was said, feared that if Stavisky were brought to book, his trial might reveal that Radical politicians were involved in his slippery financial enterprises. Filled with accusations of this sort, *L'Action Française*'s circulation briefly touched 190,000.

On 6 February, various extreme right-wing groups staged demonstrations in the Place de la Concorde and elsewhere, ostensibly to protest against the alleged corruption of the government. Railings were torn out as pikes, buses were set on fire, and an increasing number of shots were heard. When the crowd repeatedly surged towards the Chamber of Deputies, the forces of order eventually decided that baton-charges were not enough. Drastic counsels prevailed and they opened fire, raising the death-toll to fifteen and the serious injuries to several hundred.

There is little evidence to suggest that this was a broad-based plot against the regime; and certainly the situation by no means required a new broad-based cabinet to deal with it, though this was in fact what happened. Édouard Daladier's Radical ministry feebly resigned in a panic, while Gaston Doumergue was rushed into office like a fire-engine, with a ministry of all the talents. Admittedly, in a country that was fortunate enough to have been spared a lot of physical violence in the inter-war years, it was perhaps understandable that the episode should become something of a legend, both then and subsequently.

The later ramifications of the Stavisky affair provided the Press with much sensational news. One of the investigating magistrates was subsequently found tied to the railway line near Dijon, his body having been cut in three by the Paris–Dijon express. The popular speculation surrounding the incident was further confused when it was known that the son of Rasputin's secretary had been travelling to Dijon on the same train—but on his honeymoon, as it turned out. *Paris-Soir* led the general excitement by announcing that it had commissioned three well-known English detectives to investigate the mystery on the paper's behalf, one of them being 'Sir Thompson [*sic*], the man of iron'. With its characteristic sense of occasion, *Le Canard Enchaîné* riposted with the news that it had commissioned the great Chinese detective, 'Ki-san-fou, the man of straw'. Another investigating magistrate in the case was assailed by a colleague who had apparently taken leave of his senses. *Le Temps* reported him as:

breaking the windows . . . tried to strip himself naked . . . Then taking a number of banknotes from his pockets, he proceeded to tear them into small fragments . . . continued to shout 'Vive la France! Debout les morts! Take off your hats! For here is God!' Then he proceeded to sing the *Marseillaise*. In the end he was tied to a stretcher and taken off to an asylum.

As in other countries, any right-wing activity was bound to be associated in the public mind with the spread of fascism in Europe. And memories of earlier threats in the last years of the nineteenth century caused Daladier's followers among the Radicals to look once more to the parties of the Left—a tendency encouraged by the increasing restlessness that some of them felt with the conservative, deflationary policies of the ministries of the mid-1930s, with which the right wing of the Radicals was associated. The decree laws of the Laval government (7 June 1935–22 January 1936) were especially unpopular. By July 1935, the Radicals, Socialists, and Communists were holding joint rallies; and in March 1936 the Communist CGTU re-entered the womb of its mother, the CGT. The reward of co-operation came in the April–May 1936 elections, when between them the three parties of the Popular Front won an overall majority in the lower house. Out of a total Chamber membership of 618, including the 20 overseas deputies, the Communists spectacularly increased their share from 10 to 72, while the Socialist SFIO rose from 97 to 146. Their allies in the smaller left-wing groups, however, did less well, losing some 20 seats, while the Radicals dropped from 159 to 116. The Radicals had entered the contest in disarray, with a quarter of their successful candidates dissenting from the party's support for the Popular Front. The general pattern of confrontation—as it emerged from the elections in Metropolitan France—is indicated in Table 3.1, together with the distribution of votes between the various groups.

The victory was first and foremost a tribute to the election tactics of the left-wing parties who, on the second ballot, advised their voters to rally behind whichever left-wing candidate looked most capable of winning. It was this that transformed an uncertain prospect on the first ballot—where the Left polled 300,000 fewer votes than their opponents—to a resounding success on the second; but it was a success in terms of seats rather than votes. In so far as 1936

Table 3.1.

Chamber of Deputies Elections of 26 April and 3 May 1936 (Results for Metropolitan France only)

	% of vote on first ballot	Seats gained after both ballots (out of 598 seats[a])	
		Total	%
Communists	15.2	72	12.0
Socialists and allied deputies	20.8	153	25.6
Other left-wing groups		39	6.5
Radicals initially supporting the Popular Front[b]	*c.*12.2	*c.*81	*c.*13.5
Radicals opposing the Popular Front	*c.*7.9	*c.*25	*c.*4.2
Centre and right-wing groups	43.7	228	38.1
Various	0.2	0	0

Notes: The percentage of electorate casting valid votes was 82.1.

The electoral system, *scrutin d'arrondissement*, is discussed on p. 401 n. 5, and on pp. 138 and 143.

[a] There were additionally 20 overseas deputies.

[b] The dividing line within the Radical party, between supporters and opponents of the Popular Front, is hard to establish with any accuracy.

saw a significant swing in the electorate, it was inside the ranks of the Left. A substantial number of left-wing Socialist voters who had hitherto been alienated by the intransigence of the PCF were won over by the new co-operative image of Communism. Similarly, the Socialists gained votes from the Radicals—leaving the Radicals with fewer votes and seats than they had had before the election. Fundamentally, however, it was the behaviour and strategies of the parties themselves rather than the electorate that brought the Popular Front to power—especially the adhesion of the bulk of the Radical party. All of which meant that the victory of the Front was more fragile than it seemed—for what the Radical leadership had given, it could equally take away.

THE BLUM GOVERNMENT

Léon Blum's coalition government of Socialists and Radicals (4 June 1936–21 June 1937) was remarkable not only for being France's first Socialist-led government, but also for including three women, albeit in junior positions. But Blum's chances of achieving a coherent and lasting body of reforms were overshadowed by two major obstacles. The first was that each of the constituent parts of the Popular Front subscribed to it for different and partly conflicting reasons. And the second was that Blum was attempting a costly programme of reform against the background of economic depression and growing international tension. Not only was income low, but much of it would have to be spent on defence.

The divisive interests of the Popular Front parties were reflected in the fact that only the Socialists were committed to substantial social reform as an immediate duty of the incoming government; and even they were inhibited by their 20 per cent share of the overall vote—scarcely a mandate for massive change, even when supplemented by the PCF's 15 per cent. Indeed, Blum was insistent that the government should not go beyond its mandate. The Radicals were opposed to significant change and were apprehensive about the cost to the taxpayer of any but the most modest of reforms. Their prime concern remained the defence of Republican institutions against the supposed menace of the right-wing leagues. The Socialists shared their apprehension, but were less committed to the existing shape of Republican institutions. Indeed, as the next two years were to demonstrate all too forcibly, the achievement of the Socialists' reform programme was to founder on the rock of these institutions, more particularly upon the Senate, whose method of election made it a bastion of socio-economic conservatism. Yet Blum did not dare to commit his government to a restriction of the obstructive powers of the Senate—as many Socialist candidates had demanded in their electoral manifestos—or to adopt extensive decree powers that would enable him to bypass this obstruction. This would run the risk of inflaming the Radicals' protective instincts and splitting the Front.

The Communists, for their part, saw rearmament against Hitler as the first concern of a left-wing government, even if it meant diverting funds that the Socialists wished to spend on social reform. Indeed, the Communists were positively discouraging when it came to discussing a programme of social legislation, in case it alarmed the Radicals. At the same time, it is an indication of where Stalin's short-term priorities lay that in the autumn of 1936 he was reputedly investigating whether it would be possible to restore to power the former conservative Premier, Pierre Laval, in place of the diplomatically untried Léon Blum. It was also consistent with the Communists' position that they should refuse to participate in the Blum cabinet. Conscious of the difficulties of achieving significant social improvements in the current depressed state of the French economy, they were nevertheless anxious to preserve their image with the working class as the party of social transformation. It was important not to compromise this impression by participating in a cabinet which they assumed could do little. Non-participation would leave them free to criticize the government if it suited them to do so, and would also allow them greater room for manœuvre in putting pressure on Blum. Characteristically, they were later to claim that the credit for the social reforms of Blum's early months should go, not to the government, but to the working class and their Communist leaders who, by organizing strikes, had pressurized the employers into making concessions. The fact that most of the strikers did not belong to a union and that the Communists had been embarrassed by the strikes was conveniently ignored.

Like anticlericalism at the turn of the century, opposition to the right-wing leagues was the only cement that held the disparate forces of the Left together; and like the clerical issue of the *belle époque* it was arguably the least important of the major issues facing France at the time. All of which disposed Blum to proceed with caution. In any case, his own conception of what a Socialist government should do, even when unencumbered with lukewarm allies, was the product of slow and hesitant evolution. A brilliant Jewish intellectual, whose political commitment went back to the *jours héroïques* of the Dreyfus Affair, he had been a late comer to professional politics, always retaining something of his dandified youth as a successful literary critic and bright young lawyer of the Conseil d'État. Indeed, the confidential report on his background which accompanied his candidature for the Conseil d'État in 1897 remarked: 'il paraît que Monsieur Blum ne fait pas de politique'. In André Gide's opinion, he had 'trop d'intelligence et pas assez de personnalité' to be a good political leader. His integrity, however, was never in doubt—his honesty often causing his colleagues embarrassment. Long after he had embraced Socialism, with its Jaurèsian tradition of vague, humanistic Marxism, he was to describe Marx as a 'mediocre metaphysician', whose economic doctrines were constantly being outdated. Like many of his fellow Socialists, he was initially uncertain as to whether the nationalization of the economy should await the final apocalyptic victory of socialism, when workers' control would bring it about organically, or whether Socialists should embark upon it piecemeal, if and when circumstances permitted. And it was largely British and Austrian persuasion in the 1920s that inclined him towards the latter. (He believed, however, that interim nationalization of this sort would put a moral obligation on the State to compensate the former owners.) As far as the immediate tasks of government were concerned, faced with a depressed economy, he had been impressed by aspects of Roosevelt's achievements in America. Indeed, one of his critics, Joseph Caillaux, described Blum's plans as 'Rooseveltisme pour Lilliputiens'. It was symptomatic of Blum's beliefs and personality that he saw education as the true revolutionary force in society, and that he should later count among his government's foremost achievements the raising of the school-leaving age from thirteen to fourteen (2 July 1936). He also set great store by Jean Zay's attempts to reduce class size to thirty-five; and he strongly backed the abortive scheme for improving the standard of primary schoolteachers by insisting that they have the *baccalauréat* before entering training colleges.

Blum's first week in office struck the public imagination with the breadth of its achievement; and the intrinsic drama was intensified by the widespread strikes that had broken out shortly before Blum took power (p. 55). Everything that came thereafter was something of an anticlimax. The Matignon agreements, which had been wrested from the employers on 7 June, established the principle of compulsory collective bargaining and sanctioned an average pay-rise of 12.5 per cent. A few days later, legislation gave workers a fortnight's holiday with pay and reduced the official working week from forty-eight to forty hours without loss of income.

Already, however, the Popular Front programme contained the seeds of its own ruin. The wage increases varied greatly from job to job, as did the rises that followed in the next two years. Employees in department stores and other notoriously underpaid sectors saw their wages double, and by May 1938 the average national wage had increased by 47.5 per cent. Much of this was long overdue. The tragedy was that its benefits were largely undermined by the forty-hour week, which gradually came into effect, industry by industry, between November 1936 and April 1937. Even without the forty-hour week, some inflation was unavoidable, given the effect of wage increases on production costs. But it was hoped that a vigorous programme of increased production and better sales would help industry to absorb some of these costs, without passing them all on to the consumer. But what chances there were of this were destroyed by the shorter working week, which reduced productivity per man by 17 per cent and consequently put up production costs by notionally the same amount, most of which would have to be borne by the customer. The net effect of these miscalculations was that the cost of living rose 46 per cent between May 1936 and May 1938, reducing most people to the level they had been at before Blum came to power—or, indeed, lower, since civil servants were 15 per cent worse off (their pay-rises having been smaller), while pensioners were 20 per cent short. Women were also losers; their pay increases remained substantially below those of men—while for both sexes wage increases in the provinces failed to match those given in Paris, thereby increasing the traditional inequalities in France.

Very few comparable nations attempted to introduce a forty-hour week before the 1960s. Even an advanced Social Democratic country like Sweden was still working a forty-five-hour week in 1967. Although the French Socialists had included the forty-hour week in their 1932 manifesto, the continuation of the Depression had given them second thoughts, and the 1936 electoral programme merely spoke of a 'reduction in the working week', without specifying the length. That it was enacted so quickly was the result of massive pressure from below. Working-class euphoria at the electoral victory of the Popular Front had spectacularly shown itself in a series of strikes and occupations that hit industry in late May and June 1936. Beginning in the motor, aircraft, and engineering industries, they had spread in early June to the smaller enterprises and to other parts of the economy, notably the large department stores, until there was an overall involvement of nearly two million people. The basic objective was to win specific piecemeal reforms from the employers while they were still reeling from the shock of the election results. The strikers were also uneasy about whether the new government would take a sufficiently tough stance in its negotiations with management, and rightly felt that the employers were more likely to agree to a quick settlement if production were stopped.

Two-thirds of the factories and firms hit by these strikes were occupied by workers' sit-ins, but they were in no sense a workers' take-over of the type that Catalonia was to witness in the following months, when the outbreak of the Spanish Civil War triggered off the creation of workers' collectives in Barcelona and elsewhere. The purpose of the occupations was to stop production, not to take it over, the strikers accurately calculating that blacklegging would be much more difficult if the strikers were actually inside the buildings. Only in those trades where there was a danger of equipment depreciating through suspended use, or of raw materials perishing on the premises, was work continued—the prime reason being to prevent recriminations by the owners, not to demonstrate the superfluity of management.

The initial festive ambiance was well conveyed by a striker at the occupied Renault works at Billancourt:

... we arranged dances in several parts of the works. There were among us a cornet player and a violinist, and another chap had brought a concertina ... Couples of men and women, and couples of men (for there weren't enough girls to go round) danced among the vans and piles of scrap iron ... and although we had received strict orders not to damage any property, a few drunks who had got into the cars for the night were sick all over them.[2]

The bulk of the participants were non-union labour, and the union leaders themselves were mainly taken by surprise, as was the PCF and other political militants. Moreover, the highly unionized sections of the work-force, such as the railways and public services, did not participate. Significantly, the PCF leader, Maurice Thorez, was anxious to put an end to them, since they threatened to jeopardize rearmament and ran the risk of alienating the Radicals.

The forty-hour week, higher wages, holidays with pay, obligatory collective contracts, and labour's right to have the sole voice in choosing workers' delegates were prominent in the strikers' demands—and the first would not have become a reality at all were it not for the strikers' insistence. Blum had serious misgivings about it, but assumed that its depressive effect on production would initially be offset by the fact that large numbers of workers were already on short-time, while elsewhere it would create more opportunities for the unemployed. In fact, over two-thirds of the workers in those factories for which there were statistics were working a full forty-eight-hour week; and the impact of the forty-hour week on unemployment was minimal—perhaps only a 3 per cent reduction instead of the anticipated 20 per cent—since the bulk of the unemployed were unskilled and unsuited to the thriving industries that needed labour. Indeed, France suffered from a shortage of skilled labour throughout the inter-war period. Poorly equipped with labour-saving machinery, her factories needed a greater number of skilled men than those of her neighbours, yet her war losses and low birth-rate made this particularly hard to provide— added to which the Depression had curtailed the number of apprenticeships. The result was that the shorter working week did not enable the anticipated expansion of the shift system that would have kept the plant fully utilized, as well as providing employment for more men.

Shorter hours and more pay inevitably made French goods less competitive abroad. Although Blum personally favoured devaluation, he had been hesitant about doing it before consulting foreign governments, and he was also aware of the risk of alienating French investors. Even the impeccably orthodox Raymond Poincaré had been dubbed the butcher of small savers when he had devalued the franc as part of his stabilization plan in 1928. In a country where thrift was so esteemed a virtue, it was even imagined that the delayed impact of the Depression on France had something to do with the nation's sound financial habits. It was widely held in such circles that money was much more likely to be used productively if it remained in private hands, whereas if it passed to the government it would be lost in sterile expenditure. Needless to say, the idea of deficit budgeting and spending one's way out of recession was absolute anathema. Given these attitudes and the government's need to borrow, Blum's Minister of Finance, Vincent Auriol, assured the purchasers of government bonds that he was not contemplating devaluation—a pledge that later had to be redeemed with a compensating bonus.

When the inevitable came, the devaluation by an 'elastic' 25 to 35 per cent (26 September 1936) was quickly deprived of its competitive edge by the rise in French production costs following the concessions on hours and wages. In retrospect, the logical course would have been to devalue immediately on taking office, and delay the social concessions until the trade benefits of devaluation began to be reaped, with forty-five hours as the minimum working week. Increased turnover might then have enabled employers to absorb more of their costs, without raising prices so substantially. Even this, however, is to make optimistic assumptions about the competitive capacity of France's antiquated plant. But in putting the cart before the horse, Blum had been trying both to placate the investors and to get the strikers back to work.

The effect of the forty-hour week might have been less crippling if the unions had been prepared to allow its periodic suspension in key sectors of the economy. But it was understandable that they should be jealously possessive of what they had wrested with such effort from an unwilling management and government. Their suspicion was also aggravated by feet-dragging by employers in the implementation of the new concessions; and a quarter of a million workers were to come out on strike in the later part of 1936. Moreover, the failure of management and labour to achieve agreement on a system of compulsory conciliation and arbitration ultimately forced Blum and then

Chautemps to intervene and impose a system with the laws of 31 December 1936 and 4 March 1938.

However disastrous the forty-hour week, the fortnight's holiday with pay was a great psychological benefit with a minimum of economic loss. Representing a modest 3.8 per cent cut in production, its value to the worker was enhanced by a 40 per cent reduction on holiday rail-tickets—or 60 per cent on the special 'Lagrange trains' commemorating the Under-Secretary for Sport and Leisure, who readily admitted to having modelled the scheme on Hitler's Strength through Joy programme. Half a million such tickets were sold in the summer of 1936—60 per cent of the visitors interviewed on the Riviera admitting that they had never seen the sea in their lives before. Predictably, the invasion was viewed with mixed feelings by the better-off, with some of the smarter shops in Nice bearing the curt notice, 'Interdit aux congés payés'. Most families, however, could only afford to spend the fortnight with rural relatives or at home, making periodic excursions by bus or bike into the surrounding countryside.

Among its other durable achievements, the Blum government inaugurated a modest move in the direction of state control which was to receive further development at the Liberation and under Mauroy in the 1980s. Under the *ancien régime*, as in a number of other Continental countries, royal manufactures had played a significant role in developing quality goods to improve the balance of trade. More recently, the State had taken substantial minority holdings in a number of companies in the inter-war period, notably in chemicals, petrol, and air transport—generally in response to a particular problem that the firm was facing. The SFIO's leanings towards nationalization were recent and restrained (p. 54). The armaments industry, however, was a widely favoured target for take-over. The concept of state arsenals had respectable roots in a number of pre-Revolutionary monarchies, while the First World War had created considerable feeling against 'the merchants of death'. A law of 11 August 1936 empowered the State to set about their piecemeal take-over— the Brand works and the tank section of Renault being among the early acquisitions. The nationalization of the aircraft industry took the form of the creation of national companies in which the State was to hold 90 per cent or more of the shares. It would be hard to measure the effect of nationalization on these industries. The menace of Germany and the government's increase in armaments expenditure would have increased their productivity even if they had remained in private hands. Yet the fact remains that their work-force trebled and their capacity grew five times in the period 1936–9; they were also subject to fewer strikes than the private sector.

Like state arsenals, the establishment of government control over the railways was not a particular novelty in Europe. Blum was preparing a decree of expropriation just before his resignation; and it was his successor, Chautemps, who negotiated a compromise with the companies, allocating them 49 per cent of the stock in the new Société Nationale des Chemins de Fer, and leaving the State holding the other 51 per cent. State control did not cure the deficits—it merely pledged the taxpayer to picking up the bill. Yet nationalization and state support were to lay the administrative groundwork for the post-war renaissance of the French railways—even if the essential impetus came from the need to replace what had been destroyed in the war.

Back in the autumn of 1934, Blum had wanted to include the nationalization of the banks in the Popular Front programme, but opposition from the Radicals and Communists had precluded it. It was widely accepted, however, that the Banque de France needed an internal redistribution of control and influence. The changes of 24 July 1936 were largely intended to reassure the public that the government was prepared to challenge the semi-mystical beast of the 'deux cents familles', a figment of the popular imagination that ascribed vast occult power to the two hundred principal shareholders of the bank who appointed its regents and auditors. It is true that the bank had been uncooperative in the past with governments that strayed from the paths of traditional financial orthodoxy, and it was also true that the personal backing of these shareholders could be very useful to business men or politicians in need of support. But the same could be said of the leading banks in many countries, and the widening of the electoral base of the Banque de France to include all its shareholders did not substantially alter this; nor did the assumption by the government of a larger say in the appointment of its new General Council. Nevertheless, it was a symbolic victory that gave great satisfaction to many on the Left.

There was little room for satisfaction, however, in the overall economic record of the Popular Front. Industrial production fell by 4 to 5 per cent between May 1936 and May 1938—whereas it had grown by 8 per cent in the twelve months before the Front came to power. German production in 1936–8 grew by 17 per cent. On the other hand, the French trade deficit decreased during these years, thanks partly to Chautemps's further devaluation on 30 June 1937, which allowed the franc to float and helped to attenuate the deleterious effects of the domestic rise in the cost of production. Unemployment was also down—from 431,000 to 375,000, according to over-benign official figures—while short-time working had also been reduced. Cynics replied that all that had happened was that the forty-hour week had institutionalized short-time working, and that the record of neighbouring countries on reducing unemployment was far better because they had put economic recovery at the head of their priorities.

Some of the blame for the country's poor economic performance was levelled at the large-scale export of capital by private investors anxious to avoid the possibility of government taxation or semi-forced loans. While this export undoubtedly reduced the domestic funds available for industrial investment, shortage of skilled manpower, exacerbated by the forty-hour week, was the more immediate problem. The Popular Front—and with it the export of capital—did not last long enough to have had a major damaging effect on the much-needed renewal of French equipment.

A disappointment to the government was that higher wages had not created the increased demand for French manufactures that had been anticipated. Apart from the eroding effect of inflation on the workers' purchasing power, much of this short-lived improvement was spent on food rather than on clothes and household appliances—thereby intensifying the paradoxical truth that the section of society that gained most from the Popular Front was the farmers, a group that had particularly feared its advent. Agricultural prices rose between 40 and 50 per cent under Blum, while the Wheat Board, which farmers feared had been instituted to protect the consumer (15 August 1936), doubled the price of wheat and was a solid bastion of the farmer's income. Its intention was to stabilize prices in everyone's interests, the farmer's especially, but few would have predicted that by the summer of 1939 wheat would cost three times as much as it did in Britain. Moreover, peasants in debt were granted delays for repayment and offered short-term loans to tide them over their immediate difficulties—while their creditors were likewise eligible for brief loans to keep them temporarily mollified (12 July 1937).

A conversation in 1938 between Blum and his Secrétaire d'État aux Finances, Pierre Mendès France, reveals much about both men: ' "Mon petit Mendès, un gouvernement ne tombe pas sur des problèmes financiers." "Monsieur Le Président, depuis quinze ans, tous les gouvernements sont tombés de la sorte." "Mais non, mais non!" '[3]

The specific factor that brought down Blum's government in June 1937 and his later government in April 1938, was the Senate's refusal to authorize him to tackle the country's financial problems by decree. The Senate had initially been circumspect in its opposition to Blum, restricting its obstructive powers to making minor amendments to the government's legislation. What may have emboldened it to challenge Blum directly—and to challenge the majority in the lower house which had consented to the decree powers that Blum sought—was the limited success of the government's programme and the increasing tension between the Socialists and their Popular Front partners. The government's dissolution of the right-wing leagues (18 June 1936) had partly exorcized the spectre that had ensured the Radicals' commitment to the Popular Front, leaving their minds free to range over what they disliked in the government's socio-economic programme. The Communists, for their part, became increasingly critical of the government's decision not to help the Spanish Republican forces (p. 69). As for the rank and file of the Socialist militants, enthusiasm for the government was also dampened by Blum's announcement of a 'pause' in the reform programme (13 February 1937) as a result of the economic and financial problems that had beset it. Prominent among the casualties of this moratorium on further change were Blum's intentions of instituting a long-overdue National Unemployment Fund, to put France on a par with other advanced countries, and his scheme for more generous pensions. Equally humiliating for a government committed to advancing the workers' cause, was the one-day general strike organized by the CGT (18 March 1937), following the police suppression of a left-wing demonstration against Colonel de La Rocque's newly formed Parti Social Français. The police had opened fire on the crowd, killing seven and wounding several hundred (16 March 1937).

When Blum asked the Senate for decree powers in financial matters, he specifically excluded devaluation and exchange controls from their orbit, explaining that his prime intention was to introduce new taxation and force French investors to repatriate some of the capital that they had invested abroad. The Senate's refusal resulted in a new coalition government headed by the Radical, Camille Chautemps (22 June 1937–10 March 1938), and with Léon Blum as Vice-Premier, thereby preserving the outward appearance of Popular Front continuity, but leaving no one in any doubt that the reform programme was now in cold storage.

When Blum briefly became Premier again the following spring (13 March–8 April 1938), he wanted a government of national unity, with Maurice Thorez of the PCF as its left wing and Louis Marin as its right, but the conservatives were not disposed to be won over. Blum came up against the same problem, namely the Senate's reluctance to grant him decree powers. This time the powers he requested did not specifically exclude exchange controls and devaluation; and the Senate predictably refused. The Socialists dropped out of government, leaving the Radicals to form the new ministry under the Vice-Premier, Édouard Daladier (10 April 1938–20 March 1940). The Popular Front was over.

AFTERMATH

If the Popular Front was over, it would be a distortion of events to say that the demolition men now moved in—as is often alleged. Encouraged by the 'moderate' composition of Daladier's government—it contained only two Socialists who were shortly to resign—the Senate gave him the financial powers that it had specifically withheld from Blum. Nor did Daladier hesitate to use them, promptly devaluing the franc by 10 per cent in May 1938, with salutary consequences for French exports. He also started to grant industry exemptions from the sacrosanct forty-hour week, beginning with the armaments factories, and then extending them to other concerns after Paul Reynaud became Finance Minister on 1 November 1938. While the CGT was prepared to be flexible about working hours in the armaments industry, it protested against the erosion of the workers' gains elsewhere, and there was a split in union opinion between those who advocated resistance and those who urged national solidarity in the face of the worsening international situation.

The demise of the forty-hour week was undoubtedly beneficial to the French economy; and the rest of the Popular Front's statutory achievements remained largely intact, even if its wage increases had been eroded by the consequential rise in the cost of living following the institution of the forty-hour week. Certainly, the sum total of the Front's achievements was modest by comparison with the millennial expectations of many of its followers, but the error lay in expecting too much in a time of economic depression and international tension. Perhaps the Front's greatest achievement was to have come into being at all. If its record created disappointment, its existence at least demonstrated the possibility of socialist government in a largely conservative society—an encouraging thought when time was on the side of a growing industrial population.

4

The Road to Compiègne
Diplomacy and War, 1936–1940

DEFEAT at the hands of Germany was no new experience for France in 1940. The German six-week sweep to victory of May–June had its parallel in the Prussian campaign of July–September 1870. It nearly happened again in May–June 1918, after Russia's withdrawal from the war, and was only averted by the belated arrival of the Americans in force. In 1940, however, France had to wait another four years before the Americans came again.

In a celebrated broadcast to the nation just before the armistice, Marshal Pétain explained the French defeat in these terms:

Weaker than we were twenty-two years ago, we also had fewer friends. We did not have enough children, we did not have enough arms, and not enough allies—these are the reasons for our defeat ... Since our past victory, the spirit of pleasure has been stronger in us than the spirit of sacrifice. We asked for much, and we served little. We wanted to spare ourselves great efforts, and now we are face to face with adversity.

In private he was more explicit. During the German breakthrough, he angrily told Paul Baudouin: 'I cannot allow the errors of the politicians to be blamed on the army. The real culprit is Daladier. It was he who created the Front Populaire.'

While Pétain blamed left-wing governments for deflecting popular concern and national resources from military defence to hedonistic schemes for social reform, de Gaulle, in an equally celebrated broadcast on British radio on 18 June, put the blame squarely on the shoulders of the military leaders, who, unlike their German counterparts, had failed to understand the requirements of modern war.

CAUSES

If historians prefer de Gaulle's analysis to Pétain's, the issue is nevertheless complex, and needs to be looked at in stages. First, there was the failure of Britain and France to take a firm line with Hitler in the mid-1930s, when Germany was not yet in a position to resist Allied threats of armed restraint. Secondly, following the bankruptcy of appeasement, there came the failure of the Allies to take the initiative during the period of the 'Phoney War', when

Hitler was still enmeshed in the Polish campaign. And finally, when Hitler invaded France, there was the miserable débâcle of the Allies' ill-conceived and mishandled counter-strategy. Nor must the nerve, skill, and strength of Nazi Germany be overlooked amid the *mea culpas* of Anglo-French self-analysis; these would have taxed the determination and ability of the most resolute opponent.

Underlying these issues, however, was the fundamental fact that French defence plans against Germany were based on a contradiction between diplomatic and military thinking. Like most of the country, the Quai d'Orsay and the War Ministry were both haunted by memories of the First World War. They felt that the post-war years had shown that they could expect little in the way of firm support from America, Britain, or the League of Nations in the event of Germany reneging on her obligations or resuming an aggressive role. In the 1920s, the Quai d'Orsay had desperately looked for a substitute for the old Franco-Russian alliance, which for a time had kept German ambitions in check by threatening Berlin with the prospect of enemies on two fronts. Russia under the Soviets, however, was now a virtual outlaw, absorbed in her own domestic problems; and the best France could do was to construct a set of alliances with some of Germany's eastern neighbours—Poland, Czechoslovakia, Romania, and Yugoslavia. These treaties logically implied a mutual readiness to take the offensive, if need be, to protect the other members. Yet the basic principle of French military strategy and thinking became increasingly defensive.

The French general staff had learnt the lesson of the early years of the First World War, that the machine-gun, barbed wire, and other modern weapons gave enormous advantage to the defence. But they had learnt it too well, and failed to take sufficient account of the less obvious lessons of the later stages of the war. Foreign writers, notably in Britain and Germany, saw how the intelligent use of tanks and aircraft could restore the war of movement that many French generals believed lay for ever buried in the mud of Flanders. When the arch-apostle of French national security, Raymond Poincaré, devoted a fifth of the 1929 budget to defence, his Minister for War, Paul Painlevé, channelled government thinking into a massive project for fortifying the frontier with Germany. The Maginot Line, authorized by the law of 14 January 1930, reflected the conviction of Marshal Pétain and other senior survivors of the trench warfare of 1914–18 that a repetition of such an ordeal would only be bearable if the rain-filled, shell-strewn lines of the Great War were replaced by an impregnable system of concrete tunnels and turrets, where troops could be positioned and moved around in relative comfort and security. This was a concept that Pétain had vigorously advocated when he was Vice-President of the Conseil Supérieur de la Guerre (1920–31).

The Maginot Line did not of itself preclude the adoption of offensive strategies in time of war, since it could act as a springboard as well as a line of defence. But the financial stringency of the Depression years was not conducive to a vigorous parallel development of offensive and defensive weaponry. Moreover, for a country that was short of manpower and remained traumatized by recent wartime experience, a wall of guns and concrete seemed the more attractive priority. The role of the offensive would come after the enemy had battered themselves into exhaustion against the Maginot Line, at which point the French armies, like gun dogs, would sally forth and round up the crippled remnants. Clearly, such a strategy offered little comfort to France's eastern allies who, in the event of war, would be in peril of temporary occupation until such time as the French were ready to emerge from the safety of their citadel and bring them back to life.

It was the problem of Belgium, however, that would give the instruments of offence some role in the early stages of hostility in any future war with Germany. A secret military convention of September 1920 had given France permission to deploy troops in Belgium in the event of a German war of aggression. This presumed a strong, mobile force that could move rapidly into Belgium once war was declared. The Maginot Line of the 1930s was not initially intended to run along the Belgian border to the sea, since this would be embarrassing to Franco-Belgian relations: it could be misinterpreted as distrust of Belgian friendship or capability in a Franco-German war—or it might equally appear as a cynical abandonment of Belgium outside the lifebelt of French defences. Anyway, it was assumed that the thickly wooded hills of the Ardennes, masking the Luxemburg and eastern Belgian frontiers, would make a German attack unlikely in this particular sector; this was Pétain's firm conviction, clearly reiterated in 1934. In any case, the Weimar Republic had shown little inclination for foreign adventures; and neither the French nor British governments sufficiently recognized the changes in Berlin since Hitler's advent to power.

HITLER'S FOREIGN POLICY AND ANGLO-FRENCH RESPONSES

Diplomatically, the 1930s are thought of as the age of wishful thinking. Perhaps the greatest irony was that statesmen who were anxious to believe Hitler's protestations of good intentions were unwilling to believe him when his openly avowed intentions were bad. And events were to show Hitler as bad as his word: he did what he said he would do. The sheer enormity of his intentions, however, disposed peace-seeking people to hope that he was merely playing to the gallery of Germans who still smouldered with the humiliation of 1918 and its aftermath. They underestimated his determination and failed to grasp the extent to which his goals mirrored the long-term aims of preceding German governments, notably those of the late Wilhelmine Empire. From his public and private utterances in the 1920s and 1930s, it was evident that he had the same basic aims in foreign policy. He wanted to expand Germany's frontiers to include all those Germans who lay outside them—with dangerous implications for Austria, the Sudetenland, Danzig, and the German-speaking parts of Poland. But he also emphasized the need for strategic frontiers, potentially putting at risk areas that were not necessarily German. Lastly, and most ominously, he maintained that Germany should acquire new food-producing areas—an ambition which stretched as far east as the Ukraine. Closely allied with this aim was a desire to curb the growing strength of Russia—a concern that had greatly preoccupied the Wilhelmine government in its last years.

Hitler's foreign policy in the 1930s was a series of gambles, in which he hoped each time to hoodwink, rush, or openly intimidate the other powers into acquiescence—and until August 1939 his hopes seemed largely justified. He was fully aware, however, that his luck might not hold, and that Britain and France might intervene militarily against him. Moreover, there is strong evidence to suggest that by the end of 1936 he regarded rearmament as enabling Germany to take the initiative against France and Britain should they become obstructive—and that it was principally Allied appeasement that increasingly made him confident that this initiative would not be necessary.

Hitler's early gambles, such as the institution of conscription (16 March 1935) and the remilitarization of the German Rhineland (7 March 1936), which violated the Locarno Pact (1925), were passively accepted by France and Britain as regrettable but understandable assertions of German national dignity. Although these moves challenged both the Treaty of Versailles and the Locarno Pact, many observers argued that it was unreasonable to keep Germany deprived of the normal attributes of nationhood some eighteen years after the war was over. Britain accordingly decided that the Rhineland was not a 'vital British interest', and, unsure of an ally, Sarraut's Radical government was reluctant to commit France to what might be an expensive and unpopular campaign—an unpalatable prospect in a time of economic depression and with an election coming up. At the same time, Britain and France had overestimated the state of German rearmament—Britain having swallowed Hitler's claim in March 1935 that the Luftwaffe was already as big as the RAF. Yet as Hitler subsequently remarked: 'If the French had then marched into the Rhineland we would have had to withdraw with our tails between our legs, for the military resources at our disposal would have been wholly inadequate for even a moderate resistance.'

As Table 4.1 shows, French expenditure on rearmament had fallen in the early 1930s as a result of the financial problems of the Depression on the one hand, and of hopes put in disarmament talks and the Maginot Line on the other. However, in 1933–4, Daladier and his enterprising Air Minister, Pierre Cot, had drawn up plans for the expansion of land and air forces which Gaston Doumergue's government put into effect, along with naval increases, with the law of 6 July 1934. Daladier's successor as Minister for War, the elderly Marshal Pétain, showed little interest in the expansion, and was content to adopt Daladier's plan as it stood. Even less enthusiastic was Pierre Laval, who as Premier (7 June 1935–22 January 1936), proposed substantial reductions in military expenditure for 1936 in the interests of financial stability. As Minister for Foreign Affairs in the preceding cabinet, he had gone through the motions of establishing a nebulous alliance with Russia (2 May 1935), but his principal aim was to come to an eventual understanding with Hitler and Mussolini; the Russian alliance for him was little more than a short-term expedient. Indeed, contrary to Pétain's later assertion, it was precisely Daladier and the Popular Front who were most active in expanding the armed forces against the risk of German aggression.

Table 4.1.
French Defence Expenditure, 1933–1939
(Millions of Francs, at Constant 1938 Level)

	War	Navy	Air	Total
1933	12,010	4,194	2,554	19,897
1934	10,212	4,539	2,231	18,126
1935	11,180	5,075	4,035	21,507
1936	11,941	5,358	4,090	22,708
1937	13,423	5,247	4,648	24,523
1938	15,227	6,143	6,645	29,153
1939	53,668	9,897	23,904	88,584

Source: R. Frankenstein, *Le Prix du réarmement français, 1935–1939* (Paris, 1982), 304.

Blum brought back Daladier and Pierre Cot to the War and Air Ministries; and, swallowing his pacific instincts in the face of what he saw as inescapable reality, he raised the defence budget for 1937 by 10 per cent, increasing air expenditure by 14 per cent. If these were to seem very modest increases in the light of subsequent events, they were large in the context of the government's social reform programme; and, as already indicated, it was widely held in France that Hitler had as yet done nothing to justify the Popular Front's alarm. In fact, defence expenditure in the Popular Front programme rose faster than any other facet of government spending, Blum's outlay on arms substantially outweighing that on social improvement. It was also symptomatic of where responsibility was to lie for the débâcle of 1940 that a number of military and naval chiefs opposed Pierre Cot's expansion of the air force for fear that it would compete with the other services for funds. Blum's successor, Chautemps, and his Air Minister, Guy La Chambre, also achieved their much greater increases in the defence budget for 1938 (Table 4.1) in the teeth of inter-service rivalry—and of taxpayers' objections that a third of the government's revenue

was now earmarked for armaments. It is true that the strikes and nationalization of 1936–7 created a short-term plateau in aircraft production in those
years—but the subsequent leap in production owed much to the new organizational base that was created under Blum and Chautemps.

If between them they raised the air-force budget by over 62 per cent, events
in May 1940 were to show that a greater proportion of it might usefully have
been spent on dive-bombers (p. 77). Here they were the victims of the defence
chiefs, who, like the British but unlike the Germans, failed to foresee the value
of dive-bombers as army-support weapons. Accepting the verdict of the
uniformed experts, the ministers placed their money on supposedly multi-
purpose, medium-range bombers, which, apart from their supposed utility on
the battlefield, had the attraction of bringing an unambiguously offensive
dimension to what was otherwise a predominantly defensive war machine.

Blum was similarly ill-advised by the professionals on the question of tanks.
Much to his credit, in October 1936 Blum invited Colonel de Gaulle to discuss
with him his controversial views on tanks, which the latter had expressed in an
article in the *Revue politique et parlementaire* (1933) and at greater length in his
book, *Vers l'armée de métier* (1934). Whereas the prevailing view in French
official circles was that the bulk of their tanks should be scattered throughout
the army as infantry-support weapons, de Gaulle favoured the German
concept of massing large numbers together as the spearheads of major
offensive columns. The response to his views among senior military and political personnel was largely one of indifference and scepticism. Paul Reynaud was
the only major political figure to try to give substance to de Gaulle's proposals.
In March 1935, he proposed a bill in parliament instituting six armoured divisions, each of which was to consist of five hundred heavy tanks, but the bill was
lost by the end of the year. In 1934–5, the French army was still spending more
than four times as much on horse fodder as it was on petrol. Blum was personally hostile to de Gaulle's emphasis on a small, highly trained, professional
army—seeing it as a potential praetorian guard and running counter to Jaurès's
Socialist concept of the nation in arms, with its roots in the French Revolutionary tradition. But he recognized the force of de Gaulle's other arguments.
Unfortunately, Blum wrongly assumed that the High Command's professed
interest in tanks was along Gaullist lines; and, as was later to be demonstrated
so disastrously, it was the misuse of tanks, rather than a shortage of them, that
was to prove France's undoing.

Blum's concern for French security had been deepened by Belgium's
decision in 1936 to disentangle herself from the Franco-Belgian scheme of
mutual defence. Disillusioned by Anglo-French passivity over Hitler's remilitarization of the Rhineland, the Belgian government felt it more prudent to
reserve its options on the deployment of French troops in Belgium in the event
of a Franco-German war. It was understandably reluctant to appear
irrevocably committed to France in German eyes, when it was uncertain

whether France and Britain would have the will to defend Belgium if Germany invaded. This change of policy meant that French troops would have to wait for a specific invitation before crossing the frontier, even though Germany might already have initiated hostilities. This obviously reopened the question of whether the Maginot Line should be extended along the Belgian frontier to the sea—an issue that was still basically unresolved in 1939, despite a brief extension of the western end of the Line in the late 1930s. The only decisive response that the French made to the new departure in Belgian policy was to create loosely fortified zones along the Belgian frontier; but they were in no sense a continuation of the Maginot Line, and were not even envisaged as a continuous line of defence.

The remilitarization of the Rhineland saw no further major moves by Hitler until the Anschluss with Austria in March 1938. Indeed, the atmosphere was sufficiently relaxed for Blum's Foreign Minister, Yvon Delbos, to make several hesitant and inconclusive attempts to achieve a *détente* with Germany that would for the time being lessen French anxieties about the security of her eastern frontier. Historians have blamed Delbos for not using the interval to obtain a firmer alliance with Russia or to detach Mussolini from his increasing involvement with Hitler. Like other European governments, however, Delbos and Blum were uncertain of both the trustworthiness and the military capability of these powers; and, like other French cabinets, they believed that dependence on Britain was inescapable. Indeed, it was this dependence that was a factor in deciding Blum to stay out of the Spanish Civil War—although the decisive pressure came from his Radical colleagues, who, like Britain, feared that Spain would become a cockpit where Britain and France would find themselves in premature conflict with Germany and Italy. In his persistent attempts to involve Britain and France in an effective alliance against Hitler, Stalin hoped that the Spanish Civil War would create a situation where Russia, Britain, and France would find themselves supporting the Republican government, while Hitler and Mussolini backed the Nationalist insurgents—thereby inaugurating a confrontation of nations which might rapidly spill over on to the wider European stage. It was just such a scenario that Britain was determined to avoid; and although Léon Blum and many of his Socialist colleagues were anxious to help the Spanish Republic, Britain and the Radical members of the cabinet won the day.

In January 1937, the Italian Ambassador in Paris suggested that Mussolini would welcome closer relations with France—given his personal detestation of Hitler. But the price Italy set on her friendship was freedom to topple the Spanish Republic, on the understanding that Italy would persuade a victorious Franco to behave benignly towards his northern neighbour. Blum would have nothing to do with such a transaction; and in any case, it is open to question whether a Franco-Italian *entente* would have deflected Hitler from his subsequent aims. It is hard to imagine Mussolini coming to the rescue of France at any point in 1939–40. His reluctance to help Hitler before June 1940 demonstrated his extreme caution about committing Italy to war unless the outcome seemed a foregone conclusion. Nor would a French *entente* with Italy have necessarily deterred Hitler from effecting the Anschluss with Austria (12 March 1938). As Hitler predicted, the Anschluss saw no protest from Britain or France. It was gratefully remembered that the Austrian people themselves, through their elected parliament, had requested such a union in 1918, only to be refused by the victorious Allies. And although everyone, including Hitler, knew that a plebiscite on the eve of the Anschluss would probably have gone against incorporation into what was now a totalitarian Germany, the plebiscite which Hitler arranged after the event predictably ratified the *fait accompli*, thereby conveniently justifying once again the inactivity of Britain and France. Even so, it was a bitter accompaniment to Blum's second brief ministry; and his successor, Daladier, had no illusions as to the bleak prospects ahead.

Before entering politics, Daladier had been an outstanding history teacher—whose boyhood *professeur de khâgne* had been that other pillar of the Radical establishment, Édouard Herriot. Daladier embodied the familiar characteristics of so many of the leading figures of the Third Republic: intelligence, realism, and an ultimate shoulder-shrugging acceptance of the limitations of effective action under the existing system of government. With his haunted when not hunted expression, a cigarette hanging from his lower lip, he has traditionally been seen as one of 'the guilty men' of Munich. Yet the irony is that Daladier was a consistent and active exponent of the modernization of the French armed forces, and favoured firm action against Hitler over Czechoslovakia.

A few weeks after the Anschluss, Hitler was already planning an invasion of Czechoslovakia. The scheme was to engineer incidents between the Prague government and the German-speaking Sudetens that would enable Hitler to demand the annexation of the Sudetenland in the interests of its molested German inhabitants. Hitler almost certainly counted on a Czech refusal, which would give him a pretext for war in which the whole of the republic could be seized. Édouard Daladier was anxious to support Czechoslovakia. Of all the Little *Entente* countries, she was the one that it seemed most realistic to try to help. She had a large army and an important munitions industry, while her contiguity with Germany and Austria made her especially useful to France as an eastern ally. Although this could also be said to some extent of Poland, Czechoslovakia had the added attraction of being a country that Russia was prepared to help. This was not true of Poland: with six million Russians of various ethnic shades inside Polish frontiers, Russia had no interest in defending Polish integrity.

The problem was the uncertainty of British help, which Daladier considered essential to any effective deterrent to Hitler, given the rapid increase in German rearmament since 1936. In the 1920s, Austen Chamberlain had refused to be drawn into Aristide Briand's agreements with Czechoslovakia and Poland. His brother Neville was equally circumspect in 1938: Britain would help France if she were attacked by Germany as a result of an East European crisis, but France must not interpret this as a pledge of military support for French initiatives in defence of her eastern colleagues. Indeed, Chamberlain was keen to persuade Czechoslovakia to buy off Hitler by relinquishing those regions in which a majority of the inhabitants were German. Not only would this give the Western powers breathing space—conceivably 'peace in our time'—but it also accorded conveniently with Wilsonian ideals of adjusting frontiers to correspond with ethnic divisions. The principle had been temporarily subordinated at Versailles to the need to keep Germany in check—hence the transfer of the Sudetenland to the newly created Czechoslovak State—but the time had arguably come to implement Wilsonian ideals in their plenitude. Daladier understood only too well the underlying reasons for this solicitude for neglected principles; but, like Blum before him, he saw no credible alternative to going along with the policies of France's one clear ally—especially as he himself was in two minds about the wisdom of challenging Hitler on this issue. President Beneš was accordingly cajoled into giving way.

This apparent victory for Hitler was something of an embarrassment to him, since it deprived him of the *casus belli* he needed to seize the whole of Czechoslovakia. He therefore told Chamberlain that Germany must be allowed to occupy the areas in question immediately, without waiting for a proper transfer under international auspices. Hitler was right in assuming that Beneš would not accept this—but he was unpleasantly surprised to find that neither would Chamberlain or Daladier. France ordered partial mobilization on 24 September, and Britain mobilized her navy. Moreover, there was strong evidence that Russia was ready to intervene. The result was a face-saving compromise at Munich: Hitler was to be allowed to occupy the German areas, but under the brief auspices of an international commission; and he undertook at the same time to respect what was left of Czechoslovakia. Unlike Chamberlain—this 'desiccated stick' as Daladier called him—the French Premier was under no illusions about what had been achieved at Munich; on seeing the cheering crowds at the airport to greet him, he merely commented: 'Ah, les cons!'

If Daladier felt let down by Britain over Czechoslovakia, the British government for its part was inclined to regard the French as cadgers, out to ensnare Britain in their Continental entanglements. This attitude of disdainful suspicion was reflected at all levels: not untypically the British Ambassador in Paris reported home: 'As I waved farewell to the Minister of Public Works going down in the lift, I seemed to catch a faint final whiff of that sulphureous atmosphere that is commonly supposed to envelop the personage who exercised so unfortunate an influence on the declining days of the worthy Dr Faust.'[1]

Daladier's appointment of Paul Reynaud as Minister of Finance (1 November 1938) guaranteed that defence would have a high priority in government expenditure. But much of the increase in armaments in the late 1930s was financed by loans rather than through increased taxation; and Pierre Mendès France was later to criticize this back-stairs expedient, with its inflationary side-effects, as a cowardly evasion of the government's duty to confront the public with unpalatable truths.

Daladier's misgivings on the Munich settlement were to prove only too well founded in the following March. After inciting the Slovaks to challenge the authority of Prague, Hitler invaded Czechoslovakia on the pretext of restoring order, and declared the whole of Czechoslovakia 'under the protection of the German Reich' (15 March 1939). Czechoslovakia's internal dissensions were gratefully seized on by the British and French Foreign Ministers as justifications for inaction, Georges Bonnet remarking: 'the renewed rift between the Czechs and Slovaks only shows that we nearly went to war last autumn to bolster up a state that was not viable'. Yet the enormity of what Hitler had just done was far greater than his behaviour over the German-speaking areas in the previous September, when Britain and France had threatened war. The Allies had strained, albeit momentarily, at the gnat of the Sudetenland, but were now swallowing the camel of the annexation of nine million Czechs and Slovaks, the word 'protection' deceiving nobody. The difference in response stemmed not only from the convenient fiction of a Slovak bid for freedom, but also from the basic fact that Hitler had now presented them with a *fait accompli*, which could only be dislodged by an actual Anglo-French attack, since it was unlikely that Hitler would risk the international humiliation that a withdrawal under mere threat would involve. At Munich, by contrast, the occupation of the German-speaking Sudetenland had not yet taken place, and an Anglo-French threat of intervention was enough to cause Hitler to modify his plans, since the anodine demands that the embarrassed Allies made of him involved only a minor loss of face.

Chamberlain was now convinced, however, that the morale of the Western powers could not withstand another blow of this sort, and he accordingly drew up a declaration of collective security which he presented to France, Russia, and Poland. Ironically, Poland's safety was of much less interest to France and Russia than that of Czechoslovakia; and Poland was too wary of both Russia and Germany to want to commit herself to one side and risk provoking the other. The eventual outcome was that the only system of mutual security that existed in Eastern Europe was the newly signed Anglo-Polish pact of 6 April 1939 and Laval's old Franco-Soviet pact of 2 May 1935, which Georges Bonnet had foolishly—if truthfully—told Ribbentrop no longer had any effective significance (7 December 1938). In fact, Anglo-French inactivity over Czechoslovakia was to transform Bonnet's assessment into a cast-iron reality. Despairing of effective action by the Western powers, Stalin turned to the opposing strategy of an understanding with Hitler—an alternative that he had seriously borne in mind since 1935. The new Soviet Foreign Minister,

Vyacheslav Molotov (3 May 1939), was known to Hitler as an advocate of such a switch, and, sensing the change in Soviet thinking, Hitler eventually decided to involve Russia in his plans for partitioning Poland. The resultant Nazi–Soviet pact (23 August 1939) brought Britain and France face to face with the consequences of their vacillation.

Hitler still entertained hopes that Britain and France would remain true to their past pattern of behaviour and would not intervene to defend Poland. But he was perfectly prepared to accept the consequences if his hopes were unfulfilled. He knew that Britain and France had told Belgium and Luxemburg that the Allies would respect their neutrality and not set foot there unless invited. In the meantime, any Anglo-French attack on Germany would have to come from Lorraine and Alsace, and would come up against the heavy fortifications of the West Wall. In this way, Hitler was confident that the western Attack could be kept in check until Poland was defeated, when he could transfer all his forces to the western front. He could then bypass the deadlock in eastern France by violating the neutrality of Luxemburg and Belgium, and invade France from the north-east.

THE 'PHONEY WAR'

The France that was now officially at war with Germany is often presented as a divided nation, led by old men. As previous chapters have indicated, social divisions in France were in some ways less marked than in her more industrialized neighbours; and at the centre of French political life there was a solid consensus of rural and middle-class interests. There was admittedly little enthusiasm for war among them. The peasantry assumed that once again, as in 1914, they would provide the bulk of the cannon fodder, having no manufacturing expertise to guarantee them safe employment in munitions factories. Unlike a shop or an office, the operations of a farm could not easily be scaled down to suit the capabilities of a family that had lost its menfolk. The sardonic question: 'mourir pour Danzig?', was one that found echoes in all classes, far beyond the disaffected Right. Unlike 1914, there was as yet no clear indication that 'la patrie' was 'en danger'; and unlike 1914 it was France not Germany that took the initiative in declaring war—however belatedly and circumspectly. Yet if there was no enthusiasm, there was little defeatism either, and no reason to suppose that, faced with an invading German army, the French forces and civilian population would not resist as tenaciously as they had done in 1914.

If social divisions were less sharp than in some countries, political divisions on the nature of the regime and on the role of religion created disaffected minorities whose attitude to the impending war was conditioned by this disaffection. Charles Maurras expressed the view of a sizeable section of the extreme Right when he claimed that the Republic had made the double mistake of failing to build up the nation's military strength in the past, and had then challenged Hitler unnecessarily—and with inadequate resources. As Céline brutally prophesied in 1938: 'We'll disappear body and soul from this place like the Gauls ... They left us hardly twenty words of their own language. We'll be lucky if anything more than the word "merde" survives us.' Even so, once war was declared and France was invaded, most members of the extreme Right, including those who felt some admiration for the Nazi regime, fought alongside everyone else. It was essentially after the armistice that a large number of them set out to carve niches for themselves in the new Europe. At the other extreme of the political spectrum, the Communists were in a particularly embarrassing position. As a result of the Nazi–Soviet pact, the PCF was declared illegal in September 1939, and the Communist members were expelled from parliament in the following January. In the meantime, however, a quarter of the Communist deputies and senators had already resigned from the party in protest against the PCF's refusal to disown Soviet policy, and the recently reunited CGT broke off relations with the party. As for the rank and file, the bulk of PCF members participated in the war effort with everyone else, despite resentment at the proscription of the party; and, as with the extreme Right, it was only after the armistice that they started settling scores with their political opponents. Desertions were relatively rare, and Maurice Thorez was one of a very small handful who made their way to Russia to escape the embarrassment of serving against the forces of the Nazi–Soviet pact.

While it would be hard to see these divisions as a major element in the French defeat, the age of the military leaders was an undoubted factor in the reluctance of the High Command to respond flexibly to the demands of modern war. As indicated in Chapter 1, the aggravation of France's demographic problems by the First World War left her with a predominance of elderly men in senior positions, both in civil and in military life. Whereas the average age of the German military and political leaders in 1939 was fifty, in France it was sixty-seven. Admittedly, when the sixty-eight-year-old Maxime Weygand retired as Chief of General Staff in 1935, he was succeeded by Maurice Gamelin, who was a mere sixty-four. But when his suitability was put to the test in May 1940, he lost his position as Commander-in-Chief to the now seventy-three-year-old Weygand. Characteristically, when the new Prime Minister, Paul Reynaud, appointed Colonel de Gaulle (forty-nine) as temporary Brigadier-General and Under Secretary of State for War on 5 June 1940, Weygand expostulated: 'but he's a mere child!'

It was Gamelin's conviction that the Allies could not prevent Hitler from occupying Poland or any other of the small East European states. Instead of attempting an offensive rescue operation, they should wait for Hitler to turn his forces westward, and count on defeating him in what would be basically a defensive campaign. Victorious, they would then resuscitate the eastern states that Hitler had subjugated. Accordingly, the Allies merely waited for Hitler to

take the initiative against them and restricted their activities to a few perfunctory manœuvres along the Franco-German frontier. This wait-and-see
posture was encouraged by the fear that an Allied offensive might leave the
French lines of defence temporarily undermanned and exposed to the possibility of a swift German counter-attack finding a weak link. The fact that an
important part of French industry was within close striking distance of the
north-eastern frontiers was an added motive for giving defence priority over
any scheme for attack. Moreover, the assumption that France and the British
Empire had greater economic resources than Germany gave currency to the
view that time was on the side of the Allies. Even the modest step of advancing
to their intended defensive positions in Belgium was ruled out by a Belgian
declaration of neutrality on 3 September. Apart from the considerations that
had dictated the Belgians' initial caution in 1936, Brussels feared that Flemish
autonomists would oppose an anti-German war on behalf of the French.

Having defeated Poland in little more than three weeks, Hitler offered peace
to the Allies (6 October), which was duly refused. The refusal logically committed the Allies to the aim of defeating Hitler and demanding the restoration
of what he had taken, yet no initiative was taken other than cautious patrolling
of frontier territory. Bad weather in November and January caused Hitler to
postpone twice the expected attack on France. But in the meantime he invaded
Denmark and Norway on 9 April 1940, with the dual purpose of protecting the
western passage of Swedish iron-ore to Germany, and also of acquiring a string
of naval and air-bases from which to raid Britain and her Atlantic shipping.

INVASION AND ARMISTICE

The initial plans of the German High Command envisaged a right-flanking
movement in which thirty-seven divisions were to sweep across the Low
Countries into France, while a direct thrust of twenty-seven divisions would cut
through the Ardennes. It had much in common with the Schlieffen plan of
1905, modified by his successors in 1914—although the choice of the relatively
unfortified but thickly wooded Ardennes would have surprised them. They
would have been even more surprised, however, by the amendments of
6 March 1940. Following the Allied capture of the plan on 10 January, General
von Manstein, aided by General Guderian, persuaded Hitler to transfer the
main thrust of the attack from Belgium to the Ardennes, where the Allied
defences were thin. The attack would then be pushed forward to the Channel
coast, thereby cutting the Allied forces in two—one part in Belgium and the
other in the crook of the Maginot Line.

The Germans had 114 divisions ready for the campaign. The French had 94
divisions and the British 10, which, when augmented by the 22 Belgian
divisions and the 9 Dutch, gave the Allies an initial advantage and at no time

less than virtual parity—135 to the Germans' eventual 137. Even the German tanks, which were to play such a decisive role, were fewer than the Allied holdings. Accounts differ as to the precise numbers used on each side—some historians discounting the smaller and older types—but there is general agreement on the Allied superiority in numbers, fire-power, and defensive armour.[2] The average gun-calibre and armour-plating of the French tanks was of the order of 42.4 mm. and 33.2 mm., as against the German 27.3 mm. and 19.9 mm. This made for heavier tanks, however, which travelled more slowly and required more petrol than their German counterparts—and this was to be an important factor in the forthcoming campaign, where the speed and long range of the German armoured divisions enabled them to split the Allied forces in two, and then three.

In terms of aircraft, the advantage unquestionably lay with the Germans. Once again, historians offer different figures—estimates of the French machines employed in the campaign varying between 800 and 1,200, and for the British between 350 and 600. The British, in fact, kept the bulk of their 1,500 aircraft at home, ready to counter direct German attacks. Having committed virtually all of their land forces to the Continent, they wished to retain a strong defensive line of home-based fighters. The Germans used between 2,700 and 3,600 aircraft in the campaign, of which about 1,300 to 1,500 were bombers, 1,000 to 1,200 fighters, and 340 Stuka dive-bombers. Only in the sphere of fighters did the Allies have approximate parity with the Germans—even a slight advantage, according to some historians. In this, as in other aspects of the campaign, however, it was German methods rather than weapon strength that was the decisive factor in their victory.

As in Poland, the blitzkrieg method of massed armoured thrusts, closely supported by aircraft, was hard to combat except in kind—and, as it happened, the disposition of the Allied tanks did not correspond to the location of the main German assault. It was the Allied discounting of the Ardennes as the likely point of attack that proved their undoing—even more than the thinking that underlay their tactical use of tanks and aircraft. It is true that Gamelin had not formally changed his belief, enunciated in 1935 and 1937, that tanks were basically an infantry-support weapon; and 45 per cent of the French tanks used in the campaign were assigned to infantry formations. But Hitler's Polish campaign demonstrated all too plainly the validity of de Gaulle's arguments in favour of massing tanks together—and also opened de Gaulle's eyes to the tactical use of aircraft in land-battles, a factor that had had little place in his pre-war writings. In fact, some 55 per cent of the French tanks were grouped in armoured divisions on the German pattern, but there the resemblance ended. Instead of being used as the spearhead of an offensive, they were kept like a riot shield to ward off the brickbats from wherever they might come: they were little more than a Maginot Line on wheels. And since the main German attack was expected to come through Belgium north of Dinant, that was where the bulk of the armoured divisions were despatched.

The German penetration of the Ardennes on 10 May was more of a troop movement than a battle—the main problem being periodic hold-ups on the roads as a result of the sheer quantity of German vehicles moving along them. Allied troops were little in evidence in the forest, and the rivers were too small for bridge-blowing to create more than a brief delay for the Germans. Allied resistance became a reality in the open country, but with the bulk of Allied forces too far north to assist, the panzer divisions of Generals Kleist and Hoth had little difficulty in breaking through and speeding to the Channel, which they reached on 20 May.

In these and the more northerly engagements, the panzers were greatly assisted by the Luftwaffe. As in Poland, Stuka dive-bombers and low-flying fighters played the morale-shaking role that had traditionally belonged to the pre-offensive artillery barrage, with the added advantage of much longer range and the ability to respond immediately to changes in the enemies' movements. The Allied pilots had no war experience and little training in the tactical support of land forces. The Royal Air Force was an autonomous service, which saw its role as essentially the bombing of enemy economic resources and the defence of the British equivalent. The French air force, like the German, was divided between a partly autonomous reserve, enjoying some of the independence of the RAF, while the rest were assigned to service with various army formations. Given the nature of the Allied strategy, this assignment was mainly defensive in character, consisting largely of reconnaissance and the interception of enemy raiders. The low-level attack attempted by the French bombers in military engagements lacked the accuracy and intimidating effect of the German Stukas—an important defect, since a large force of French dive-bombers could have inflicted considerable damage on the German armour and troop-carriers as they proceeded along the open roads of Belgium and northern France. The importance of air-power was also impressively demonstrated further north, where the German attack on Holland and northern Belgium on 10 May was remarkable for the incisive use of paratroops, which not only opened up advance objectives for the land forces but created a widespread feeling of insecurity in the Allied ranks.

While the westward dash of Hoth and Kleist was cutting the Allied forces in two, Guderian's Ardennes panzers had turned south and were isolating the garrisons of the Maginot Line from the rest of the French-based army—thereby completing a tripartite division of the Allied forces. Viewed as a whole, the remainder of the campaign was now merely a matter of rolling up the northern and eastern sectors, and then pursuing the remainder southward. Indeed, the German victory in the north might have been absolute, if Hitler had not been worried about his advanced detachments being cut off from the rest. He therefore held back for a fateful two days, with the result that the British

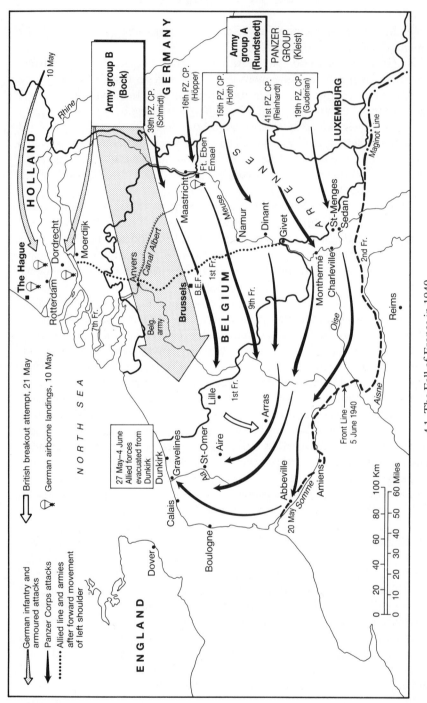

ENGLAND

Dover

NORTH SEA

The Hague

Rotterdam

Dordrecht

Moerdijk

HOLLAND

10 May

Rhine

Army group B (Bock)

39th PZ. CP. (Schmidt)

16th PZ. CP. (Höpper)

GERMANY

Ft. Eben Emael

Maastricht

7th Fr.

Anvers

Canal Albert

Belg. army

Brussels

B.E.F.

1st Fr.

BELGIUM

Meuse

Namur

Dinant

Givet

9th Fr.

15th PZ. CP. (Hoth)

Army group A (Rundstedt)

PANZER GROUP (Kleist)

41st PZ. CP. (Reinhardt)

19th PZ. CP. (Guderian)

ARDENNES

St-Menges

Sedan

Monthermé

Charleville

2nd Fr.

LUXEMBURG

Maginot Line

Oise

Reims

Aisne

Lille

1st Fr.

Arras

St-Omer

Aire

Gravelines

Calais

Boulogne

Dunkirk

Abbeville

20 May

Somme

Amiens

Front Line 5 June 1940

27 May–4 June Allied forces evacuated from Dunkirk

British breakout attempt, 21 May

German airborne landings, 10 May

German infantry and armoured attacks

Panzer Corps attacks

Allied line and armies after forward movement of left shoulder

| 0 | 20 | 40 | 60 | 80 | 100 Km |
| 0 | 10 | 20 | 30 | 40 | 50 | 60 Miles |

4.1 The Fall of France in 1940

Source: Brian Liddell Hart, *A History of the Second World War* (London, 1970), 64.

were able to evacuate 338,000 men from Dunkirk (27 May–4 June). He nevertheless assumed that once France had capitulated Britain would come to terms; and by 14 June the Germans were in Paris.

The sheer speed of the German campaign minimized the number of casualties on both sides; the total French service losses were 92,000 killed and some 200,000 wounded—a modest toll by the appalling standards of the First World War, where the British lost more than half that number on the first day of the Somme offensive. The impact on civilian life was on the surface more striking. Perhaps six to seven million French people took to the roads to escape the advancing Germans; and four months later half of them were still refugees from home.[3] The Red Cross alone returned 90,000 lost children to their parents following this migration—not counting those who were reunited or rescued by other bodies. Official organization of the exodus was minimal: the Versailles *mairie* reputedly posted a simple notice: 'Ordre d'évacuation. La mairie invite tout le monde à fuir.' Three-quarters of the Paris population had left by the time the Germans entered, while many northern towns were virtually deserted: only 800 of Chartres's 23,000 inhabitants stayed behind and only 30 of Troyes's 58,000. In those areas bordering Germany and Belgium, the Germans forbade the refugees to return, with the result that a year later the *département* of the Ardennes was still without two-fifths of its inhabitants. Conversely, a southern city like Bordeaux grew from 258,000 to 800,000 in a matter of days, posing enormous problems of feeding and accommodation.

In the midst of the débâcle, the extrovert but circumspect Mussolini felt the issue sufficiently settled for him to slide off the fence and declare war on the Allies (10 June). The new French Commander-in-Chief, General Weygand, was convinced that an armistice was essential; and on 16 June, Paul Reynaud relinquished the premiership to the Vice-Premier, Marshal Pétain, who was firmly resolved on making peace. There is no statistical evidence as to how the mass of the population viewed the prospect of an armistice. But all the indicators suggest that most people saw it as the only realistic option open to France. Memories of 1914–18, combined with the experience of the last few weeks, militated against accepting further destruction and loss of life in a war whose outcome seemed already settled. Few expected Britain to remain in the conflict for long. Like Napoleon, Hitler would have been content—at least in the short run—to establish a *modus vivendi* with Britain, as he demonstrated on 19 July when he publicly offered her peace. The terms he had in mind would in fact have deprived Britain of only a few of her possessions, notably her mandate over Iraq and her control over Egypt. Churchill had only come to office on 10 May, and there had been little time or opportunity for Continental observers to appreciate the changed character of British leadership. Indeed, the 'desertion at Dunkirk', where British captains had initially refused to take French soldiers on board while there were still British troops ashore, seemed to many Frenchmen to be in the worst traditions of British selfishness. Churchill's rejection of

Hitler's terms on 3 August would have surprised them—but all that still lay in the future.

One man who was not surprised by Churchill's determination was de Gaulle. As Reynaud's newly appointed Under Secretary of State for War (5 June), he had met Churchill several times in the week before Reynaud's resignation and was convinced of his steadfastness. When Pétain's intentions to sue for peace became evident, de Gaulle returned immediately to London (17 June), where Churchill put the BBC at his disposal. In a broadcast of 18 June, in which he invited Frenchmen to join him in London, he denounced the new leaders in France whose tactical incompetence had lost the recent campaign and who were now suing for peace. When Weygand ordered him to return to France, de Gaulle wrote back urging him to raise the banner of resistance in North Africa.

It had seemed to a determined minority that a retreat to North Africa was preferable to an armistice. Nearly a thousand French pilots had already flown their machines there to escape enemy capture, and a sizeable section of the Mediterranean fleet was at anchor at Mers-el-Kebir. The German fleet was weak by Anglo-French standards and in no position to impede a major transference of French troops across the Mediterranean. Reynaud's cabinet had discussed the merits of moving the government out of France, but Pétain had warned (13 June) that a cabinet in exile might lose all control over the population. At the same meeting, Weygand had alleged that the Communists had taken over Paris and that Jacques Duclos was presiding in the Élysée. It was all too easy for rumours of this kind to find fertile ground in minds that were living in the double shadow of the Popular Front and the Nazi–Soviet pact.

The Pétain government (16 June–12 July) that took over from Reynaud's was a broad-based cabinet—including two Socialists—in which commitment to a cessation of fighting was the main common factor. On 17 June, Pétain broadcast to a relieved population that he was seeking an armistice. A cabinet decision on 19 June to transfer parliament and part of the government to Morocco did not represent a change of policy: it was merely an attempt to ensure the safety of government pending negotiations. It was in any case overtaken by events and changing attitudes, and only twenty-nine deputies and a senator made the journey—to be apprehended three weeks later as moral deserters.

Hitler's terms were relatively lenient. He did not wish to commit the men and resources that a complete take-over of the country would entail, nor did he wish to strengthen the hand of the few who were urging the government to refuse an armistice and take up a stand in London or North Africa. He nevertheless insisted that the railway carriage in which the 1918 armistice had been signed should be brought out from the transport museum in the château of Compiègne, and set up in the forest clearing where Germany had admitted defeat in 1918. The armistice which was signed there (22 June) divided France

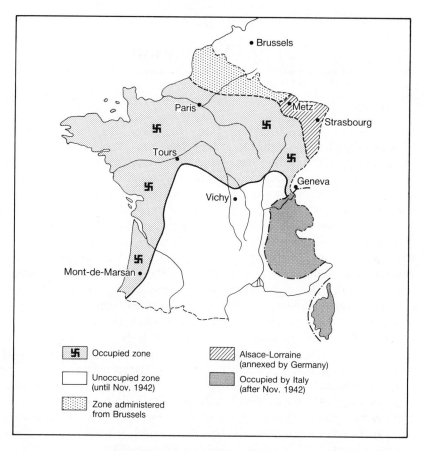

4.2 The Divisions of Occupied France

Source Colin Dyer, *Population and Society in Twentieth-century France* (London, 1978), 110.

into two parts: an Occupied Zone, including all the northern half of France plus the Atlantic coast, and an Unoccupied Zone, consisting of the south-eastern third of France. The French government—which was shortly to take up new headquarters in Vichy—was theoretically to govern both parts; but its role in the Occupied Zone would be directly subordinate to the wishes of Germany.

5

The Occupation

IT is not easy to compare the Occupation years with other sombre periods in French history. The Allied occupation that followed the defeat of Napoleon was brief and made little impact on the daily lives of the bulk of the population. The same might be said of the German occupation of 1871–3—despite the Paris Commune and the change of regime. More comparable with the 1940s was the German take-over of north-eastern France in the First World War; but the close proximity of continuous fighting on a horrendous scale, in which an un-defeated France was managing to hold the line, gave the occupation a more unsettled character of fear mingled with hope. The 1940s, by contrast, saw not only the complete defeat and occupation of the nation, but the degradation of its government and people to being the spectators, when not the active accomplices, of Nazi policies. The refusal of a small minority to accept this situation only served to underline the humiliation of the rest. The choices that were open to them seemed to many far less clear-cut at the time than they did after the Liberation, when the anger of active resisters was augmented by the censorious observations of Anglo-American commentators, who conveniently forgot that Anglo-American inactivity in the 1930s had been a major factor in encouraging Hitler to try his luck. Between the opposing poles of collaboration and resistance there was the huge expanse of middle ground where the vast majority of people lived out these years—many, especially in the remoter areas of the south-east, scarcely setting eyes on the occupying forces, even after 1942 when the distinction between the two zones was abolished.

For the mass of the population, attitudes were contingent not only on their personal circumstances but also on changes in German policy and fortunes abroad. Chronological landmarks that had a profound effect on individual choices included Hitler's invasion of Russia in June 1941; the abolition of the Unoccupied Zone in November 1942; the institution of compulsory labour service in February 1943; the turn of the tide of war in 1943; and the Anglo-American invasion of Normandy in June 1944. These were all developments that deeply affected people's calculations as to where their best interests lay.[1]

In the immediate aftermath of defeat, however, stunned acceptance was the attitude of the overwhelming majority. Most people assumed that Britain would be forced to come to terms with Hitler; and it was not until Russia and America were brought into the war in 1941 that the hope arose that Germany would not necessarily be left as undisputed master of Europe. In these circumstances, and with over six million people on the road, it was easy enough for Marshal Pétain to obtain the full powers he sought, in the name of stabilizing the situation and establishing a *modus vivendi* with Germany. These were granted by a joint sitting of Chamber and Senate on 10 July, with only eighty dissentient voices, most of them Socialist and Radical. Although not clearly appreciated at the time, the regime was transformed overnight from an ultra-parliamentary democracy into the most autocratic scheme of government that France had had since the eighteenth century. Even Napoleon had had a legislature of sorts. What permitted this transformation with such a minimum of fuss was the widespread assumption that the new arrangements were purely provisional, pending the return to normality and the framing of a new constitution.

Acceptance was likewise greatly aided by the fact that the newly created Chef de l'État Français was Pétain, for whom there was widespread sympathy and respect in most sections of political opinion. Blum had called him France's 'noblest and most humane soldier'—the man who was not only 'the victor of Verdun' but the constant champion of the minimizing of hardship to the common soldier. Indeed, his urging of an armistice in 1940 was seen by many as a confirmation of his realism and refusal to sacrifice the well-being of the French people to the vain glory of a fight to the finish. The subsequent cult and adulation of Pétain under Vichy was carefully fostered by the architects of the regime, in the knowledge that he was the government's prime claim to respectability.

Not that there was much reluctance among eminent contemporaries to join in the praise. Paul Claudel's notorious 'Ode to Pétain' of 1940 was only to be matched by his equally nauseous 'Ode to de Gaulle' in 1944—both reflecting widely and sincerely felt emotions of the time rather than mere time-serving. Perhaps the best-known of these overblown tributes was Georges Gérard's *Pater noster*: 'Our father who stands before us, thy name be glorified, thy kingdom come, thy will be done on earth so that we may live. Give us our daily bread, though we give nothing in return. Give once more life unto France. Lead us not into false hope nor into deceit, but deliver us from evil, O Marshal.'

Yet, side by side with the prestige, there remained the fragile reality of the eighty-four-year-old man. His propensity for falling asleep during meetings had led Georges Mandel to refer to him as 'le conquistador' (le con qui se dort), in ostensible tribute to his former services as Ambassador to General Franco. His table-talk was no more scintillating than that of most octogenarians: 'Bouthillier, parle-moi des Chinois. Ils sont jaunes, n'est-ce pas? Je n'aime pas les jaunes.'[2] Yet, despite the frailties of age, he remained politically sharp and exerted a real influence on the conduct of affairs, within the limitations placed on him by the German occupation. He was far from being the tool of his deputy, Pierre Laval; he recognized Laval's political utility and endeavoured to put up with the clouds of cigarette smoke with which Laval enveloped him every time he brought Pétain a document for perusal.

Anglo-American cartoonists of the war years portrayed Laval as a Charles Addams creation, a toad-like figure with gin-trap teeth beneath a clipped moustache. In appearance and outlook he could have been a degenerate brother of Aristide Briand, former Premier and apostle of peace in the 1920s. Having followed a similar primrose path from being a young Socialist lawyer to becoming an easygoing independent of no particular allegiance, Laval combined a genuine commitment to international amity with an eclectic opportunism when it came to methods of achieving it. Like many politicians of the time, he had pinned considerable hope on Mussolini as a bulwark against German expansion in the mid-1930s, and regretted the effects of the Abyssinian occupation and the Spanish Civil War on Western European attitudes towards him. He had a sharper eye for personal advantage than Briand, and amassed considerable assets in the regional press and radio. Having already been Prime Minister four times, but still only in his fifties, he had the necessary experience and stamina to direct affairs under Pétain—while his pragmatism and dexterity seemed to fit him for the tasks ahead. He had little sympathy, however, for the *bien pensant* paternalism favoured by many of the Marshal's early cronies, and personally regarded the National Revolution as a high-minded anachronism, however useful in distracting the public from the awfulness of the country's predicament. For Laval, the main business of government in the foreseeable future was to establish a tolerable working relationship with Germany that would enable France to survive with as much sovereignty as possible until the end of the Anglo-German war—after which an independent France could take up the threads of its past. A man without firm convictions, he had no long-term blueprint for France, believing that such planning was too often upstaged by unpredictable changes of circumstance. Although he was technically only Pétain's Deputy Prime Minister, the role was virtually an acting premiership, given Pétain's age and time-consuming duties as Head of State.

The choice of the small spa-town of Vichy as the seat of government stemmed from its centrality, its proximity to the frontier between the two zones, and its abundance of hotel accommodation that could be turned into ministerial offices. It also had the virtue of not being the political fief of any prominent Republican figure.

France continued in a constitutional limbo. Although neither of the houses of parliament had been formally abolished, laws merely consisted of semi-monarchical edicts: 'We, Philippe Pétain, Marshal of France, Head of the French State, decree', etc. It is true that a National Council of *notables* was instituted on 24 January 1941, with committees that discussed the framing of a new constitution; but nothing came of their deliberations, other than a loose, unrealized proposal for a parliament consisting mainly of nominated members. Unencumbered by a legislature, the government could also count on the loyal support of the civil service. It was not just a question of bureaucrats safeguarding their salaries. France had to be saved from chaos; and few *fonction-*

naires foresaw the dark compromises and shifty stratagems that were to be the daily fare of government in the years to come. However, the terms of the armistice specifically obliged the administrative authorities in the Occupied Zone to 'collaborate faithfully' with the Germans, and this was quickly to create conflicts of conscience.

The bizarre life of Vichy—a Whitehall transported to Cheltenham or Tunbridge Wells—has been described often enough. The early riser might see the cleaners of the Ministry of Marine swabbing the decks of the Hôtel Helder, as though it were a ship of the line. The climax of the day for Vichy's retired residents was the eagerly awaited appearance of the Marshal on the steps of the Hôtel du Parc, prior to a *petite tournée* among the seats and chirping sparrows of the gravel concourse, where he would nod benignly to the elderly *receveuse* lurking beneath the trees to catch the odd sly stroller who might seek to steal a few minutes unpaid-for rest on one of her wrought-iron chairs. And in the evening, for the privileged, there were the dinner parties, at which one might meet what Richard Cobb has called

the very *fine fleur* of collaboration . . . the lugubrious, sad, yellow-faced Georges Bonnet, the papal nuncio, always clothed by Jeanne Lanvin, who used to make his beautiful silk soutanes, the minister of education, wearing makeup, his eyes running with henna in the heat, the Spanish ambassador . . . decked out like a . . . Philip II galleon, and covered in braid and medals, so that he could hardly sit down.[3]

If in some ways Vichy represented a conservative backlash against the Popular Front, it was not one that personally benefited the pre-war parliamentary leaders of conservatism. Their very associations with the old system made them *personae non gratae* in Vichy governmental circles. As soon as the armistice was signed, Pétain got rid of seven parliamentary figures from his government (12 July 1944), and in his subsequent cabinets there were to be only four ministers and four Secretaries of State who had sat in parliament. Prominent among these was Pierre-Étienne Flandin, who viewed the Vichy regime as a temporary Roman-style dictatorship; and in his brief ministry (14 December 1940–9 February 1941) he was more deferential to public opinion than other Vichy leaders. Eighteen of Vichy's overall total of thirty-five ministers and seven of its eighteen Secretaries of State might be termed experts and professionals, a pattern that was new to Republican France, and was not to be seen again until the Fifth Republic. Not only were products of the *grandes écoles* more evident in the cabinet, but the Third Republic's practice of seconding Inspecteurs de Finances and members of the Conseil d'État to important posts in the various ministries was considerably extended. It reached its apogee under the Darlan ministry (10 February 1941–17 April 1942), when the problems of meeting German economic demands necessitated expert planning at top level. With Laval's next spell of office, however, in April 1942, France rapidly entered the ice-age of subservience to Germany, when the strain on conscience put a premium on loyalty to Laval rather than on straight-forward expertise—and the cabinet's membership reflected the fact.

Dealing with the Germans was the overriding task of government during the Occupation. It was the issue which determined all other policies and the issue on which ministers rose and fell. Compared to the occupation of northern France in the First World War, the behaviour of the German forces was initially conceded to be 'correct'. It was the German invasion of Russia in June 1941 that introduced the savage phase in German repression. Not only did it see the Communists entering into the Resistance, bringing with them a programme of assassinations—'à chacun son boche'—and intensified sabotage, but the Russian campaign put a severe strain on German manpower, limiting the forces that it could retain in France and inclining the Germans to resort increasingly to terror to maintain a grip on the country. In the meantime, however, a victorious Germany, whose military activity was restricted to minor campaigns in south-eastern Europe, the Near East, and North Africa, could afford to affect an outward attitude of condescending propriety, even if the enormous economic burdens placed on France revealed the reality of the merciless exploitation that lay beneath.

Laval established an early rapport with the German Ambassador, Otto Abetz, who was widely—if mistakenly—regarded in Parisian society as a Greek among Romans. Despite his pre-war cultural connections with France and his deep interest in French intellectual issues, Abetz was essentially trying to make the French recognize the superior merits of the Nazi system. And if he brought to his task a finesse and understanding that was a refreshing contrast to his masters in Berlin, he remained the willing instrument of a system of ruthless imposition and repression.

Initially, Germany had little interest in Laval's various offers of unofficial help against Britain, notably the services of some two hundred 'volunteer' pilots. What changed the German position, however, was news on 28 August 1940 of de Gaulle's success in obtaining the support of French Equatorial Africa and the fear of West Africa following suit. Vichy's repulse of the ill-fated Anglo-Gaullist naval attack on Dakar in late September underlined the advantages of treating France as an unofficial ally rather than a conquered country—at least when it came to the defence of her overseas empire. More-over, the indefinite postponement of the attempt to subjugate Britain by direct attack, and Hitler's increasing interest in an attack on Russia, made it all the more desirable to establish some working relationship with France to secure his rear. Even so, Hitler did not seek to make France a formal and active ally. Not only was there little chance of persuading Vichy to take such a step, but it would have committed Germany to helping with the protection of the French empire, and would have tied Hitler's hands in the ultimate share-out of spoils at the end of the war. A co-victorious France could scarcely be deprived of her former possessions, nor be exploited in the meantime. And Hitler was well aware of Mussolini's desire to obtain Corsica, Nice, and Tunisia. Although the meeting at Montoire between Hitler and Pétain on 24 October 1940 resulted in nothing more specific than a vague commitment to collaboration, the hand-shake between the two leaders startled opinion and symbolized the direction that affairs were now taking.

However, Laval's close relations with Abetz and his negligence in keeping his colleagues and Pétain closely informed of his dealings created irritation and suspicion in Vichy. So much so, that Pétain removed him from the vice-premiership on 13 December and put him under temporary house arrest—much to the anger of Abetz, who had found him a congenial colleague. Laval's successor, Flandin, did not enjoy Abetz's confidence, and he was rapidly replaced by Admiral Darlan, whose Anglophobia was a strong commendation. The ill-judged British attack on the French fleet at Mers-el-Kebir on 3–4 July 1940, to prevent a possible German seizure, had helped to foster a strong current of anti-British feeling; and this was exacerbated by the Anglo-Gaullist occupation of Syria in June–July 1941, following the installation of German air-bases there.

It was at this time, however, that Hitler's invasion of Russia triggered off the French Communists' campaign of violence against the occupying forces. The Germans retaliated by shooting over a hundred hostages selected from imprisoned Communists and common-law offenders. Pétain's response was to offer himself as a hostage; and the shootings were temporarily suspended—but only on the understanding that Vichy would co-operate in preventing acts of violence by the Resistance. It was to be Vichy's growing dilemma that in order to preserve such shreds of independence as it still possessed, it preferred to be the responsible executant of Germany's repressive policy—even to the point of choosing the Communist hostages who were to be shot.

French internal politics continued to be dominated by developments abroad. After Hitler declared war on the United States (11 December 1941), Darlan was obliged to drop his long-term schemes for a closer military understanding with Germany. The vulnerability of the French West Indies to American attack constrained him to a policy of delicate neutrality—which in turn made him less attentive to the Germans. The outcome was that German pressure brought Laval back to office on 18 April 1942—this time with the title of Head of the Government, leaving Pétain simply as Head of State. Laval continued Darlan's policy of wary deference towards America, while publicly expressing hopes for 'the victory of Germany' in Europe as the best guarantee against 'the triumph of Bolshevism' (22 June 1942). Both Pétain and Laval hoped that the success of German arms in Europe would oblige America to come to a negotiable peace with Germany, when France would emerge with a privileged position in the new Europe, together with Italy and Spain. Indeed, Laval was personally in favour of going further than Pétain and joining Germany in war against the Allies, provided that Germany recognized France's pre-war status and overseas possessions. Laval optimistically hoped that the Allied landings in North-West Africa on 8 November 1942 would make this an attractive offer; but Hitler preferred freedom of German manœuvre to the questionable value of a formal French alliance—as was shown by his occupation of the whole of France in the following weeks.

Hitler feared that the Allied foothold in the Maghreb might eventually be used as a springboard for an invasion of southern France; hence his decision to bring German forces into the Unoccupied Zone. Hitler gave Pétain advance notice of his intention—and an escape to North-West Africa would have been possible had Pétain been so minded. Indeed, de Gaulle was later to say that Pétain could have entered Paris at the Liberation riding a white horse, had he chosen to leave at that juncture. The fiction of a Vichy sovereignty under German supervision was now extended from the Occupied Zone to the whole of France; and the token French Armistice Army was dissolved, which Hitler justified on the grounds of the defection of the French forces in North-West Africa. To prevent the Germans from attempting to seize the French fleet, its crews took the heart-breaking step of scuttling it (27 November 1942). While there were strong arguments in favour of this sacrifice, it nevertheless deprived Vichy of a significant bargaining counter—leaving France with little else to barter for concessions other than the remnants of empire not already occupied by the British.

Laval, however, was not a man to let a crisis pass without picking up some advantage—if only for himself. He utilized it to extract from Pétain full powers to issue legislation without Pétain's signature (17 November 1942)—assuring the Germans that this would enable him to pursue a more Germanophile policy. Hitler, for his part, made it clear to Pétain (29 April 1943) that he would not tolerate any interference with Laval's tenure of the premiership. When de Gaulle convened a Consultative Assembly in Algiers on 3 November 1943, Pétain pathetically planned to counter it with a public announcement declaring Laval to have been dismissed, and promising that power would revert to parliament in the event of his death. But the Germans sat on the scheme immediately, and appointed Cecil von Renthe-Fink to vet all legislation and policy contemplated by Vichy (18 December 1943). Realizing that Laval was in need of tough committed supporters, they also bullied Pétain into accepting Joseph Darnand (of Milice notoriety) as Secretary-General for the Maintenance of Order (1 January 1944), and Marcel Déat as Minister of Labour (16 March 1944). By then, however, everyone in office recognized that the fortunes of war had irrevocably turned against Germany, and that it was now a matter of judging the right moment to make a personal get-away from the corporate responsibilities of the Vichy gamble. What held the higher echelons of government together, until well after the Allied invasion of northern France in June 1944, was uncertainty about how they would be treated by the Allies if they defected to the winning side. The former Minister of Industrial Production, Pierre Pucheu, had earlier made his way to North Africa, only to find himself facing a firing-squad. Apart from thought-provoking examples of this kind, there remained the force of inertia. Traditional habits of professional loyalty were also buttressed by the calculation that it was probably safer to stay put and thereby demonstrate to the victors the sincerity of one's faith in the legitimacy of Vichy—and the fact that the government had been at pains to remain neutral. Indeed, the effectiveness of this tactic was largely demonstrated at the Liberation, when retribution was to be relatively mild, except for a handful of key figures.

THE NATIONAL REVOLUTION

If the Occupation was a period of savage constraint, there were many French people—by no means confined to the Right—who also saw it as an opportunity to set about tasks that had been beyond the powers of the Third Republic, with its succession of weak ministries constantly at the mercy of an impregnable parliament. While none of the twenty-four pension bills introduced by the Popular Front and its successors ever survived the *via dolorosa* of parliamentary discussion, Vichy instituted an improved pensions scheme by simple decree (14 March 1941). There was consequently considerable initial enthusiasm among 'moderate' reformers and socially minded Catholics for the progressive potential of Vichy. The Right, however, saw the break with the Third Republic in redemptive, if not to say retaliatory, terms. This was the opportunity to rebuild society and government on the old certainties and disciplines of the traditional moral order, the neglect of which had allegedly contributed to the military defeat of 1940; and Pétain himself preferred the term 'National Renovation' to 'National Revolution' for the programme of changes that Vichy introduced.

The propertied classes, who had been unnerved by the Popular Front, felt benignly disposed to these changes, without necessarily subscribing to all of them. Support, however, was more specific among three sections of French opinion, each of which favoured certain aspects of the programme while entertaining reservations about others. After the Church and the various right-wing movements that found a niche in Vichy, there were the many proponents of a vigorous government of national solidarity, in which traditional class antagonisms would be subordinated to a common determination to get things done. Indeed, both Vichy and its intellectual counterparts in the Resistance were anxious to re-create or discover the sense of community which each side felt had been lost in France—hence Vichy's attention to the family and to youth organizations. Symptomatically, the Catholic poet Charles Péguy, who had been killed in 1914, became a banner of both Vichy and the Resistance, reflecting the complex amalgam of attitudes that each side represented and, indeed, the complexities of Péguy himself. Yet, despite the strong influence of various strands of Catholic thought, and despite the presence of devotees of Action Française in several influential government posts in the early months of Vichy, the National Revolution arguably owed more to nineteenth-century Positivist ideals of a technocratic society. The relative strengths of these differing currents altered with the passage of time; and although the traditionalists, who extolled hierarchy and the tried-and-true decencies of the past, were much in evidence in the first six months of the regime, they increasingly lost influence to the technocrats, especially when Darlan took over government in February 1941.

The initial character of the National Revolution was reflected in the replacement of 'Liberté, Égalité, Fraternité' with 'Travail, Famille, Patrie'. Pétain declared that family rights 'take precedence over the state as well as over the rights of individuals'; and, like inter-war governments, Vichy was concerned to increase the birth-rate, especially in view of recent war losses and the exile from family life of 1,800,000 potential fathers in prisoner-of-war camps. Building on the precedent of Daladier's Code de la Famille of July 1939, family allowances were increased in April 1941, and mothers who stayed at home were given a substantial grant, notionally equivalent to the official average wage in the locality. This was a potentially important factor for women whose husbands had been killed or taken prisoner. Yet these increases and accompanying tax concessions came nowhere near meeting the cost of rearing children in a period of rapid price-rises; and the standard of living of a family with four children was only half that of a childless couple—and more like a third, if the childless wife went out to work. Having a large family only became an arguable advantage in the later stages of the Occupation, when it reduced a father's liability for labour service in Germany.

Yet the intriguing fact remains that the Occupation years did see a growth in the birth-rate. The explanation would seem to lie mainly in the impact of the post-1918 baby-boom on the number of marriageable people living in France—the beginnings of which were already observable before the Vichy government came to power. There are also the familiar claims that the curfew, plus fuel and lighting shortages, made bed a congenial refuge in the long winter evenings. Yet before too much is made of the peculiar circumstances of Occupied France, it is worth remembering that non-belligerent countries such as Switzerland and Sweden also experienced a rise in the birth-rate during the war period. Not surprisingly, the number of illegitimate births in France rose by over a quarter during this period of shifting populations and transitory attachments—although it still remained below

10 per cent of the total birth-rate. Despite the fact that German servicemen were not permitted to marry the inhabitants of occupied countries, 80,000 women in the Occupied Zone alone were claiming children's benefits from the German military authorities by the middle of 1943. And many were to remember with some wistfulness the peculiarly German aroma of leather and aftershave that was added to the familiar blend of sweat, Gauloises, and garlic that made up the Metro smell.[4] There were perhaps even some who naïvely saw themselves re-enacting Jacques Feyder's popular film, *La Kermesse héroïque* (1935), where the womenfolk of an occupied country sought in the foreign soldiers a welcome change from the stale routines of their defeated husbands.

Running parallel with the 'carrot' aspect of Vichy's natalist policies was the 'stick' of prohibitive legislation. A law of 11 October 1940 attempted to limit female employment; but the demands of German labour service were later to oblige Laval to draft women into French factories to replace the missing men. With an eye to the future, primary education for girls was given a more domestic and less vocational content than that of boys (15 August 1941), while respect for family ties was supposedly strengthened by forbidding divorce during the first three years of marriage (2 April 1941). The most notorious piece of stick was yet to come, however—the guillotining of a laundress on 29 July 1943 for effecting twenty-six abortions.

The protection of the family had its counterpart in the organization of youth. In a message to schools of 1 January 1944, Pétain said: 'ce qu'il faut à la France, à notre cher pays, ce ne sont pas des intelligences, ce sont des caractères'. De Gaulle was later to make similar remarks in the 1960s, and it has been a familiar criticism of state schools under the Third Republic that they attached too exclusive an importance to academic achievement and not enough to character formation—an opinion by no means confined to the champions of Catholic private schools. Vichy's youth programme was essentially an attempt to compensate for these alleged deficiencies, rather than a semi-fascist-style attempt to enlist youth in a monist vision of society and the State, despite the programme's emphasis on self-discipline, public spirit, and national loyalty. It was also an attempt to provide occupation and interest for young people at a loose end, whom the economic dislocation of 1940, and the abolition of all but a token army, had left in something of a limbo—until the unexpected prolongation of the war and the demands of the German economy remorselessly wiped out the problem.

Youth affairs were put in the hands of Georges Lamiraud, a paternalistic Catholic of the Marshal Lyautey stamp, while sport was entrusted to the tennis champion, Jean Borotra, who was a keen Anglophile and admirer of the British scout movement. The Baden-Powell spirit was much in evidence in the Compagnons de France for adolescents, and in the compulsory Chantiers de la Jeunesse for young men of draft age (30 July 1940). The Chantiers involved eight months' labour service, supplemented with civic and moral instruction, in which individual vigour and corporate effort were extolled, the inspiration ranging from Emmanuel Mounier's Catholic Personalism to the basic *nostra* of Action Française. Its members were exhorted to cultivate *le regard clair et franc*, 'that gaze through which one may read your soul, and for this there is but one means: be pure ... be a man ... be French'.[5]

The instructors in these various youth organizations were trained in special *écoles de cadres*, of which the best-known was the École Nationale des Cadres d'Uriage near Grenoble, where the influence of Mounier's Personalism was very strong. Its staff included not only Mounier himself—whose haunted face suggested an amalgam of Camus and Fernandel—but Hubert Beuve-Méry, future founder and editor of *Le Monde*. Mounier had long had an enthusiastic admirer in the German Ambassador, Otto Abetz, whose Catholic upbringing and early social democratic sympathies had drawn him to Personalism, in which he saw the possible basis of a new Western order, civilizing those aspects of National Socialism which his cultural fastidiousness found uncongenial. His masters, however, were suspicious of all French youth movements—seeing them as schools for patriotism—and had already banned their existence in the Occupied Zone. Uriage itself was dissolved within a few weeks of the German take-over of the Unoccupied Zone, and the Chantiers were eventually wound up in June 1944. Mounier himself had already been imprisoned in the summer of 1941 for his critical attitude towards Vichy; and his periodical, *L'Esprit*, was suppressed, after enjoying wide initial favour in Vichy circles.

In the field of education itself, the government's changes were mainly confined to favouring its friends and weakening its enemies. Pétain had long inveighed against what he saw as the demoralizing influence of the primary-school *instituteurs*, with their anticlericalism and alleged pacifist sympathies. Their training colleges—'these anticlerical seminaries'—were replaced with Instituts de Formation Professionnelle (18 September 1940), which would only take candidates who had obtained the *baccalauréat* through the secondary-school system. Moreover, the reintroduction of fees into the higher classes of secondary schools (15 August 1941) would help to ensure that intending schoolteachers would be socially and politically *comme il faut*. When it came to friends, Catholic private schools were the main beneficiaries, for Vichy had no doubt that it could count on the loyal support of the Church.

THE CHURCH AND VICHY

It is easy enough to compile an unedifying anthology of Catholic statements, recognizing in Vichy a liberation of France from the anticlerical traditions of the Third Republic.[6] The Papal Nuncio spoke of 'the Pétain miracle', and an embarrassingly large number of senior churchmen echoed his sentiments.

Admittedly, they were in part a desperate search for consolation in the disaster that had struck France, and few Catholics positively welcomed what had happened since 1939. Indeed, some of these early panegyrics of Pétain came from ecclesiastics who were later prominent in their denunciation of Vichy's subservience to Germany and its anti-Semitic policies. Yet there is no doubt that the majority of committed Catholics, of both left- and right-wing sympathies, began with favourable expectations of Vichy. They welcomed its benevolence towards the Church, and those who had been active in the 1930s in the Jeunesse Agricole Chrétienne were attracted by Vichy's talk of regenerating agriculture.

Members of religious orders were now formally allowed to teach once more in France (3 September 1940), a measure which admittedly did no more than recognize what had long since been a *fait accompli*; and religious congregations were at long last put on a legal par with other private associations (8 April 1942). More controversially, religious instruction was briefly reintroduced as an optional subject into state schools (6 January 1941); but this was soon altered to time off for optional instruction outside school (10 March 1941)—to the private relief of some senior churchmen, who had feared an anticlerical backlash. What the bishops were much more keen to obtain was some form of financial assistance from the State for private Catholic schools. But, despite persistent lobbying, all they got was a temporary annual grant, which was supposedly intended for schools with particular problems (2 November 1941). Admittedly, the government permitted sympathetic municipalities to give help, if they were so minded; but this only became a reality in areas where Catholicism was a strong force in the local population. Even so, these various tokens of government favour encouraged more Catholic parents to resort to the private sector, with the result that its share of the primary-school population rose from 18 to 23 per cent by 1943.

In the meantime, the Church had sanctified the Marshal's twenty-one-year-old marriage to a divorcee. Madame Pétain's first marriage was declared void by the Vatican, and the happy couple were legitimately united by the Archbishop of Paris in a church wedding—albeit with a stand-in for the Marshal, after the fashion of *Romanoff and Juliet*.

Ecclesiastical confidence, however, began to wane under Darlan, and then much more markedly during Laval's second government. Neither minister was sympathetic to the Church, and the clergy were uncomfortably conscious of their rapidly declining influence. At the same time, there was growing concern over Vichy's increasing subservience to Germany. Anti-Nazi feeling was strong among committed Catholics, especially after the wholesale purging of the Church in Alsace-Lorraine, where Strasbourg Cathedral was turned into a war museum and the clergy put under tight surveillance. The Nuncio protested to Pétain in August 1942 about France's involvement in Germany's anti-Semitic programme, and a few months later a political spokesman for the bishops privately advised Pétain to resign rather than become the mere tool of German policies. But, despite these misgivings, episcopal loyalty to Pétain remained strong; and even as late as February 1944, the assembly of cardinals and archbishops denounced Resistance activities as 'terrorism'.

However, this formal acceptance of the legitimacy of Vichy and the duty of obedience to it did not prevent individual bishops from denouncing the immorality of the government's treatment of Jews; and several of them finished the Occupation in prison camps. Censure of German policies and Vichy compliance was also a major factor in the deportation of over a thousand priests, a fifth of whom died in captivity. Prominent among Vichy's critics was the Archbishop of Toulouse, Mgr Saliège, whose physical infirmities narrowly saved him from imprisonment when the Gestapo came to fetch him in June 1944.

THE ECONOMY

The solid, honest France that Vichy claimed to be rescuing from the double tyranny of the secular *idéologues* of the Third Republic and the urban *idéologues* of the Popular Front, had its feet firmly planted in the soil. Pétain reminded France that 'La terre, elle, ne ment pas', and that France would always remain 'une nation essentiellement agricole'. He punctuated his indefatigable tours of the provinces with appreciative speeches, extolling the 'beaux produits de chez nous' and the 'fruits de la bonne terre de France', not forgetting the 'belles vaches' and 'braves chevaux de chez nous'.[7] This language had admittedly been the stock-in-trade of Ministers of Agriculture since the nineteenth century—*Madame Bovary* contains a celebrated pastiche—but it had never been so dominant a theme in the perorations of a Head of State. All of which reflected the undeniable reality of a fertile country wth a large agricultural population. But, as outlined in Chapter 1, much of its potential remained unexploited because the farms were so small and it was difficult to apply modern methods to them. Yet, as Gordon Wright has argued, Vichy's agricultural reforms 'sought not to alter France's traditional agrarian system, but to protect it against erosion'.[8]

It is still difficult to assess the potential worth of Vichy's agricultural measures. They took place in a period of such disruption and difficulty—with wheat production down by 20 per cent and potatoes down by over 40—that it is impossible to disentangle the consequences of government policy from those of the underlying circumstances. Their principal architect was Pierre Caziot, Minister of Agriculture from July 1940 until April 1942, when his inconvenient Germanophobia decided Laval to replace him with Jacques Le Roy Ladurie— an ironical choice in view of Le Roy Ladurie's later resignation over the labour-draft law in September 1942. The immediate problem of French agriculture was the labour shortage caused by the large number of draft-age peasants in prisoner-of-war camps. The government made forlorn attempts to tempt back to the land families that had earlier pulled up their roots and migrated to the towns. But, despite the bait of subsidies, only 1,561 families applied for them, and of these nearly a third failed to make the change successfully. Perhaps the most immediately effective of Caziot's reforms was the government loan scheme to improve rural housing, which resulted in the repair and modernization of 100,000 farm dwellings and cottages. And there was some attempt to address the basic problem of the small size and scattered nature of French holdings. Caziot's law of March 1941 eased some of the legal obstacles that prevented a farmer from bequeathing his farm intact to one child, and it likewise facilitated the consolidation of scattered holdings by mutual exchange between neighbouring farmers. Unlike the old law of 27 November 1918 (p. 5), it reduced the required assent of local farmers to a simple majority, and it also permitted the State to take the initiative in proposing consolidation in a particular village. However, the scheme hardly had time to operate in the Vichy period, and it was only under the Fourth and Fifth Republics that it slowly began to bear fruit.

Agriculture, like the whole of the French economy, laboured under the burden of German demands. Despite food shortages in France as a result of reduced productivity and the requirements of the resident Germans, 15 per cent of French agricultural output was going directly to Germany by 1943.

It is not easy to make a clear distinction between the punitive tribute paid by France as a defeated nation and the increased exports to Germany resulting from German contracts and the wartime closure of alternative foreign markets. The basic obligation to pay for the upkeep of the German occupying forces took 58 per cent of the government's total revenue—the enormity of the burden also reflecting a 40 per cent adjustment of the exchange-rate in Germany's favour. This was far in excess of what French taxation could provide, and could only be met by heavy borrowing from the Banque de France. But the new exchange-rate gave all commercial dealings with Germany a quasi-punitive character. By 1943, the Germans were taking 40 per cent of France's total industrial output, including 85 per cent of her vehicle production. The Vichy government initially tried to exclude war material from German contracts; but the eagerness of French business men to find economic outlets in a period of restricted domestic opportunities quickly overrode these inhibitions. After all, German contracts guaranteed regular supplies of raw materials and prompt payment for the finished product, while the refusal of a German offer could result in economic strangulation and the possible imprisonment of the owner—as happened to Marcel Dassault. Nor were other sectors of industry immune from these double-edged offers. With the Russian campaign making more and more demands on German resources, 1943 found Albert Speer turning increasingly to France for consumer goods as well. Indeed, when all of the occupiers' demands were taken into consideration, over a third of the French labour force was directly serving German needs.[9]

The pressures of supplying Germany were nevertheless a major factor in accelerating the rationalization of French industry that took place in these years. The increase in state control of the economy is often presented as one of Vichy's positive achievements. Yet it was largely an intensification of a trend that had already been set in motion in the mid-1930s. It served a double purpose under Vichy. Until the total occupation of November 1942, it was hoped that it might enable France to survive as a respected element in Hitler's New Order; but more immediately, it would gear the French economy to meeting the heavy demands of the Germans.

Comités d'organisation were set up on 16 August 1940 to pre-empt any move by Germany to take over the running of the economy. Some were largely in the hands of employers' organizations, while in areas where employers were less well organized, civil servants often played the leading role. These bodies were to survive the Liberation, playing an important role in post-war planning. As is so often the case in wartime, government control of the economy was greatly strengthened by its control of resources; and in the Occupied Zone, inefficient firms quickly found themselves starved of fuel and raw materials by the Germans.

It was Darlan's government of technocrats, however, that saw the most coherent attempt at a government economic strategy.[10] Characteristically, Darlan's first Minister of Production was a former head of a steel *comptoir*, Pierre Pucheu, while his new planning agency, the Délégation Générale à l'Équipement National (DGEN), was master-minded by a former director of Renault motors, François Lehideux. As Darlan said to a critic on Pétain's staff: 'That's better than the soft-cheeked altar-boys who form your entourage! We don't want any generals or seminarians, but sharp young fellows who will come to terms with Fritz and cook up a good soup for us.' And indeed, most of the men that Vichy appointed to key economic posts were in their early forties.

Among the DGEN's prime tasks was the elaboration of a ten-year plan (1942) and then a short-term Tranche de Démarrage (1944) for the post-war period. The ten-year plan envisaged a major role for private enterprise and the free market; and, unlike the Fourth Republic's plans, it did not set production targets. It was essentially a broad-front strategy, without the Fourth Republic's emphasis on key sectors. It envisaged national investment in industrial equipment, and was a counterpart to the law of 17 December 1941 permitting the government to close down inefficient firms. However, the DGEN's inability to breach the individual ministries' control over resources severely weakened its effectiveness as a co-ordinator of the economy; and Laval's return to power in 1942 saw the plan fall into disfavour. The disappearance of Lehideux and other committed backers from the government guaranteed that ministerial jealousies would henceforth prevent the organization from realizing its prime objectives.

The Tranche de Démarrage of 1944 was to fare somewhat better, in that parts of it were to be taken over by the de Gaulle government, its greater concern for social equality giving it some affinity with the Resistance programme. It gave agriculture a less favoured place than the ten-year plan, and shared the concern of Jean Monnet's team for the modernization of industrial plant and a more vigorous lead from government. It also deplored what it called the 'general Malthusianism' of the inter-war period. However, it did not envisage nationalizations, and it lacked the *esprit de consultation* of Monnet's schemes.

As with agriculture, however, the fruits of Vichy's planning elude the statistician, in that whatever advantages government policy may have conferred, they were swallowed up in the dislocation of war. Despite the sticks and carrots of German demands, French industrial production in 1941 was only 65 per cent of what it had been in 1938, while in the first half of 1944, it fell to a mere 44 per cent of what it had been in 1941. Apart from the obvious effects of loss of manpower to Germany, labour at home was overworked and undernourished, using run-down equipment; and the increasing disaffection of the work-force killed any inclination to show initiative or work harder than was necessary to escape disciplinary action. Moreover, French industry had to make do with three-fifths of its normal coal supplies—and even that compared well with the domestic consumer, who was down to little more than a third by 1944. By contrast, growth did take place in relatively new industries geared to the German war effort, such as aluminium, half of whose production went to the Luftwaffe. At the same time, the Occupation could bring unexpected, indirect stimuli to industry: for example, the otherwise disruptive division of France into two zones was an influential factor in encouraging industrial initiatives in southern France.

Vichy's economic planning had its dark, repressive aspect, as was reflected in the dissolution of trade unions, which, together with employers' associations such as the CGPF, were branded as inimical to 'energetic and effective authority'. Léon Jouhaux and the CGT initially volunteered to enter into an employer–worker organization; but the government was bent on creating new mixed bodies. Henceforth, trade unions were allowed to exist only at a local level (9 November 1940). The Labour Charter of 4 October 1941 envisaged the setting up of twenty-nine professional families for the various sections of the economy, each of which would embrace both management and workers. But only one such family was fully operative by 1944; and in the meantime the prohibition of strikes and lock-outs, together with the mandatory acceptance of arbitration, emasculated any bargaining power the workers might have had. Wage negotiations and other work-related issues were treated after a fashion in a multi-tiered structure of social committees—of which the basic company unit was partly to inspire the *comités d'entreprise* of the Liberation era.

The reality of working conditions in these years swung violently with the vicissitudes of the economy and the pressure of German demands. Although the Popular Front's forty-hour law remained on the statute book, Vichy followed Daladier's and Reynaud's example and authorized longer hours by decree. By 1944 it averaged 46.2 hours, and was considerably higher in the key industrial sectors. In the matter of wages, Otto Abetz declared in July 1942 that French industrial levels must be kept below their German equivalents to encourage volunteers to work in the Reich. And indeed, real wages in France dropped by over 25 per cent between 1938 and 1943 as a result of official prices doubling but wages rising by less than half.

When black-market prices were taken into the equation, however, the drop was much more marked. Bread was rationed as early as September 1940, and most other comestibles, clothes, shoes, and tobacco followed suit by the autumn of 1941. People who through honesty or poverty stuck to the official rations, had to make do with 1,200 calories a day by 1943—instead of the 1,700 which are currently considered to be the minimum necessary to health. Consequently, a large number of townsfolk had recourse to the black market and rural connections. Indeed, small country railway stations in the Paris basin saw their passenger traffic increase twenty-fold in the general search for food. Black-market food prices varied between two to eight times the standard rate, with butter especially expensive. Alfred Sauvy argues that those worst affected by the food shortages were the young, whose weight loss and vitamin deficiencies showed a marked aggravation during the Occupation—whereas weight loss among the middle-aged made for greater health and longevity in a number of cases.[11] Not only was wine rationing introduced, but 'jours sans', when no wine could be officially bought, became part of the secular liturgical year. If deaths through tuberculosis went up by 10 per cent, those through alcoholism went down; and, coincidentally or not, the death-rate in the six north-western *départements* most noted for their high alcoholism, fell on average by 17.5 per cent during the Occupation, despite the other privations of war.

REPRESSION AND PERSECUTION

However harsh living conditions were for those who could not afford black-market prices, the aspects of the Occupation that have made the deepest impression on the imagination of later generations are its repression of critics and the persecution of the Jews. Any government constrained to deal with an occupying power is immediately faced with the problem of involvement in the maintenance of order. The Germans had only some 10,000 police of all branches in France, and were therefore heavily reliant on the 100,000 or more French police to maintain their authority. It was, of course, open to Vichy to refuse all co-operation and to pay the inevitable price: loss of any say in the running of the country, and loss of any power to protect those whom it wished to protect. And many have argued that loss of autonomy would have been a smaller price to pay than the indignities that France suffered as a result of Vichy choosing collaboration.

Yet the history of repression in Occupied France was not just a question of carrying out German orders. Vichy had its own *chats à fouetter*. The role of Freemasonry in the anticlerical policies of the Third Republic made it an obvious target in Vichy's settling of scores with the Republican tradition.[12] Freemasons were excluded from the public services and Masonry was proscribed along with other 'secret societies' (13–19 August 1940). Bernard Fay, the obsessively anti-Masonic director of the Bibliothèque Nationale, was busily engaged in combing the Masonic archives for unsavoury evidence. Pétain assured him: 'La Francmaçonnerie est la principale responsable de nos malheurs; c'est elle qui a menti aux Français et qui leur a donné l'habitude du mensonge ... qui nous a amenés où nous sommes' (15 January 1943). Lists of Masons were compiled, and 14,600 names were published in the *Journal Officiel*. This itself engendered its own corrupt traffic—a suitable sum in the right quarter would save one's name from appearing, or so it was claimed. But Laval had too many political friends in Masonic circles to sympathize with the campaign, and its prescriptions were irregularly observed. So much so, that Action Française devotees denounced Laval as a sly protector of Freemasons— an imprudent accusation that contributed to the government counter-purge of Maurrassians when Laval returned to office in April 1942.

The Protestants, on the other hand, were never seriously at risk in these years. Despite the traditional enmity of the extreme Right towards them, their established position in high finance and the senior civil service preserved them from serious harassment. The Jews were another matter.[13] Anti-Semitism had been a significant factor in French politics since the 1880s, when Jews had been made the scapegoats for the economic recession of the late nineteenth century as well as that of the 1930s. The arrival of refugees from Tsarist pogroms was followed by those from Hitler's Germany, until by 1939 the Jewish population in France was 300,000, half of whom were not French citizens. Although the armistice of 1940 obliged France to repatriate refugees, Hitler had little interest at this time in taking back Jews. Quite the reverse: Germany immediately deposited 3,000 Jews from French Alsace in the Unoccupied Zone, and 6,000 west German Jews were shortly to follow, despite French protests. Moreover, Jews in the Occupied Zone who sought asylum elsewhere were forbidden by the Germans to return.

There was initially no attempt by Germany to persuade Vichy to embark on an anti-Semitic programme. And yet the Vichy government did so on its own initiative. Apart from the repressive measures affecting immigrant Jews, its Statute of Jews of 3 October 1940 also debarred French Jews from elected public bodies and from posts of responsibility in the public service, as well as from teaching and the news media. And, lest the pure essence of Vichy be contaminated, Jews were not permitted to enter the *département* in which Vichy was situated; there was to be no influx of Jewish tailors, hairdressers, entertainers, and property dealers to develop and enliven the future seat of government—Vichy being all the drearier for their absence. It has been suggested that these early, unsolicited measures may also have been intended to impress the Germans. By pursuing policies that were thought to be congenial to the Nazis, the government may possibly have hoped to keep German interference in French affairs to a minimum. Yet one must recognize the existence of a strong indigenous anti-Semitism in Vichy circles, even if neither the Marshal nor his successive Deputy Premiers were personally regarded as committed anti-Semites.

It has been further argued that the ideology of French anti-Semitism was less overtly racist than that of Germany—the issue of cultural identity being its mainspring, or so its militants claimed. (The enormous contribution of Jews to French culture, science, and scholarship was either studiously avoided, or dismissed with the comment that Jews, from Offenbach to Proust, had infected France either with decadent self-indulgence or decadent self-doubt.) It is often pointed out that Vichy did not introduce any discriminatory legislation against Blacks. Pétain's first Minister of Colonies was the Martiniquais, Henry Lémery, a close confidant of the Marshal; and it was German insistence, not French volition, which excluded Senegalese regiments from the Armistice Army. On the other hand, the official Vichy definition of a Jew was specifically racist—as a Vatican spokesman pointed out. It was Rome's view that a Jew ceased to be a Jew on conversion to Christianity.

Nor was there any denying that all three of Vichy's successive Commissaires Généraux for Jewish Affairs were noted anti-Semites: Xavier Vallat; the unambiguously racist Louis Darquier de Pellepoix; and Charles Mercier du Paty de Clam, son of the officer who had originally arrested Dreyfus. The summer of 1941 saw further occupations closed to Jews, and the extension to the Unoccupied Zone of a thinly disguised system for confiscating Jewish businesses. Quotas were put on the proportion of Jews allowed into the professions, universities, and even secondary schools in June 1942; and in the Occupied Zone, Jews were made to wear a yellow star (28 May 1942)—a measure that Pétain refused to permit in the Unoccupied Zone, even after the German unification of November 1942.

Eminent Jews were accorded exemptions from these various prohibitions and indignities; but many of the beneficiaries refused to accept them out of solidarity for their fellow Jews. There were already signs of the horrors to come. Many of the Jewish refugees from Central Europe were, technically speaking, German nationals, since their homelands had been annexed by Germany. With the co-operation of the French police, several thousand foreign Jews had already been rounded up and interned in the Occupied Zone in the summer of 1941; and May and June 1942 saw the regular deportation of such Jews to

Auschwitz in accordance with the extermination policy which had been planned at the Wannsee Conference in the previous January. The real turning-point came with Himmler's demand on 11 June 1942 for 100,000 Jews from France—including the Unoccupied Zone—and Laval's agreement to send 10,000 foreign Jews from the Unoccupied Zone, provided that French Jews in the Occupied Zone were spared—except in the event of a shortfall. By the following February, however, Laval was investigating a possible trade-off of French Jews for a guarantee of French territory at the peace settlement—even though in principle he wished to confine the deportation to foreign Jews, since the arrest of long-established neighbours and colleagues would represent a greater test of public loyalty than previous issues had done.

When, and to what degree, Vichy realized what was happening to deported Jews remains a matter of controversy. Certainly, French Jewish leaders warned Laval in August 1942 that 'hundreds of thousands of Jews' were being 'methodically exterminated' in eastern Europe; but it was tempting for Vichy to dismiss such reports as enemy propaganda. In fact, over 70,000 Jews were deported from France—a third of them French citizens—and only 2,500 survived. It has been argued that Laval's attempts to save French Jews as distinct from foreign Jews may have prevented the numbers of those deported from being even greater than they were; and it is often pointed out that France lost only 26 per cent of her Jews, including foreigners, compared with 55 per cent for Belgium and 86 per cent in Holland, where admittedly their concentration in Amsterdam made them particularly vulnerable. On the other hand, the involvement of the French forces of order in locating foreign Jews and putting them in internment camps undoubtedly enabled the Germans to deport far more than they could have done if they had been entirely reliant on their own forces. Gypsies were likewise herded into French concentration camps by the Vichy authorities, and many of these were later sent eastwards by the Germans to the death camps.

If the fate of the Jews remains Vichy's greatest crime in the eyes of historians, it must be confessed that it ultimately put less of a strain on public loyalty than the government's acceptance of labour conscription for work in Germany (STO). Indeed, there was considerable resentment at first that Jews were exempt from it, and many Gentiles naïvely assumed that the deportation of Jews to camps was a belated rectification of an unjust anomaly.

Of the 1,800,000 Frenchmen who had been in prisoner-of-war camps since 1940, roughly half were based in Germany; and of these, half had been put to work in factories and another third on farms. But this was not the skilled and semi-skilled labour that the German munition factories were seeking. And although by 1943 some 250,000 'suitable' prisoners of war had been offered and had accepted civilian status to join the German work-force, this still did not meet the economy's needs, despite the exploitation of Polish and Russian labour. Nor did the *relève* scheme of June 1942, by which Germany agreed to repatriate one French prisoner of war for every three skilled workers who volunteered for work in Germany. There consequently followed a gradual move to conscription between August 1942 and February 1943, when the Service du Travail Obligatoire systematically embarked on the enlistment of whole age-groups.

Estimations of the number of Frenchmen sent to work in Germany vary from 650,000 upwards. The majority were lodged in camps and worked an eleven or twelve-hour day—as many of their grandfathers had done in the French factories of the 1870s. In addition, there were also some 44,000 French women drafted to Germany. France, in fact, was second only to Poland as a source of labour for Germany, and was second to none in supplying skilled workers. Indeed, it was only the larger number of Russian and Polish female conscripts that prevented France from heading the overall list of deported labour: 3.3 per cent of the total French population found themselves working in Germany under one scheme or another, compared with 3.4 per cent of the much smaller Belgian population and 3 per cent of the Dutch.

The productivity of French workers in Germany was almost as high as that of their German counterparts—and much higher than that of Dutch or Danish deported labour, whose levels were only 50 to 70 per cent of the German equivalent. It has been suggested that the lower standard of machinery and nutrition in wartime France may have made working in Germany less of an ordeal for French conscripts than for the Dutch and Danish. Whatever the explanation, more than 10,000 French workers and prisoners of war chose to stay in Germany after the war.

VERDICTS

Even the sternest critic would concede that Pétain and Laval devoted a lot of time and considerable skill to trying to minimize the impact of German demands on France. But this leaves unanswered whether the moral price they paid in becoming the reluctant accessories to the horrendous aspects of German occupation outweighed the advantages they obtained for France. Laval himself claimed that the price he paid was a personal one. If he had not undertaken the ungrateful role of dealing with the Germans, someone else would have done—probably less successfully—whereas the concessions he extracted from the Germans were enjoyed by the tens of thousands of Frenchmen whom his haggling saved from labour service in Germany, and by the individual Jews whom his backstage dealings saved from the fate that he could not prevent befalling others. To objections that his policies strengthened rather than weakened Germany, his reply was that Bolshevik Russia was a greater danger to the democratic world, and that National Socialism was merely a temporary set-back in Germany's slow evolution towards a peaceable democracy.

In a declaration said to have been drawn up by Maurras's colleague, Henri Massis, in August 1944, Pétain justified his own rule by maintaining: 'If I could not be your sword, I tried to be your shield'. And at his trial he claimed: 'While General de Gaulle carried on the struggle outside our frontiers, I prepared the way for Liberation by preserving France, suffering but alive.' Laval likewise asserted at his trial that he had enabled France to 'avoid the worst', and that he had acted as a 'screen' between the people and the Germans.

Historians are still a long way from being able to test Robert Paxton's contention that a France directly administered by the Germans would not have fared any worse than it did under the mediating filter of Vichy. Apart from the speculative nature of the question, the experience of other occupied countries is still insufficiently documented to make comparisons easy. What is clear, is that the degree of exploitation depended more on German attitudes towards the conquered people than on the constitutional form of political control. Those countries with some ethnic affinity to the Germans fared best, probably because Germany was pursuing long-term goals of close association or, in some cases, assimilation, and the advantages of co-operation with Germany needed to be demonstrated to their inhabitants. There was no hope or desire of assimilating France, and therefore little to be gained from moderating the level of exploitation. German provisional plans for the post-war settlement showed an independent France (shorn of the traditionally disputed territories such as Alsace-Lorraine), apparently lying outside the free-trade zone mapped out by Germany. It is not surprising, therefore, that Norway and Denmark should appear to have come off much better than France; and that Holland and even Belgium (with its Flemish element) seem to have done marginally better, despite their lacking a Vichy-style element of independence. It has been estimated that France provided Germany with more food than any other occupied country, including Poland, with the consequence that nutritional levels in France were probably the lowest in Occupied Western Europe—with the one exception of Italy, which was a prolonged theatre of war.

If the debate continues today with less bitterness and a more resigned acceptance of the limits of human nature than it did in 1944–5, it continues nevertheless.

Public Attitudes

In 1944, however, the moral distinction between the roles of witness, judge, and accused were much more blurred. There was a sense in which everyone who had lived through the Occupation was vicariously under scrutiny. The attitude of the man in the street to Vichy and the Germans during the Occupation had been as much a matter of temperament as of class or political allegiance. Although there were unquestionably more resisters on the Left than on the Right—given the nature of Nazi Germany and the ethos of Vichy—it would be a gross oversimplification to see the pattern of resistance and collaboration as corresponding to these traditional divisions. Similarly, the distinction between Vichy and the Germans was also crucial, acceptance of one frequently being combined with bitter hostility to the other. Most people passively accepted the Vichy regime as attempting to make the best of a bad job, given the apparent military supremacy of Germany. Active resistance was initially confined to a handful of individuals who were so outraged at the defeat of France that they felt compelled to assert their revulsion by militant subversion. There were royalists and men of the extreme Right who opposed both Germany and Vichy because they saw them as the agent and the accomplice of France's humiliation, despite a certain sympathy for one or other of the types of government they represented. The bulk of the population, however, continued to keep their heads down, and tried not to become involved with either side.

What gradually changed this situation was the worsening of conditions in France following the occupation of the whole country in November 1942. The mass conscription of able-bodied Frenchmen for compulsory work in Germany alienated many people from a government no longer able to protect them; and the next nine months saw 150,000 go into hiding to avoid conscription, while more than 300,000 were to do so in the first half of 1944. Clandestinity, however, was not resistance, even if a sizeable number eventually found themselves in the Maquis formations which came into being in significant numbers in the winter of 1943–4. Resistance, after all, was to embrace no more than 1 per cent of the adult population. But it represented disaffection; and when concealment and nourishment became more difficult in the winter months, it was more tempting to accept the material support of subversive organizations. Moreover, with the turn of the tide of war in 1943, the solid support for Vichy that had dominated France in the first two years of the Occupation was clearly melting at the edges. Nevertheless, it took the Allied invasion of northern France in June 1944 to provide the decisive impulse, when the question of being on the winning side was posed at last in all its immediacy.

The Right

For the professional politicians of the Right, the Occupation years had much to offer. Yet Vichy and Paris attracted markedly different types of clientele. Most of the leaders of inter-war neo-fascism gravitated towards Paris, where the German-subsidized news media offered broad opportunities of rewarding employment. But few figures of the extreme Right reached ministerial rank at Vichy, although a number of them were given administrative duties appropriate to their hates or enthusiasms. Jacques Doriot suffered the triple disadvantage of seeming too chauvinistic for the Germans, too extreme for respectable Vichyites, and too tainted by his Communist past for conservatives in general. Laval nevertheless regarded him as a potential rival, and was at pains to turn Abetz against him—even to the point of trying to persuade the Germans to dissolve Doriot's PPF, albeit unsuccessfully. The result was that Doriot had to content himself with organizing the Légion des Volontaires contre le Bolchévisme, which attracted over 12,500 enthusiasts, many of whom were to die in the snows of Russia.

The only prominent inter-war *ligueur* to achieve ministerial office under Vichy was Marcel Déat, who was valued by Laval and the Germans as a counterweight to Doriot. Yet, at the end of the day, the quasi-fascist organizations were too divorced from the main currents of French opinion to be of more than limited interest to the Germans. Only Joseph Darnand's Milice, founded in January 1943, attracted their support, but largely as an instrument of anti-Resistance repression. The 45,000 volunteers who joined the Milice embodied a mixture of motives, a number of which were straightforwardly political. Some sought to avoid German labour service, or wanted to obtain a more privileged existence in a time of acute shortage; others saw the Milice as a bastion against anarchy. But the most disreputable found it an opportunity to settle scores or to indulge sadistic inclinations in its torture chambers in the Rue Lauriston, where the Gestapo methods practised in the Avenue Foch found ready imitators.

The collaborationist Right also had its literary lions, notably Pierre Drieu La Rochelle (now entrusted with the *Nouvelle Revue Française*), Louis-Ferdinand Céline, and Robert Brasillach of *Je suis partout*. They represented, in Stanley Hoffmann's words, 'a romanticism—elitist, sentimental, attractive in its nostalgia for rejuvenation and action, and repulsive in its pose of virility, ... its *fin de race* character, and above all its vindictiveness. These were men to whom the petty, compressed, depressed, and querulous France of the 1930s was anathema.'[14] Brasillach and Drieu La Rochelle had both adopted a patriotic stance in 1939–40. But Brasillach saw the Occupation as an opportunity for national regeneration, while Drieu La Rochelle saw it as the beginnings of a united and renovated Europe. Both were to become progressively disillusioned with the German leaven in 1943—as, indeed, had Céline. Drieu La Rochelle eventually committed suicide in April 1945, while Brasillach was executed at the Liberation.

Collaborationist circles in Paris also had their fair share of drifters, as Stanley Hoffmann and Richard Cobb have described: 'at ease nowhere, too cynical for intellectual daydreams, too undisciplined for the gangs, accumulating only grievances against all established orthodoxies'.[15] There were likewise those 'who had made a mess of their careers in the 1930s ... had been caught with their hands in the till, had been condemned for minor sexual offences ... and ... would naturally welcome the opportunity to make a fresh start and to avenge themselves on a social system and on a political regime that, as they would see it, had let them down'.[16] These and others made up a 'rather seedy army of frayed fanatics in suits shining at the seats and at the elbows and stained under the armpits'. But, semi-pathetically, there were also 'the collaborationists ... who often could not even find anyone to collaborate with', for 'the Germans did not keep absolutely open house'.[17]

Vichy, on the other hand, attracted a very different segment of the Right. Charles Maurras was too old and deaf to play a significant personal role at Vichy, but several of his Action Française disciples held influential posts in Pétain's entourage. The Germans were understandably suspicious of *Action Française*, with its celebration of France's role in the world, and the paper ceased to appear in the Vichy zone after the German take-over of November 1942. Indeed, Maurras himself came close to arrest in 1943. Similar suspicion was displayed towards that other veteran of the inter-war Right, Colonel de La Rocque, who was treated coolly by Vichy as well as by the Germans. As for the royalist pretender, the Comte de Paris, he alternately offered his services to Pétain in August 1942 and then to the Allies a few months later, but neither side was interested in a restoration.

As for the wider category of propertied conservatives, it has been argued that the neo-corporatist sentiments of Vichy aroused their hopes of exercising an effective influence on government economic policy.[18] The experience of the Popular Front had shaken their confidence in the ability of their parliamentary representatives to defend their interests, while on the other hand, they felt little enthusiasm for the quasi-fascist movements of the extreme Right, which tended to appeal to the *petits bourgeois* and the artisans—and to certain sections of the professional middle class. Vichy, on the other hand, appeared to offer a state structure which eschewed both the uncertainty of parliamentary politics and the *étatism* of fascism, and promised instead a privileged place to economic interests. Conservatives likewise found it consoling that Vichy was tending to replace the professional politicians of the Third Republic with respectable apolitical elements, drawn from the civil service and local *notables*. But many of the corporatist plans of Vichy remained little more than blueprints; and in any case, the exigencies of the Germans always took priority over whatever proposals French business interests wished to promote.

The Left

The Radical and Socialist parties did not have an official party line on how to behave towards either the Germans or Vichy. The Communists, by contrast, gave a firm line on Vichy—and, after June 1941, on the Germans as well. This not only reflected the much more disciplined nature of the PCF, but also its greater success in keeping a tangible organization going during the Occupation.

The broad clientele of pre-war Radicalism contained within it a split between those who accepted the ethos of Vichy and those who did not. Many of the peasantry were attracted by the 'back-to-the-land' aspects of Vichy, but others were suspicious of its initial clerical overtones—as indeed were the archetypal *démocrates laïques* of the party, who saw Vichy's anti-Masonic measures as a direct attack on the secular conscience of the nation.

Until the Communists belatedly joined the Resistance in the summer of 1941, the Socialists were the most clearly identifiable element in the amorphous collection of groups and individuals that made up the active opposition to the Germans, and to Vichy. Trade-union militants bitterly denounced the labour legislation of Vichy and the part played in it by their former colleague, René Belin. The only other prominent Socialists to accept office from Vichy were Paul Faure and Charles Spinasse, a former Popular Front Minister of the National Economy. As for Léon Blum, he and five of his Popular Front ministers were arrested in September 1940 and were later put on trial by Vichy for their alleged responsibility for the French defeat of 1940. But even the Riom court found it impossible to make the charge stick.

Turning to the Communists, their record in the 1939–41 period was not an easy one for apologists to defend. The bulk of the French Communists were acutely embarrassed by the Nazi–Soviet pact of August 1939, and remained so until Hitler's invasion of Russia in June 1941. Even so, their leaders tried hard to come to terms with the Germans in the Occupied Zone, and Marcel Cachin publicly encouraged the workers of the Paris region to fraternize with the German troops. Nor were they above shopping some of their former Radical and Socialist allies to the occupying authorities. On the other hand, the Communists were consistently hostile to Vichy. Not only was the whole ethos anathema to them, but its intensification of the anti-Communist measures taken by the Daladier government encouraged them to build a highly organized clandestine network in the southern zone, which was to prove very valuable once active resistance got under way in subsequent years.

6

Resistance and Liberation

THE Resistance in France played three roles: it was part of the campaign against Germany, it was a forum for planning a new France, and it was an instrument for re-creating French self-respect. The last of these was arguably its greatest achievement, far outweighing the material damage that its limited resources could inflict on the occupying forces. The information it supplied to the RAF, the Allied air-crews whom it led to safety, the acts of sabotage that harassed German communications and morale, all these represented a tiny percentage of the factors that brought the Occupation to an end. Even the leadership of the provisional government that was to administer France at the Liberation was largely determined by other forces: it was primarily the result of distant decisions in London, Algiers, and New York, even if the predilections of the internal Resistance were among the many factors taken into consideration. But the heroism of the Resistance created a source of inspiration that helped to sustain the hopes of many during the humiliations of both the Occupation and the post-war chaos. It gave to the names of its clandestine networks and newspapers a legendary quality that long outlived their material achievements. There survived a salutary myth of the Resistance, just as there grew up a spurious counterpart that extended the mantle of membership to large numbers of people who only abandoned Vichy when Vichy abandoned them, months after the Anglo-American victory was a certainty.

The term, 'the Resistance', may suggest a disciplined organization which sympathizers could formally join, like the Territorial Army or the Rotarians. In reality, however, it was as fragmented and diffuse as 'the Mafia' or 'the Enlightenment', since its clandestine nature inevitably made contact difficult between the various groups that independently came into being at various stages of the Occupation. Nor was mutual contact always sought or desired; political differences often created suspicion and distrust between them, particularly between Communist-dominated groups and those with a significant right-wing membership. On the other hand, Resistance groups were equally a meeting-ground for men of otherwise conflicting political loyalties, and were a prime source of hope that post-war France might likewise find Communists, Catholics, and other former antagonists working together to improve society.

In 1943, Jacques Soustelle estimated the active Resistance as representing no more than 1 per cent of the adult population. For what it is worth, 170,000 were officially given veteran status as 'Resistance volunteers' after the war, while 100,000 were reported to have been killed while engaged in Resistance activity. Yet the hard-core resisters, those who were engaged in activities of major risk, would probably not have exceeded 45,000, while at the other end of the scale, the number of sympathizers prepared to take the risk of reading Resistance literature may have totalled as many as two million. Romantics have tried to enlist many others among the angels, such as the newspaper-vendor who allegedly said: 'Voici ta feuille, grand con' to a regular German client throughout the Occupation; but similar greetings were being thrown at po-faced officials from the British Council a decade later.

The early months of the Occupation, when defeat seemed absolute, saw few active resisters: only fanatics and misfits with nothing to lose, as they subsequently enjoyed describing themselves. 'We only had on our side Jews, negroes, hunchbacks, cripples, failures and cuckolds', de Gaulle later remarked to Mendès France. Discreet landmarks in the dark months of November and December 1940 included the surreptitious launch of Henri Frenay's newspaper, *Combat*, and Robert Lacoste's and Christian Pineau's *Libération*; the same period also saw the foundation of a short-lived Resistance group at the Musée de l'Homme in Paris, where the anthropologist, Jacques Soustelle, first set foot on the tortuous political path that subsequently made him de Gaulle's most powerful aide and then his most bitter critic.

Resistance in the Occupied Zone had a visible enemy and was specifically anti-German, which facilitated the coming together of men of widely differing political persuasions. In the south, however, where Vichy was the only legal authority, opposition tended to be dominated by men of the Left whose political sympathies inclined them against the paternal conservatism of the regime.[1] Thus, Emmanuel d'Astier de La Vigerie's Libération-Sud specifically sought to bring together Socialists, Communists, and Catholic trade-unionists in mutual opposition to Vichy's policies, while the eclectic Franc-Tireur became the focal point of the unorthodox Left. On the other hand, Libération-Nord in the Occupied Zone grew from the joint enterprise of Socialist and Catholic trade-unionists, and was aimed directly at the German forces—a strategy not easily accepted by the Communists before June 1941. Spanning both zones was Front National, which became increasingly Communist-dominated after June 1941, but which continued to enjoy the support of leading Catholic resisters such as Georges Bidault of Combat, as well as prominent representatives of other persuasions. Indeed, its senior committees included François Mauriac, Jacques Debû-Bridel of the Conseil National de la Résistance, and Max André of the MRP, as well as a couple of priests and a Protestant pastor. At the same time, there existed more homogenous Resistance groups, corresponding to particular segments of political and religious opinion, such as the Socialist Libérer et Fédérer and the Catholic Témoignage Chrétien in the south, or the conservative Organisation Civile et Militaire in the north.

Links with de Gaulle's movement in London were initially tenuous or non-existent. The main architect of closer contact was Jean Moulin, a former prefect who became closely involved in the activities of Combat, a Lyon-based organization with a strong Catholic following. Moulin played a leading role in securing for de Gaulle the allegiance of the Conseil National de la Résistance, conceived by the Socialists in 1942, and which rapidly embraced a wide spectrum of Resistance opinion. He was also prominent in the creation of the Mouvements Unis de la Résistance the following year, drawing together Combat, Libération-Sud, and Franc-Tireur, with de Gaulle as their acknowledged leader.

Arms and money from London and Algiers tended to go to MUR and its associated networks, and not into the Communist-dominated Front National and its subsidiaries. While Resistance groups were glad enough of the help provided by the British SOE, it was not without its attendant problems. The archetypal case was the British agent who injured a leg in a parachute-landing and had to be taken to hospital—with all the risks of self-revelation under the anaesthetic. Equally remembered was the British expert in French peasant customs who was dropped into France in ethnic peasant costume, his *folklorique* disguise causing initial amusement and then acute embarrassment among his hosts entrusted with his safety.

Moulin's subsequent capture, torture, and death at the hands of the Gestapo in the summer of 1943 was one of the great tragedies of the Resistance. It was made more bitter by the suspicion that he had been betrayed by a Resistance faction hostile to his policies—which had their critics even among those who admired his courage and organizational flair. Among others, Henri Frenay of Combat objected to Moulin's attempts to effect a Gaullist take-over of the Resistance, and he also regretted Moulin's concessions to the Communists, which he saw as a mistaken attempt to create a spurious underground Popular Front.

The Métro-platform assassinations inaugurated by the Communists in 1941 had long raised questions as to whether the alleged benefits were worth the savage reprisals and the general tightening of German rule that followed them. De Gaulle openly condemned them as an expensive and fruitlessly premature activity; but many groups of varying political persuasion refused to accept de Gaulle's insistence on obedience in matters of method and policy. At the same time, considerable tension often existed between the relatively well-disciplined Resistance groups organized by ex-officers, and the free-wheeling *francs-tireurs* who put too much faith in the ability of enthusiasm to triumph in the end. The distinction between Resistance and plain crime became harder to discern, as the problem of staying alive encouraged theft from the non-combattant population. There were cases of summary executions of 'criminal' resisters by more disciplined groups, while the political rivalry between Communists and other groups led to further settling of scores.

De Gaulle's efforts to win control of the internal Resistance were only a part of his larger attempt to be accepted by Britain and America as the provisional government of France. His London-based Free French National Committee of 24 September 1941 claimed to be 'the provisional guardian of the national patrimony'; but these assertions, if aesthetically appealing, were constitutionally and diplomatically questionable. There was no doubt that the last parliament of the Third Republic had delegated its authority to Pétain, and de Gaulle based his position on the argument that Pétain had morally abdicated when he allowed the French government to become the tributary of Germany. De Gaulle had nevertheless found an early admirer in Churchill, despite the latter's oft-quoted postscript that the hardest cross he had had to bear was the cross of Lorraine. De Gaulle's truculence, as he himself later explained, was a reflection of his weakness: it was his means of obtaining attention. He suspected that Britain wanted to eradicate the French from Syria, and that South Africa had designs on Madagascar. Until November 1942, however, it was in Britain's interests not to close off all possibility of negotiation with Vichy; and although France had broken off diplomatic relations with Britain on 8 July 1940, the first two years of the regime saw several unofficial if abortive attempts at contact. It was therefore quite logical for Britain's recognition of de Gaulle to remain restricted to the definition accorded on 28 June 1940—leader of 'all Free Frenchmen who rally to him in support of the Allied cause', a mere tautology in de Gaulle's opinion.

His following in London was pitifully small and lacking in senior personnel, the overall total rising from 7,000 in July 1940 to some 35,000 by the end of the year. But of the 18,500 naval personnel who found themselves in Britain in the summer, all but 250 of them opted to go back to conquered France when the opportunity was offered to them—a telling indication of how strong the feeling was that the war was virtually over. De Gaulle's followers came from the Right rather than the Left, and his later rallying of support overseas augmented them with a significant number of imperially minded *colons* and administrators. By contrast, the left-wing Pierre Cot, architect of French aerial rearmament, was positively rebuffed by de Gaulle when he attempted to join him. The result was that London contained a fair number of Third Republican exiles who regarded de Gaulle as an arrogant, would-be dictator; and although the French service of the BBC was engaged in all-day broadcasts to the French, de Gaulle's Free French organization was allowed only two five-minute programmes a day.

Furthermore, America's involvement in the war in December 1941 brought a further complication to Anglo-Gaullist relations, since America, like Russia, continued to have diplomatic representation at Vichy. Roosevelt's vision of the post-war world was very different from that of either Britain or de Gaulle. He saw little place in it for the great colonial empires of the pre-war era, and was later to speak of the desirability of putting parts of the French empire under United Nations control. Apart from its ideological aspects, Roosevelt's scheme would open up the French empire to American commercial exploitation. Such a vision was much more likely to be accepted, however reluctantly, by French politicians who had accepted the defeat of 1940, than by the Gaullists who had not. The Americans, moreover, were under the illusion that Vichy was divided between a neutralist wing, represented by Pétain and Weygand—with whom a post-war settlement was possible—and the collaborationist wing of Laval, which had to be undermined if possible in the interests of the neutralists.

Herein lay part of the attractiveness to America of Darlan's defection in North Africa. The Anglo-American invasion of North-West Africa on 8 November 1942—without consulting de Gaulle—brought about a bizarre switch of allegiances on the part of Admiral Darlan. Commander-in-Chief of Vichy's armed services since April 1942, he personally took charge of the defence of this strategic corner of the empire for six days, and then he not only accepted a cease-fire but agreed to come over to the Anglo-American side, when the Allies threatened to take over government of the Maghreb. Darlan thereupon became High Commissioner of 'l'État français', justifying his action to his subordinates on the unlikely grounds that Pétain secretly approved of his conduct. In fact, Pétain had just appointed the French commander in Morocco, General Auguste Noguès, as 'sole representative in North Africa of the Marshal, Head of State'; but Noguès, on his own initiative, entrusted these powers to Darlan two days later.

When this congenial situation was extinguished by the assassination of Darlan on Christmas Eve 1942, the Americans transferred their favour not to de Gaulle, who was still in London, but to General Henri Giraud, who was made High Commissioner. Giraud had escaped from a German prison in April 1942; and, after declaring his loyalty to Pétain, he had been secretly brought to North Africa by a British submarine in November, as part of the general Allied strategy of keeping North Africa independent of Axis influence.

North-West Africa now became an alternative focus of loyalty for those inhabitants of Vichy France who were increasingly perturbed about the direction events were taking in the *métropole*. Those who transferred their allegiance included ex-Vichy ministers, such as Flandin and Pucheu, and senior civil servants such as Maurice Couve de Murville. For six months the Maghreb continued to be ruled in the spirit of the National Revolution—but with the Allies, not the Germans, as the spectre at the feast. Marshal Pétain's portrait continued to hang in all government offices, and a systematic campaign of expulsions and arrests was conducted against suspected Gaullists.

These were uneasy months of self-doubt for de Gaulle, during which he privately wondered whether Giraud was better placed to represent France. Even so, Britain remained attracted by the fact that the Gaullists were the only French group to have committed themselves to reconstructing the Anglo-French alliance, while at the same time de Gaulle persuasively portrayed himself as a sound bulwark against the possibility of a Communist take-over in post-war France—a consideration that counterbalanced British suspicions that he was a potential dictator. Moreover, Giraud's anti-democratic views and pro-Vichy (if anti-German) sentiments were fast becoming an embarrassment to the Americans. In the meantime, de Gaulle was enjoying a growing esteem among the French-based Resistance, including the Communists, and was particularly well regarded by the Russians. The outcome was that an uneasy merger was effected on 3 June 1943 between de Gaulle's London-based Free French National Committee and Giraud's North African administration—the new body taking the title of the Comité Français de Libération Nationale. While few of its members initially regarded themselves as personal partisans of either de Gaulle or Giraud, Giraud's influence rapidly waned, until de Gaulle finally succeeded in manœuvring him off the committee altogether in April 1944.

Despite Anglo-American reluctance to follow the Russian example and formally recognize the committee as 'the representative of the state interests of the French Republic', de Gaulle gave it a more governmental character in the autumn of 1943 by bringing in former figures of the Third Republic, such as Henri Queuille and Pierre Mendès France, as well as members of the French-based Resistance, such as Emmanuel d'Astier de La Vigerie, François de Menthon, and Henri Frenay. Although the committee was not privy to the preparations for the Normandy landings in June 1944, the Commander-in-Chief, General Eisenhower, believed that the conquered areas would have to be civilly administered by the committee—and he acted accordingly. To pre-empt any alternative, the committee declared itself the provisional government of the French Republic three days before D-Day; but neither Britain nor America made formal recognition of the change until 23 October, when most of France was free of German occupation.

In France itself the later part of 1943 had seen the development of an important new dimension in the internal Resistance—the growth of guerrilla-style armed groups, who took their name from the scrub they inhabited, the Maquis. The larger size of their units enabled them to undertake the more substantial tasks of blowing up communications and installations, as well as mounting attacks on small German detachments. Apart from STO evasion and a general sense of the turning tide of war, they were strengthened by pre-war Republican refugees from Spain, who brought with them experience of the Spanish Civil War and a determination not to be interned or extradited. But even in January 1944, the *maquisards* were still thought to be no more than 30,000, mainly in southern France. The London headquarters of the Forces Françaises de l'Intérieur—incorporating the Maquis and the Communist-dominated Francs-Tireurs et Partisans Français—had little understanding of the principles of guerrilla warfare, and made a number of major errors, none greater than the proclamation of a general insurrection on 6 June 1944, which resulted in an extravagant loss of life in uncoordinated action, even if it did help to delay German units joining the main theatres of war.

6 June was the date set for 'Operation Overlord', the gigantic Second Front in France that the Russians had been demanding for so long, and which was

now launched against the Caen–Cotentin coastline. The success of Anglo-American action against U-boats in 1943 had enabled America to build up her manpower in Britain to 1,300,000 by May 1944, thereby making Britain the springboard for the main attack. The Germans had only fifty-eight divisions in France, of which a mere twenty-four were of a comparable size to what the Allies were launching against them. Furthermore, the Germans were slow to reach the battle-area in strength, partly because Hitler insisted that all moves should be referred to him, and partly because the Germans, still fearing an attack in the Pas-de-Calais, did not wish to commit too much of their strength to western France. It was this uncertainty that enabled the Allied spearhead of five divisions to establish a secure landing-place for further forces, despite the superior German numbers that were available for a firm counter-thrust, had the Germans chosen to launch one.

A major factor in the initial success of the Allies was their overwhelming superiority in the air, thereby reversing the situation in 1940—11,000 planes against the 500 comparable aircraft that the Germans had at their disposal in the campaign area. Paris fell to the Allies on 25 August, and three weeks later American forces had reached the German border. In the meantime, a Franco-American force under General Alexander Patch invaded south-east France on 15 August and started to make its way up the Rhône Valley, attracting to it large numbers of Resistance fighters. By 12 September, this 'Operation Anvil' had already met with the southern fringe of Operation Overlord less than 150 miles from Paris; and by the end of the year most of the country was liberated.

The French contribution to the liberation of their own country was severely limited by the distribution of French forces in the early summer of 1944. Of the 400,000 outside Vichy France, only half were armed, 150,000 were scattered on garrison duty in various parts of the empire, and 120,000 were fighting in Italy. Those in Italy made an important contribution to Operation Anvil in southern France, but the invasion of northern France was almost entirely devoid of official French military participation. Even General Leclerc's celebrated Second Armoured Division, which was to be accorded a key role in the liberation of Paris, had to be shipped from Algeria to England. So bitter was de Gaulle at the minimal role and consultation afforded the French, that twenty years later he refused to attend the commemorative ceremonies of the Normandy landings. Even the FFI, which played an important role in disrupting communications, had only enough arms for half its members, with the result that the losses it was able to inflict on the Germans amounted to only about 2 per cent of the enemy's casualties.

The Allied invasion of France confronted both Vichy and de Gaulle with the problem of contriving a transfer of power which would respect order and legitimacy, yet would leave the balance of moral dignity with oneself rather than with the other side. De Gaulle simply refused to receive the emissary whom Pétain sent to negotiate 'the avoidance of civil war', while Laval for his part attempted to set in motion a plan for reconvening the National Assembly (12 August 1944). The Germans, however, forced the Vichy leadership to withdraw beyond the range of the advancing Allies; and, protesting at their removal, Pétain and Laval both refused to exercise their functions—leaving France in a state of political vacuum until the Gaullists arrived. After years of waiting, Jacques Doriot had at long last the brief chance of heading a government in exile, but with no means of making his will felt in France. Like Aguirre, adrift on a raft of squabbling monkeys, his time was spent in bitter quarrels with his fellow refugees, until he was eventually killed in mysterious circumstances— allegedly by an Allied bomb hitting his car.

The failure of Laval's scheme to recall parliament removed a potential threat to de Gaulle's authority. Another threat, more imagined than real, was the spectre of a grass-roots social revolution, carried out by left-wing Resistance groups at the level of local government. In fact, most Resistance groups, including the Communists, were well aware of the importance of initial French unity, as long as the destiny of the country depended so much on foreign factors. Even so, the saga of the liberation of Paris gave rise to the myth of a social revolution nipped in the bud by reactionary Gaullists.

As the Allied forces neared Paris, Henri Rol-Tanguy and other Communist Resistance leaders in the capital launched an insurrection on 19 August, which rapidly occupied the Hôtel de Ville and other key buildings. Their position was very insecure, however; and, fearing massive German reprisals, the more cautious Resistance leaders, including the Gaullists, Alexandre Parodi, and Jacques Chaban-Delmas, negotiated a temporary truce with the German garrison commander, General von Choltitz, with the help of the Swedish Consul-General (20 August). But the cease-fire was imperfectly observed on both sides, and the Resistance leaders, including the Gaullists, decided to resume fighting. Communists, both then and subsequently, have suggested that the truce was a desperate attempt by the Gaullists and propertied classes of Paris to prevent a left-wing take-over of the capital reminiscent of the Paris Commune of 1871—and that it was only the expected arrival of Allied forces that decided them to resume the fight, the inference being that the regular troops could be relied on to prevent the wild men of the Left taking power. Such accusations are hard to substantiate; but it is nevertheless clear that de Gaulle was at pains to keep the Conseil National de la Résistance well away from the levers of power in the days to follow.

It had been Eisenhower's intention to outflank and surround Paris; but the news of the Paris insurrection decided him to order General Leclerc's Second Armoured Division to enter the capital—an imaginative step which did much to bolster French morale. Choltitz, who, to his eternal credit, ignored Hitler's order to set fire to Paris, surrendered to Leclerc on 25 August; and de Gaulle was able to make a triumphal entry on the same day, claiming that Paris was 'libérée par son peuple, avec le concours des armées de la France'. If this was ungenerous to the Allies, it reflected de Gaulle's acute awareness of the need to restore French self-respect. He then made it clear that the CNR as such would have no part in the future State, even if some of its members were offered government posts in their own right. Bidault, in fact, was the only member to accept what he was offered—the Quai d'Orsay. When the CNR came—uninvited—to take part in the triumphal procession down the Champs Élysées, de Gaulle kept Bidault firmly behind him. Whenever Bidault quickened his pace to draw level, his plump thighs chafing in the hot weather, he received a curt: 'Monsieur, un peu en arrière, s'il vous plaît.' Similarly, when Bidault and the CNR had earlier invited de Gaulle to follow tradition and proclaim the new Republic from the balcony of the Hôtel de Ville, de Gaulle refused, pointing out that the Republic had never ceased to exist as long as he had held the banner of resistance in London and Algiers. Besides, like Frederick William IV, he had no intention of seeming to pick up a crown from the gutter.

September 1944 allegedly found de Gaulle contemplating early retirement: 'It is necessary to disappear, France may again have need of a pure image. I must leave her that image. If Joan of Arc had married, she would no longer be Joan of Arc.' Whatever the truth of the matter, he formed a new government of 'national unanimity' on 9 September 1944, two-thirds of it drawn from the former Algiers provisional government, and a third from the French-based Resistance. Its political span embraced two Communists, four Socialists, three MRP (Catholic democrats), three Radicals, and a conservative, together with nine unaffiliated figures, of whom most were followers of de Gaulle—Gaullism as such having no party until 1947. Connoisseurs of irony were subsequently to relish the inclusion of Robert Lacoste (Industry) and later of Jacques Soustelle (Information), who were to play such contrasting roles in the events that brought de Gaulle to power fourteen years later.

The assertion of government control over the provinces proved less difficult than de Gaulle had anticipated. As the Germans retreated, local government had been taken over by Resistance bodies, the *comités départementaux de libération*, which de Gaulle feared might attempt to become similar to soviets, arrogating to themselves the function of keeping a watching brief on the local echelons of the public administration. He therefore despatched *commissaires de la République* into the provinces, to ensure that the newly appointed prefects were not obstructed by left-wing militants with visions of a grass-roots social revolution; and he also incorporated the FFI into the regular army (23 September 1944) and dissolved the paramilitary *milices patriotiques* (28 October). To his relief, the Communists were co-operative in this and other potentially explosive matters.

The tasks confronting government, both short-term and long-term, were dominated by the need to re-create an economic base for the country's material recovery. As indicated in Table 6.1, the total loss of population sustained by France as a direct and indirect result of the war was just under one and a half million—which was admittedly fairly modest compared with over three million in the First World War. But, added to the bereavement of families and the long-term loss of these people to the labour force, there were the problems of repatriated labour-service conscripts and prisoners of war returning to France. Although 71,000 individuals had already escaped from captivity, the rest, emerging *en masse*, presented problems of employment in a dislocated economy. Moreover, it was claimed that 60 per cent of married men returning from STO found their marriages had broken up, and a large majority of engagements had likewise fallen through.

Equally unwelcoming were the material circumstances to which these bewildered exiles returned. Less than half the rail network was serviceable; and after German requisitioning and the destruction of the 1944 campaigns, the average age of surviving French machine tools was at least twenty-five years, creating a situation in which a given task required eight times the number of workers to perform it as it did in America. Horses were down by a third (1938–44)—a blow to butchers as well as to transport—and when hostilities ended in May 1945, imports were running at five times the level of exports. The franc had fallen to a sixth of its value between 1939 and 1946, with debtors as the only beneficiaries—the government being among their number. Serving the national debt was now costing only 12 per cent of government revenue in 1946, instead of twice this in 1939, despite the large increase in the capital borrowed.

The hard winter of 1944–5 and the cold spring of 1945 exacerbated shortages, and infant mortality in Paris rose 40 per cent higher than the previous winter. Black-market prices were three to four times the legal retail level, while even produce sold discreetly to friends and neighbours fetched double the official price. With the wheat harvest down by a third in 1945—and only half the pre-war average—bread rationing remained in force until the autumn of 1949, with one brief respite. Indeed, bread consumption in 1947–8 was a mere 40 per cent of what it had been in the 1930s, despite the fact that the food harvests of 1946 had temporarily eased matters and allowed most farmers to regain their pre-war levels of income. Even so, supplying the black market remained their prime method of finding the wherewithal to buy new agricultural machinery.

The post-war years were to be a period of dynamic change in French economic attitudes and practice. How much this was due to government planning is a matter of continuing debate. Certainly, the Liberation era marked a new departure with the establishment of 'un premier plan d'ensemble pour la

Table 6.1.
French Loss of Population Resulting from the
Second World War (000)

Civilians	
Victims of bombardment	65
Killed directly by the fighting	65
Shot or massacred	30
Deportees, racial	70–100
Deportees, political	60
French workers in Germany	40
TOTAL	360
Military	
Battle of France, May–June 1940	92
Forces Françaises Libres	58
Forces Françaises de l'Intérieur	20
Alsatians and Lorrains in the Wehrmacht	40
Dead in captivity	40
TOTAL	250
Migration and natural losses	
French and foreign emigrants	300
Frenchmen remaining in Germany	20
Excess deaths	297
Reduced births	201
TOTAL	818
TOTAL LOSSES	1,428

Principal source: Colin Dyer, *Population and Society in Twentieth-century France* (London, 1978), 127.

modernisation et l'équipement économique' of France and her empire, which was substantially beyond the restricted attempts at state planning envisaged by Vichy (p. 96). Established by a decree of 3 January 1946, without consulting the Constituent Assembly, it was one of several examples of how the most durable achievements of the post-war period were brought about without recourse to the hurdle-race of parliamentary approval. The plan was initially created by Mendès France (23 May 1944), but it was Jean Monnet who persuaded de Gaulle of the need to institute a Commissariat Général du Plan.

The small team that Monnet assembled in the autumn of 1945, and the proposals which they submitted to de Gaulle on 4 December, did not seek to challenge the free economy. Indeed, the dominant feature of the French planning system was that it should be a series of recommendations not directives to producers, and that its appeal should lie primarily in the information and forecasts that it proposed, giving firms some sort of yardstick by which they could measure their own achievement and potential. Persuasion would principally lie in the force of the arguments and evidence, but it would also be strengthened by the direction and quantity of the help that the commissariat could give to those firms prepared to gear their activities to the commissariat's plans. It would be a matter of the carrot, rather than the stick; and with the advent of substantial American funds, the carrot in the government's hands became increasingly attractive.

A brandy-merchant by trade and a pragmatist by temperament, Monnet had observed how private industry in America and Britain had been galvanized behind the war effort without losing its basic independence; and he was convinced that this could be an example for peacetime reconstruction and the pursuit of prosperity. He had subsequently become the CFLN's *Commissaire* for armaments, reconstruction, and supply in 1943. His Commissariat Général du Plan (CGP) employed no more than thirty *chargés de missions*, essentially making use of existing public and private bodies—stimulating them into asking new questions of their material, and proposing new areas for investigation. Very rapidly its modest headquarters in the Rue de Martignac became the venue of wide-ranging meetings of *patrons*, trade-unionists, and civil servants. Among its key men were Étienne Hirsch, an engineer of the École des Mines who had been responsible for armaments and supply with the Free French; and it was his name that was eventually to be given to the Commissariat Général's second plan of 1954–7. Another close colleague was Robert Marjolin, former right-hand man to Mendès France at the Ministry for Economic Affairs. However, the results of their labours were primarily to be felt in subsequent years, rather than in the immediate post-Liberation era (pp. 185–95).

However important the task of economic regeneration, it was equally essential in de Gaulle's view to restore to the nation a sense of self-respect. Much would be at stake in the post-war settlement, and it was vital that France should be accepted as an equal partner in the victorious team that would be dictating terms. Both for the nation's own morale, and in order to present a united French front to other countries, the internal split between the former supporters of Vichy and the Resistance had to be healed. And to assure France of her rightful share of influence and compensation, she had to be seen by her allies as making a major contribution to the defeat of Germany.

It is commonly alleged that de Gaulle's prime interest was international affairs, and that domestic policies always came second to this. It was equally the case, however, that de Gaulle saw a vigorous foreign policy as a means of restoring unity and self-respect to a divided, demoralized nation. In a famous passage at the opening of his *Mémoires de guerre* he was to write: 'France is only truly herself when she is in the front rank ... only great enterprises are able to

compensate for the ferment of dispersion that her people carries within itself.'
Not only was the pursuit of 'great enterprises' a unifying force that helped the
nation to forget its differences, but the choice of the international stage as the
scene for such enterprise helped to underline what Frenchmen had in common
by bringing them into confrontation with 'the foreigner'. It ran parallel with de
Gaulle's search for a constitution that would enable government to base itself
upon what united Frenchmen rather than on the selfish, sectional concerns that
were all too effectively represented by the parliamentary parties, with their
endless capacity for prolonging and deepening the divisions of French society.
Permeating these concerns was de Gaulle's professed belief in the importance
of psychological factors. For de Gaulle, a people's capacity for self-fulfilment
was essentially the outcome of an attitude of mind, and in his view most politicians failed to take into account the importance of appeals to the collective
imagination and creative instinct.

While all this needs to be recognized, it must also be conceded that de
Gaulle's personal aptitudes and instincts found their greatest fulfilment in the
sphere of national security and international relations. As a soldier and the
leader of a government (many would say a faction) in exile, de Gaulle's experience and skills lay in the defence of French interests against those of other
nations. Those spheres in which he claimed no particular competence, he
tended to leave to the ministers concerned, except when wider issues were
involved.

De Gaulle was often accused of entertaining illusions of grandeur on the
French behalf. In fact, he understood the cold reality of France's vulnerable
position in the mid-1940s better than most statesmen; and his policies, like
those in the 1960s, were based on a bleak assessment of the effort that would be
needed for France to preserve any freedom of manœuvre in the world of the
superpowers. Admittedly, the status of a second-rank power was a hard one for
the French to accept, in that France, like Britain, had played a leading role in
European affairs between the wars, partly as a result of the abstention or
ostracism of the three most powerful countries. America had withdrawn from
the political scene in Europe immediately after the First World War, while
Soviet Russia and vanquished Germany were semi-pariahs in European
politics. Britain and France had therefore found themselves playing a disproportionately prominent role in inter-war affairs. The re-entry of America
and Russia on to the European stage in the Second World War cut Britain and
France down to a size commensurate with their limited strength.

The broad consistency of principle underlying de Gaulle's foreign policy in
the 1940s and 1960s tended, paradoxically, both to emphasize and to obscure
the strong element of affinity that it had with the policies of other French
governments. Bitter experience had taught France that in the last resort she
could only rely on herself. The brief history of the League of Nations had
demonstrated the limited willingness and capability of international bodies to

defend individual countries against aggressive neighbours, while the behaviour of Britain and America in the 1930s and 1940s inclined the French to conclude that alliances with particular countries were no stronger than the material self-interest that brought them about. De Gaulle remembered, too, that Churchill had rejected the offer he had made in November 1944 of close co-operation with Britain, and that Britain had preferred to develop its special relationship with America. America was not the obvious patron for France. Marshall Aid, welcome and influential when it came, had not yet been devised; and many Frenchmen assumed that once the immediate problems of the aftermath of war had been settled, American self-interest would reassert itself, as it had after the First World War.

Having been invaded by Germany three times in seventy years, French pre-occupations were initially centred on preventing a recrudescence of German expansionism. Russia, likewise a victim of Germany in the last two wars, was also determined to keep Germany weak; and there was therefore a strong affinity of interests on this point. The old Franco-Russian alliance of the *belle époque* had reflected this interest, and many considered that it could have been re-created as an effective deterrent to Hitler in the 1930s, had not Britain and a succession of vacillating ministers in France turned deaf ears to Stalin's overtures. At the same time, de Gaulle was less susceptible than America or Britain to the ideological aspects of what they considered to be the Russian threat. The nation-state loomed large in de Gaulle's concept of human affairs; and while he had very little liking for Communism, he saw Russia essentially as a national entity, motivated by national self-interest rather than by ideological Messianism. As such, Russia was no more nor less to be trusted than the other superpower, America.

De Gaulle had hoped for a profitable working arrangement with Britain, which, as an impoverished imperial power overshadowed by the Russo-American giants, laboured under similar difficulties. Indeed, it was to be an enduring tenet of de Gaulle's foreign policy that the second-rank powers should work together to preserve their freedom of manœuvre. Britain's preference for a close relationship with America was, in de Gaulle's view, a betrayal of diplomatic class loyalty—and one which he was to bring up against Britain during her unsuccessful attempts to join the EEC in the 1960s.

It was against this background of frustrated hopes that de Gaulle suffered what he saw as a series of calculated snubs to French pride. He felt that the French contribution to the defeat of Germany had been consistently hamstrung by Anglo-American parsimony in supplying his troops with adequate weapons. He was particularly hurt at not being invited either to the Yalta or Potsdam meetings of the Big Three in January and July 1945, despite the compensation of receiving a permanent seat for France on the Security Council of the United Nations with a right of veto (May 1945). Nor was he mollified by the promise of membership of the Council of Foreign Ministers in August 1945.

The offer was overshadowed by the nuclear destruction of Hiroshima and Nagasaki a few days later, which served to emphasize still further the power gap between America and France, as well as to demonstrate the closeness of the Anglo-American relationship. It correspondingly increased the French fellow-feeling for Russia, who had also been kept in ignorance of what the Anglo-Saxons were up to. Nor did it dispose de Gaulle to respond calmly to Truman's suggestion that he should drop the Communist ministers from his cabinet.

It was in the same month of August 1945 that American talk of acquiring a string of military bases across the world fuelled French fears that some of these might be at the expense of the French empire. De Gaulle was already smarting from events in Syria, where the British had ordered French troops to be returned to barracks (31 May) and had then assumed responsibility for keeping order there. With these affronts to French prestige overseas, de Gaulle was determined to safeguard France's position in Europe. He unsuccessfully pressed Truman for control of the left bank of the Rhine, and vainly argued that the administration of the Ruhr should be an international responsibility, given the importance of coal to the regeneration of a war-stricken Europe. Even appeals to Russia were fruitless, since the Soviet government had no desire to weaken still further a German economy that it intended to exploit for its own advantage. Nor was Stalin won over when de Gaulle offered to back Russia's claim to the Oder–Neisse line as Germany's eastern boundary—which would irrevocably commit the Poles to dependence on Russia in the event of a resurrection of German irredentism.

At the end of the day, France had to be content with a 20 per cent share of German reparations (December 1945) and eventual responsibility for an Allied Control Zone, consisting of the Sarre, the Palatinate, and the south-western corner of Germany, together with a section of Berlin. When France complained on 24 March 1946 that German coal supplies to her were much less than expected, her Anglo-American allies replied that German economic recovery must come first, otherwise Russia might seek to extend her control westwards. France's worst suspicions were confirmed on 5 September 1946, when America made clear her commitment to the creation of a German national government and her final rejection of all of France's outstanding demands.

It is true that de Gaulle's resignation in January 1946 had deprived France of the force of his personality in these later negotiations. And it perhaps did not help France's case that his successor, Félix Gouin, retained Georges Bidault at the Quai d'Orsay. *Le Canard Enchaîné*'s 'Georges Bidet' was an abrasive, squat caricature of Maurice Chevalier, with his square, protruding lower lip—'such a dear little man', as Ernest Bevin testily called him. But his toughness and Resistance record commended him as a dogged champion of French interests—even to those, like de Gaulle, who disliked him personally, and to many on the Left who regarded him as a detestable reactionary.

But whoever held France's cards at the negotiating table was perpetually handicapped by her need for material aid. It was hard to play tough with America in the wake of the Blum–Byrnes agreement of 28 May 1946, in which the United States magnanimously cancelled outstanding French war loans and allowed the money to be redeployed in the battle for French economic recovery. There were, moreover, further American loans to come—but at a price. Import restrictions on American goods were to be lifted; and French cinemas and book-stalls were flooded with a growing stream of American films and magazines, much to de Gaulle's disgust. Whatever the French *mission civilisatrice* overseas, it was now under heavy pressure at home; and it is doubtful whether de Gaulle's 'certaine idée de la France' had ever envisaged the sad spectacle of the luminaries of the Deux Magots making a cult of Tweety-Pie and other Fred Quimby creations that were to become the inevitable curtain-raisers at the major cinemas of the Fourth Republic.

7
Retribution and the New Jerusalem
The CNR Charter

THE Liberation government was expected not only to find solutions to gigantic, material problems, but also to fulfil the far-reaching promises of the Conseil National de la Résistance in its charter of 15 March 1944. De Gaulle disliked redeeming cheques he had not signed. But his government was basically drawn from the two main strands of the Resistance; and he was well aware that he partly owed his triumph over Giraud and other Allied counter-scenarios to the endorsement of the CNR. Not only did this make it awkward for him to postpone payment, but the CNR's pledges were enthusiastically supported by a wide segment of popular feeling and political opinion. Even so, de Gaulle preferred to talk about 'renovation' rather than 'revolution' (as proclaimed by left-wing elements in the Resistance)—a distinction that perhaps caused the incarcerated Pétain to give a wan smile of fellow-feeling as he remembered his own semantic scruples in 1940–1.

LES DIEUX ONT SOIF

High on the list of the charter's pledges was the promise of retribution to those 'who have come to terms with the enemy or have been actively associated with the policy of collaborationist governments'. Many saw the post-Liberation purges as a quasi-sacramental necessity. Édouard Herriot, who was normally wary of firebrands, nevertheless spoke in the old Jacobin tradition of the Radicals when he proclaimed: 'c'est par un bain de sang que la France devra passer d'abord'. De Gaulle, however, was deeply conscious of the debilitating divisions in French society and of the need to present a united front to a hard, competitive outside world. Reprisals, in his opinion, should be kept to a minimum.

Many resisters, of course, had taken justice into their own hands. 5,234 people were listed as summarily executed before the Allied landings, and 4,439 were thought to have been executed subsequently, of whom only 1,325 were given some semblance of a trial. More recent local evidence suggests that the sum total of unofficial killings may have been considerably higher. The official trials, on the other hand, which continued until 1951, sentenced 2,853 persons to death, and a further 3,910 *in absentia*, but only 767 executions were actually carried out.

The main show trials were not all that they might have been. The eleven-day trial of Laval (4–15 October 1945) rapidly degenerated into an unedifying affair in which he was shouted at by the jury, and his execution was especially gruesome, since he was carried before the firing-squad in a state of semi-collapse, after a long session of stomach-pumping following an unsuccessful attempt at suicide. Pétain, after a three-week trial (23 July–15 August 1945), was symbolically sentenced to death; and then, as anticipated, the sentence was commuted to life imprisonment, where he died six years later. The thirst for justice had many grotesque aspects. A former *milicien*, who had subsequently lost both legs fighting with the Liberation forces, was nevertheless condemned to death; and the macabre spectacle of his being lifted out to face the rifles was only averted by the threatened resignation of the Garde des Sceaux.

A further 38,266 people were sentenced to imprisonment; but all but 1,570 of them were released by October 1952, and only nineteen remained in gaol when the Fourth Republic came to an end in 1958. In proportion to the size of the population, these figures were modest compared with neighbouring countries. Whereas only 94 out of every 100,000 inhabitants were sentenced to prison in France, the corresponding proportions elsewhere were 633 for Norway, 596 for Belgium, 419 for Holland, and 374 for Denmark.

Leading members of the Resistance voiced misgivings about the haphazard nature of Liberation justice, with charges and penalties often depending on the prevailing attitudes in the region—or on the material resources and contacts of the accused. In effect, it was the poorer sections of society that were the hardest hit. The richer defendants employed able lawyers, while the most severely viewed forms of collaboration—membership of the Milice and other para-military bodies—tended to concern those whose modest circumstances had made them more susceptible to the material attractions of enrolment. And, indeed, it was ironic that three-quarters of the judges dispensing retribution had themselves served the Vichy regime, whose courts had punished 135,000 people for alleged political offences. Procureur-Général Mornet of the High Court, who sent Laval and other Vichy figures to the firing-squad, had himself narrowly missed being Vichy's public prosecutor at the Riom trials (p. 107).

A further 49,723 people were sentenced to national degradation, but only a quarter of them still remained under sanction in 1952. Nearly 50,000 government employees were investigated at the Liberation, of whom 11,343 were sacked or subjected to lesser punishment, representing a mere 1.3 per cent of the total force. And of those dismissed, many were to be readmitted in the next five years. Moreover, the amnesty laws of January 1951 and 6 August 1953 allowed many of the remaining victims of sanctions to recover their former employment or at least to receive a commensurate pension.

This was a mild purge compared to Vichy's sacking of 35,000 civil servants and demotion of 15,000 servicemen. Even the shake-out of the police was relatively modest, despite the role that they had been obliged to play during the Occupation, when the pursuit of Resistance figures engaged in sabotage or assassination was part of the business of keeping law and order. De Gaulle was very conscious, however, of the need to keep a strong police force at the government's disposal during the hazardous uncertainty of the Liberation period, when some feared a Communist coup and others feared the unbridled settling of scores by private, trigger-happy hit groups. Like the police, the Church could not justifiably complain of vindictive reprisals, in that the ecclesiastical purge was much milder than de Gaulle and Bidault would have wished. At the end of the day, the Vatican agreed to remove an archbishop and three bishops, whereas Bidault had a list of a dozen for dismissal.

The CNR charter had also promised the confiscation of black marketeers' property and the taxation of profits made during the Occupation. Yet many firms that had prospered on German contracts continued to grow fat on reconstruction work; and it was more or less inevitable that the demands of the postwar economy and the need to create employment should override the calls for atonement. Even so, the early Liberation saw some short-term, grass-roots take-overs of factories, notably in Lyon and Marseille, where *commissaires de la République* temporarily authorized workers' occupations of factories whose owners had disappeared or who were popularly considered guilty of collaboration. From the government's point of view, this enabled production to continue; but none of these *ad hoc* situations survived the restoration of normal peacetime conditions.

Newspapers were the sector most severely hit by official sanctions; 482, mostly local, lost all of their property, the evidence of their collaboration being plain to all in black and white. Like journalists, entertainers and musicians were also the victims of the public nature of their occupations—Maurice Chevalier and Alfred Cortot being among many performers who were interrogated for having been heard on the German broadcasting station in Paris. Teachers were another sector vulnerable to accusation, unguarded remarks in the class-room leaving memories that were all too easily recalled at the Liberation. *Lycée* teachers of German were especially at risk, particularly those who had sought to perfect their pronunciation by conversing with soldiers billeted in the locality.

The extent of private profit made during the Occupation is, of its nature, impossible to estimate; and the grey area between profiteering and earning one's livelihood was a large and confusing one. There was endless debate as to whether a manufacturer who lacked the courage to refuse a German order was necessarily more guilty than a shopkeeper or bar-owner who did not dare turn away members of the occupying forces. Apologists pointed out that Marcel Dassault served several hard years in a German concentration camp for refusing to manufacture aircraft for the Germans, and that such fortitude was not for the ordinary run of mankind. Critics of industrial profiteering replied that it was in the nature of war to leave no middle ground between heroism and cowardice. The ordinary soldier going over the top into a hail of machine-gun fire had a simple choice between a posthumous mention on a list of honour, or court-martial for cowardice and perhaps the firing-squad; why should the wealthy be pardoned for taking the unheroic path? Many felt that men like Louis Renault were lucky not to suffer more than the confiscation of their property.

When it came to the black market, estimates are no easier to make. Paul Bodin's famous survey in *Combat* (January 1947) claimed that the illicit dealings of recent years had created a new bourgeoisie which had acquired some 30 per cent of the real estate of its predecessors. Alongside them were the more modest pickings of the peasantry and small shopkeepers. He estimated that over 20 per cent of the working population were supplementing their incomes by illicit racketeering of one sort or another.

Prominent among those who were determined to deprive the profiteers of their pickings was Pierre Mendès France. He had been the Finance *Commissaire* of the Algiers government; but de Gaulle's new cabinet of September 1944 found him transferred to the Ministry for Economic Affairs, with Finance eventually entrusted to René Pleven. Mendès France argued that as long as profiteers and the wealthy in general could afford to pay more on the black market, price controls would be circumvented, shortages in the shops would continue, and the ordinary citizens on a state-controlled wage would continue to suffer. His solution to this problem was monetary reform on the lines that Belgium adopted a few weeks later. By recalling existing money and issuing a new currency in exchange, hoarders and profiteers would have to reveal the extent of their wealth, and would have to justify it before receiving new currency. Those afraid to do so would find their savings valueless in the face of the new currency; and in this way a fresh, 'honest' start would be made to the distribution of wealth within the country. But the permanent officials of the Ministry of Finance and the Banque de France opposed the scheme as likely to create ill will in a nation anxious to forget the deprivations and tyranny of the past four years. There were doubtless many whose profits and savings long predated the Occupation, who had no desire to be obliged to account for their wealth and perhaps become liable for long arrears of tax. This category included numerous peasants who had briefly risen above their habitual tight circumstances as a result of the high prices paid for food by unofficial purchasers—and who in many cases intended to use the money to replace antiquated equipment or pay off debts. Numerically, such people were an important segment of the electorate and needed to be handled carefully—or so it was argued.

Pleven preferred a return to the traditional soft options of the Third Republic. He launched a couple of loans and chanced his arm with a minor capital levy, spread over four years, which gave the profiteers plenty of time to disperse and conceal their winnings. When he introduced the new currency scheme in April 1945, it was merely in the shape of an informal note-for-note exchange over the counter, with no effective check on the total number of notes presented by each individual during the eleven-day period of exchange (4– 15 June 1945). The unscrupulous were thereby able to clothe their ill-gotten gains in the shining apparel of the new currency.

Mendès France had already offered his resignation in January 1945—to have it refused by de Gaulle with, it is said, the disconcerting comment: 'aren't all the experts against you?'. But in April Mendès resigned definitively. Rapid inflation was the inevitable consequence of the government's easygoing financial policy. The index of retail prices in January 1946 was to stand at 481 in relation to 1938's 100; by the end of the year, it woud be 865.

SOCIAL REFORM AND NATIONALIZATION

Running parallel to the CNR's pledge to adjudicate and punish the past was its promise to transform the future. Like the Popular Front, the Liberation era is seen in the mythology of the extreme Left as a monstrous missed opportunity: a period when the working class were ready to assume power but were betrayed by their accredited leaders—Blum in the 1930s and Thorez in the 1940s. Such an interpretation overlooks what was economically feasible in these periods, and ignores the publicly declared aims of the left-wing parties when they assumed office. Far from encouraging millennial expectations in the working class, they were at pains to point out the limitations of what could be hoped for. Even so, it is arguable that the reforms of the brief Liberation era were a more significant advance on the past than Labour's creditable six-year record in post-war Britain.

De Gaulle personally had no strong views on social reform. Various of his statements between 1941 and 1943 indicate a readiness to accept the national-ization of key sectors of the economy and the extension of state social security. Yet neither the SFIO nor the PCF was deeply committed to state ownership in the period before the last trump sounded and working-class rule became a reality (p. 54). In similar fashion, state intervention in the matter of wages and industrial disputes was arguably in the *étatist* logic of earlier regimes, and as likely to appeal to a nationalist like de Gaulle as to left-wing politicians who would continue to be somewhat wary of state intervention as long as it remained in the hands of the propertied classes. More enthusiasm, however, was felt in left-wing circles for the development of social security, where the element of distributing wealth from the rich to the poor was more evident.

A strong influence on the social programme of the CNR was the CGT plan of the mid-1930s; and the decision to provide the charter with a post-war programme of social reform was principally a Socialist initiative.[1] Léon Blum had favoured the idea while in prison in 1942—the experience of the Popular Front having convinced him of the utility of mutually agreed declarations of intent when it came to reminding disparate allies of their common objectives. And he informed de Gaulle of his plan when he pledged his support to the General in November 1942. It has also been argued that the enthusiasm of the London-based Socialists was indirectly stoked by the discussions surrounding the Beveridge Report of 1942.

The Communists, by contrast, were initially hostile to pledges that could create divisions in the Resistance and distract attention from the immediate task of expelling the Germans—an attitude that had obvious affinities with their position in the Popular Front era (p. 53). Support for the Socialists' programme came mainly from the CGT, the CFTC, and the Christian democrats (the later MRP)—and also from the heterogenous Resistance groups where the non-Communist Left and the Catholics were most numerous: Libération-Nord and Libération-Sud, Franc-Tireur, and Combat. At the same time, the unedifying role of big business in Occupied France had sapped the opposition of the Communists to nationalization, and caused conservatives to recognize, however unwillingly, the necessity of propitiating public indignation by the sacrifice of part of industry on the altar of state control. Even the arch-capitalist of the Communist imagination, François de Wendel of the Comité des Forges, conceded the need for gestures of this sort, mindful perhaps of the Communists' proposal to the CNR to 'send before the firing-squad the traitors of the Comité des Forges, the Comité des Houillères, the Kuhlmann Trust, the big banks, etc.'.

Even so, it was essentially the Socialists and the CGT who persistently waved the charter in the faces of the Liberation ministers and the Constituent Assembly—the CGT forming the Délégation des Gauches as a broad-based ginger group for this purpose. With representatives of the PCF, the SFIO, and the Ligue des Droits de l'Homme, the Délégation had a whiff of the old-style anticlericalism of its namesake in the Combes era; and, feeling cold-shouldered, the MRP issued its own manifesto on 24 September 1944, calling for the nationalizations listed in the CNR charter.

Paradoxically, however, the cause of social reform benefited greatly from the delay in the emergence of representative institutions (pp. 138–42). The programme of the Popular Front had ultimately foundered on the opposition of the Senate; and the Socialist measures in the 1950s were to encounter obstruction in the lower as well as the upper house. While it is true that the two houses elected in 1946 both had majorities that were pledged to reform, it is hard to know whether the social measures of the Liberation would have become law so painlessly had they had to run the traditional gauntlet of both houses. As it was, the first series of measures were enacted by ordinance of the Provisional Government, while the second were passed by the unicameral Constituent Assembly of October 1945–June 1946.

Given the strength of the Resistance parties in the Assembly (p. 139), a smooth passage was not in doubt. But it was all the smoother for being a single rather than a double run—with no Senate to worry about—and was also facilitated by the fact that the Constituent Assembly saw itself as an *ad hoc* body, committed to laying the foundations of the future rather than concerned with the day-to-day critique of government and with inter-party jockeying for advantage. There was consequently less danger of the basic content of the reforms being damaged by party tactics.

The charter had aggressively promised 'the setting up of a true economic and social democracy, entailing the eviction of the great economic and financial feudalities', brave words that understandably aroused the apprehension of the industrial bourgeoisie. Characteristically, the textile manufacturer and conservative politician, Joseph Laniel, who had reluctantly endorsed the CNR programme in March 1944 in the interests of Resistance unity, now felt free to oppose it on orthodox, free-economy grounds. The charter specifically stipulated 'the return to the nation of the great monopolies in the means of production, the sources of energy, mineral wealth, insurance companies, and the large banks'. Consequently, an ordinance of 13 December 1944 set up the state-controlled Houillères du Nord et du Pas-de-Calais, with the trade unions occupying a third of the seats on the consultative committee. The remaining coalfields were nationalized by an unopposed vote of the Constituent Assembly on 26 April 1946—the opposition having expended its energies on the more contentious issues in the nationalization programme. Thus, sixty-four of its members had opposed the nationalization of gas and electricity when the Constituent Assembly comfortably passed the measure on 28 March 1946.

The same group voted against the take-over of the insurance companies on 24 April 1946—and had to content itself with the sour satisfaction that, whereas the Communists had been in favour of nationalizing the lot, the Socialist Minister of Finance, André Philip, had cut the shopping-list to 60 per cent. Thirty-five of the hard-core conservatives had likewise tried to save the banks from a similar fate on 2 December 1945, when the Banque de France and the four largest clearing banks, the Crédit Lyonnais, the Société Générale, the Comptoir National d'Escompte, and the Banque Nationale pour le Commerce et l'Industrie came under state control.

The formal case made against the banks was their excessive caution in the past and their attempts to influence policy by varying their readiness to lend to governments—the Popular Front being a conspicuous victim in the 1930s. The former shareholders were indemnified—if rather less generously than they had initially been led to expect—and the Banque d'Algérie was added to the list on 26 April 1946. While the nationalization of the banks gave the State a tighter control over credit, it did not substantially increase its ability to channel the direction of investment. The Communists and Socialists had also wanted to nationalize the two leading merchant banks, the Banque de Paris et des Pays

Bas (Paribas) and the Banque de l'Union Parisienne. But this was left to the Mauroy government to bring about three decades later—when it was to provoke much the same hostility from business circles.

Conservative opposition to other extensions of state control had been largely pre-empted by the use of ordinances months before the Constituent Assembly was elected. Admittedly, it would have required considerable daring or insensitivity to fight the cause of private enterprise over the empire of Louis Renault, whose goods and factories were nationalized without compensation on 16 January 1945, following his heavy record of supplying the Germans with armoured and other vehicles. In fact, the Renault works were to continue to operate with the virtual independence of a private company. The Socialist Minister for Industry, Robert Lacoste, would have liked to install trade-unionists as management; but opposition from both the government and the firm itself resulted in the appointment of an archetypal industrialist, Pierre Lefaucheux, who proceeded to ignore the pleas of the Commissariat Général du Plan for heavy lorries, and launched the 4cv family car in 1946–7 with new mass-production methods.

A further ordinance of 29 May 1945 transferred to the State the shares of Gnome et Rhône aero-engines—likewise guilty of supplying Germany. Under its new title, SNECMA, the company retained its existing structure and the former shareholders were compensated. In the less sensitive realm of commercial air transport, Air France and Air Bleu were nationalized on 26 June 1945—the two companies eventually being amalgamated as Air France in 1948. This was a logical consequence of the wartime state control that had been established in September 1944. The same logic underlay the state acquisition of a majority shareholding in the merchant shipping fleet, with compensation for the former shareholders (18 December 1944). On the other hand, it was trade-union pressure and the demands of the Parisian *comité de libération* which led to the permanent municipalization of Paris transport (RATP)—the source of sinister posters in the 1950s: 'Pendant que le Métro sommeille, les autobus circulent', which vied in chilling appeal with the health service's stark announcements of 'Journées de Sang', suggesting a re-run of the September massacres rather than the benign visit of a transfusion team.

While the virtues of public ownership were to be a matter of bitter debate in the next forty years, only hard-pressed employers were likely to contest the immediate benefit of the reorganization of the social security system that was decreed on 4 and 19 October 1945 in an attempt to meet the CNR charter's demand for 'a complete plan of social security ... with control over it assured for representatives of beneficiaries and of the state'. Henceforth, 6 per cent of the worker's wage was deducted from his pay, to which the employer added a sum equivalent to 10 per cent, and the joint amount was then paid into a *caisse primaire départementale*, which was part of a national hierarchy of social security funds. In fact, social security contributions were calculated only on the lower portion of workers' wages, as were benefits. Thus, although the state pension at sixty years of age was notionally equivalent to 20 per cent of the worker's wage at his period of maximum earnings, it was 20 per cent of only a basic proportion of his wage. In practice, however, most workers did not retire until they were sixty-five, and their pension was correspondingly higher (p. 207).

Some nine million workers and their families were now covered for illness, incapacity, and old age—while, with effect from October 1946, the employer bore the cost of insuring them against death and accident at work. Although 80 per cent of medical expenses were reimbursed to the patient, the system was less attractive than the British health service because it involved a considerable initial payment and then tedious queueing to recover the 80 per cent. The Constituent Assembly extended the social security system to include most sections of the population (22 May 1946), but old-age pensions took somewhat longer to reach a similar proportion, with the relevant measures dating from September 1946 and January 1948. Even so, some categories of non-salaried persons had to wait until August 1967 before being covered by the main provisions of the system.

On the contentious matter of who controlled the funds of the social security system, a law of July 1946 entrusted them to elected representatives of the unions—much to the chagrin of the MRP and the Confédération Générale des Cadres, who objected to the overly working-class complexion of this arrangement. They were to obtain their revenge, however, three years later, when they succeeded in withdrawing their favourite benefit, family allowances, from the jurisdiction of the *caisses* (law of 21 February 1949). Certainly, the sums involved in family allowances were large (pp. 209–10). Employers could also take satisfaction in the exclusion of unemployment benefit from the competence of the *caisses*. This remained entirely under the control of the State, thereby assuaging employers' fears that worker-controlled unemployment benefits could become a potential source of covert strike pay.

A major demand of the CNR charter was 'the participation of the workers in the ordering of the economy'. Accordingly, an ordinance of 22 February 1945 established *comités d'entreprise* in firms employing a hundred or more workers, a threshold that was lowered to fifty by the Constituent Assembly on 16 May of the following year. Largely a development of the Popular Front's *délégués du personnel* and Vichy's *comités sociaux*, they met once a month; and, in addition to the traditional concerns of welfare and working conditions, they were entitled to examine the firm's accounts. Lacking mandatory powers, however, their role was largely consultative; and many firms took little notice of their recommendations, except in day-to-day matters. Committee membership was initially based on competing lists of candidates proposed by the various unions, but non-union labour acquired increasing representation in the years to come. Even so, the committees were a pale shadow of what the charter had envisaged—and were to remain so, despite the efforts of Socialists in the 1980s to give them bite.

Alexander Werth was later to claim that 'The whole subsequent history of the Fourth Republic was largely a struggle between those who wished to see the CNR charter applied, and those who were determined to ignore it...'. Foreseeing such a possibility, the charter had stated that: 'The representatives of Resistance proclaim that they have decided to remain united after liberation' in order to ensure its fulfilment. Yet it was recognized that the concept of a formal Resistance party as such was an impossibility. Given the differing preoccupations of the Communists, Socialists, and the MRP, for a Resistance party to survive it would either have had to be so relaxed in its discipline as to be nothing more than a loose coalition—little different from the Tripartism of the Liberation era—or, if it had tried to impose unity on the basis of a few agreed common policies, it would rapidly have become the scene of continual struggles for dominance among the constituent elements, with a series of secessions by the disillusioned. Moreover, de Gaulle was the only obvious leader for such a party. But he was too hostile to the traditional system of parliamentary party politics to make close co-operation with the three main Resistance parties a workable proposition. It was only while France's constitutional options were still open, and de Gaulle's preferences still among those options, that he was prepared to head governments composed of these parties.

The Union Démocratique et Socialiste de la Résistance was a forlorn attempt to keep alive the solidarity of the Resistance years; but without the unifying factor of German oppression, it retained only a handful of supporters, Pleven and Mitterrand being among the few names that either carried, or were to carry, significant weight. Losing the bulk of its socially progressive members to the SFIO, the rump was to gravitate rapidly towards the Radicals, eventually forming an easygoing federation, the Rassemblement des Gauches Républicaines.

THE COMMUNISTS

If a Resistance party was no more than a vain dream, a working alliance of Resistance parties seemed for a while to be a possibility, with even the Communists prepared to continue co-operating with their erstwhile colleagues. Stalin was anxious for France to be united behind de Gaulle, not only as long as there was a German menace, but as long as France could be useful to him in the diplomatic complexities that lay ahead. Russia's concerns were focussed essentially on the territories that lay between Russia and Germany, the high road of previous Western attacks on Russia; and, given his wish to secure Allied acceptance of Soviet predominance in Eastern Europe, he had no desire to provoke unnecessary hostility by encouraging Communist take-overs in territories that did not immediately interest him. Accordingly, Thorez exhorted his followers to 'make war; create a great French army; work like blazes, rapidly rebuild industry' (27 November 1944); and he likewise roundly condemned strikes and unreasonable wage demands (21 January 1945). Moreover, the PCF strongly supported French demands for the internationalization of the Ruhr and French control of the Saar. Indeed, while serving in the French cabinets of 1946, Thorez was to be in open dispute with Stalin over the need to put the Saar under French administration.

In reality, the resort to a revolutionary seizure of power was never seriously considered by the PCF in the Liberation era—despite subsequent claims by André Marty that a golden opportunity had been missed (pp. 115–16). A Communist insurrection would have been a chancy business. The recent fortunes of the Left in Spain were not encouraging; and it could not be guaranteed that the work-force in any of the large factories would be prepared to answer such a summons. The police were loyal to the government, the bulk of them having served under Vichy, with long experience of hunting Communists; and they were too grateful to have survived the first purges of the Liberation to risk themselves in a left-wing coup, for which few of them would have felt any sympathy.

The French Communists counted on increasing their popularity by exploiting their war record and by adopting a co-operative stance in the creation of the new France. They hoped to win over the rank and file of the Socialist party, while remaining outwardly on good terms with the Socialist leadership. They would seek to drive a wedge between the Socialists and the MRP by resurrecting the clerical issue; and they would likewise use the clerical and constitutional issues to win *ad hoc* support from the Radicals against the MRP and de Gaulle.

At the same time, the PCF hoped to infiltrate the administrative apparatus as deeply as possible. The Liberation's modest purging of Vichyite personnel, and the fact that Communists held ministerial portfolios, afforded them a favourable opportunity for such an attempt. It was therefore not surprising that a myth should rapidly arise among journalists that the Communists were colonizing the civil service. The Communist-held ministries undoubtedly saw a considerable influx of Communists—just as ministries held by other parties saw the appointment of men whose political views were congenial to the minister. But even the personal *cabinets* of the Communist ministers were not wholly Communist in composition.

It is true that the newly formed riot police, the CRS, included a number of Communist officers; and when he was Minister of the Interior in 1947–8, Jules Moch was to complain of the large number of Communists in its rank and file. But this situation had largely arisen from the attempts of de Gaulle's first Minister of the Interior, Adrien Tixier, to neutralize the Communist element in the Francs-Tireurs et Partisans Français by giving them disciplined employment under his watchful eye.

Whatever suspicions were harboured about the loyalty of the Communists, the Socialists were initially regarded by many as the linchpin of Resistance idealism and interparty co-operation, just as they had been a force for unity under the Occupation. While Blum was in prison, Daniel Mayer had clandestinely kept the Socialist party going; and he had carefully explored the possibilities of forming a post-war workers' party with the Communists. But Blum was implacably opposed to the notion, and Communist behaviour in Eastern Europe quickly destroyed such enthusiasm as there had been for it among the Socialists.

The MRP, on the other hand, raised Socialist doubts of a different kind. The allegiance of its early members to the ideals of the charter seemed safe enough, whatever misgivings might be felt about particular individuals such as Bidault. But it was its post-Liberation adherents that made many Socialists shudder.

Formally founded on 26 November 1944, the MRP set itself the dual aim of reconciling the working classes and the Church, and of reconciling the Church with the Republic. It drew on a tradition embodied in a variety of pre-war organizations, notably Catholic youth movements such as Jeunesse Étudiante Chrétienne, Jeunesse Agricole Chrétienne, and Jeunesse Ouvrière Chrétienne. Its basic principles owed much to the formulations of a young disciple of Emmanuel Mounier, Gilbert Dru, who had been shot by the Gestapo in July 1944; and these were later developed by such kindred spirits as the philosopher, Étienne Gilson, who was a regular contributor to MRP publications.

Pluralism came easily to its way of thinking—given its confessional rather than class basis, and given the division of its loyalties between an international Church and the French nation. The pluralism of the MRP distinguished it markedly from the Jacobin centralist traditions of the Socialists, Radicals, and Gaullists, as well as from the equally centralist concepts of the PCF. On the other hand, it claimed to eschew the individualism of the conservatives. This, however, was an attitude more characteristic of its founders and the left wing of its devotees, rather than the rank-and-file electors that were increasingly drawn to it in 1945 and 1946. With the traditional conservative parties discredited by the Vichy experience, many conservative voters began to look to the MRP as the least offensive of the Resistance parties. Indeed, it is commonly claimed that perhaps as much as three-quarters of the votes cast for the MRP in October 1945 came from the Right—leaving it vulnerable to taunts that it was simply a 'Machine pour Ramasser les Pétainistes'.

This influx rapidly created tension between the relatively progressive aims of the wartime leadership and the conservative outlook of its new adherents, who were committed to little else than defence of the Church and their own economic interests. Initially, the leadership was prepared and able to keep to its path, despite the deflecting demands of the growing conservative grass roots. But the rehabilitation and creation of rival right-wing parties, especially after

1946, was to oblige the MRP to take more heed of its conservative supporters, for fear of losing them to the new groups.

But all this seemed no more than an ugly possibility in the early months of 1945, and the MRP still had sufficient factors in its favour to make committed Republicans overlook its clerical connections. Although largely Catholic, it was not exclusively so; and prominent among its early parliamentary members was the archetypal trio of the after-dinner clerical joke, a Protestant, a Jew, and an atheist. The Resistance record of such MRP leaders as Georges Bidault, Maurice Schumann, François de Menthon, and Pierre-Henri Teitgen, assured the party considerable credit in secular circles, while it also profited from what were popularly—if inaccurately—assumed to be de Gaulle's sympathies for the MRP. De Gaulle's Catholicism, the fact that he had not yet created a party of his own, and the strong presence of the MRP in de Gaulle's government furthered this assumption. But de Gaulle considered most of the MRP leadership to be 'dangerously incapable', and his creation of the RPF in 1947 was to confront the MRP with its most formidable rival.

Paradoxically, the influence of the MRP on Catholic voters was that much greater as a result of the discredit into which the bishops had fallen, following their complaisant attitude to Vichy. Like the conservative parties, the episcopacy kept a low profile at the Liberation, waiting for the clouds of opprobrium to disperse before attempting to reassert their traditional leadership over the faithful. This moral vacuum of authority gave laymen a louder voice in Catholic debate, just as it was to be a major factor in the rich diversity that characterized Catholic writing and thought in post-war France. The Vatican, too, emerged from the Second World War with its image tarnished in the eyes of the victors, with the result that it was initially hesitant to exercise a strong controlling influence over the more venturesome propositions of French Catholic thought. Pius XII (1939–58) was deeply disliked in Liberation France. His diplomatic record in the 1930s and his circumspect behaviour in the war were not easily forgotten, while his face, suggestive of a sinister brain surgeon, did not lend itself to the proliferation of mantelpiece photographs that made his successors such a familiar feature of *bien pensant* households. It needed the development of the East–West conflict in international affairs, and the reassuring domestic context of a strong Christian Democrat government in Italy, for the Vatican to take up more vigorously its customary watching brief over the intellectual life of the Church. But in the meantime the Catholic Left flourished.

8

The Constitution of the Fourth Republic

AFTER the social 'New Deal' of the CNR charter, there still remained the fundamental matter of the constitutional framework for post-war France. The charter was understandably imprecise in its constitutional demands, given the need for relative unity and the genuine uncertainty as to what type of constitution would gain acceptance at the Liberation. But it emphasized the need for a 'democracy combining effective control exercised by popular representatives with continuity of action in government'. In other words, there was to be no sinking back into the old habits of the Third Republic with its unstable ministries and inability to pursue vigorous policies. And it was assumed that the constitution of the Third Republic would have to be changed accordingly. Unfortunately, the suppression of democratic liberties and institutions during the Occupation intensified traditional suspicions of anything suggesting a strong executive; and the desire to safeguard democracy was to blind many politicians to the equal need for effective government. There was consequently little enthusiasm on the Left for the semi-presidential schemes favoured by a sizeable section of the Gaullists.

Yet, whatever systems were proposed, few doubted that women would at long last have the predominant voice in saying 'yes' or 'no' to them. The Algiers Committee had pledged itself on 21 April 1944 to include women in the electoral base of the new Republic; and, given their longer life-span, they were a majority of the population. Paradoxically, female suffrage had not featured in the CNR charter—largely because of the opposition of the Radicals and a sizeable section of the CGT, who were still mesmerized by the traditional fear of 'les curés au pouvoir'. But no one seriously questioned that the Liberation would see it become a reality. Whether women would act as a new force in politics, or largely follow their husbands' lead, remained to be seen. To later claims that four-fifths of them voted the same way as their husbands, feminists deftly responded that the truth of the matter was that the men voted as their wives wished them to. Others tactfully pointed out that marriage was essentially a growing together, and that explanations were superfluous.

The female vote was a major factor in de Gaulle's wish to put the future constitution to a referendum. And the matter rapidly became a trial of strength between de Gaulle and the Communists. The Communists expected that the forthcoming Constituent Assembly would have a sizeable majority of left-wing enthusiasts, and hoped that it would vote for a unicameral Parliament with extensive powers over the executive—reminiscent of the Convention of 1792 or the Paris Commune of 1871. But they were well aware that the electorate might be suspicious of a proposal that was so resonant of the Revolutionary tradition. For their part, the Socialists and the MRP supported de Gaulle's call for a referendum, mainly because they feared that the Communists might win a larger number of seats in the Constituent Assembly than in fact they did—and, in such an eventuality, the referendum would be a useful safety net should the Communists attempt to push through an unacceptable constitution. The Radicals, on the other hand, opposed the idea as reminiscent of plebiscitary dictatorship, and argued in favour of the Constituent Assembly simply becoming a new Chamber of Deputies under the old Third Republican constitution—which had served the Radicals so well in the past.

In fact there were two referendums—one on the mandate to the Constituent Assembly (21 October 1945), and the other on the constitution it produced (5 May 1946). In the first referendum, not only did 96 per cent of the voters demand a new constitution rather than a return to the Third Republic, but two-thirds of them insisted that the document drawn up by the Constituent Assembly should itself be put to a referendum. De Gaulle had won his point, and events were to justify his persistence—at least in the short run.

THE CONSTITUENT ASSEMBLIES AND DE GAULLE'S RESIGNATION

Since the main purpose of the Constituent Assembly was to frame a constitution rather than to support a government, the prime concern of the architects of the electoral system was to create a representative body that would faithfully reflect the diversity of opinion in the country, rather than provide a strong majority that would ensure strong and effective government. The elections of 21 October 1945 were therefore based on a *départemental* system of proportional representation (*scrutin de liste*), which was broadly similar to the electoral system of 1919–27. In principle, de Gaulle favoured a system that would strengthen the representation of whichever party was the first past the post, since it would improve the chance of a clear, decisive majority emerging in the Assembly. And, from this point of view, the pre-war *scrutin d'arrondissement* of single-member constituencies was preferable to the currently favoured *scrutin de liste*. Alternatively, his constitutional adviser, Michel Debré, envisaged a two-ballot *scrutin de liste* that would enable the leading list in a *département* to scoop the pool—provided it obtained an absolute majority on the first ballot or a relative majority on the second. But de Gaulle feared that in the context of Liberation France the party first past the post might well be the Communists; and so he wearily accepted the one-ballot *scrutin de liste* that the consultative committee seemed to prefer. At least it was better than a nation-wide system of proportional representation, whereby even the tiniest parties got a toe-hold in the assembly—as had notoriously happened under the Weimar Republic.

The basis of the system was that electors voted for party lists of candidates presented in the *département* as a whole, rather than for individual candidates representing single-member constituencies. The net effect was to favour the disciplined parties that could present attractive lists in a large number of *départements*, while the losers were the more easygoing parties that had traditionally rested on the capacity of their individual members to cultivate grass-roots support in the intimacy of the pre-war constituencies. The Communists, the Socialists, and the MRP came into the first category, while the Radicals and conservatives came into the second. And their respective fortunes in the Constituent Assembly election of 21 October 1945 partly reflected this fact.

But the electoral system was not the main reason why the results were as they were. The Communists, Socialists, and MRP were the parties most associated with the Resistance and New Deal attitudes, whereas the Radicals and conservatives not only had a chequered Vichy record, but were seen as embodying the bad habits of the pre-war era. There were doubtless large sections of the middle class and peasantry who still felt closest to the Radicals and conservatives—but it was unfashionable and, in some areas, positively dangerous to campaign vigorously on their behalf.

While de Gaulle could take some pleasure in the victory of the Resistance parties, he gloomily assumed that their constitution would fail to provide France with a strong executive. It seemed to him increasingly plain that the swiftest way to salvation was to distance himself temporarily from the downward drift of French political discussion by resigning. In his estimation, the next ministry, whatever it was, would make such a mess of things that there would be a widespread demand for his return—which he would then make conditional on the politicians accepting a constitution to his own liking. This bleak assessment of what he should do was intensified when Léon Blum turned down the General's request to hold the fort in the meantime—Blum pleading that his health was not up to occupying the premiership at this critical juncture (November 1945).

De Gaulle's obsessive worry that government would revert to being the plaything of the parties was reflected in his increasingly irascible treatment of his ministers. In a blistering tirade to the cabinet on 28 December he declared: 'Instead of obeying me, the ministers kow-tow to their parties. One cannot serve both France and one's party at the same time.' And he roundly berated Bidault for turning up late to the meeting. The dumbfounded Bidault had just got married an hour earlier amid the devotional elegance of Saint-Philippe-du-Roule, only to find himself bundled unceremoniously into an official car outside the church at the behest of his irate master.

As it happened, the pretext de Gaulle was seeking for resignation came his way only three days later, with Socialist demands in the Assembly for a 20 per cent reduction in military expenditure. The announcement of his departure was dramatic in its suddenness and decisiveness; as he remarked to one of his entourage: 'Il faut être pittoresque dans ses actes. En partant sans me retourner, j'emporte avec moi mon mystère.'

Astonished by his resignation, his ministers were left in a mingled state of elation and dismay, depending on their party allegiances. The Communists saw this as their chance. When de Gaulle had reshuffled his cabinet in November 1945, he had brought in a couple more Communists, including Thorez; but he had impressed on the latter that there could be no question of the Communists having either Foreign Affairs or Defence—where the Communists' Moscow connection made them unacceptable as guardians of French national interests. Nor would he let them have the Ministry of the Interior, with its control of the police and elections. Still smarting from this demonstration of distrust, Thorez now invited the Socialists to join the PCF in a government under his own leadership. But the Socialists, like the other parties, were too aware of what was going on in Eastern Europe to be tempted by the offer, and insisted that the MRP should be included as countervailing ballast. Faced with this proposal, the MRP in its turn mumbled misgivings, indicating that there could be no question of serving under a Communist Premier. It seemed that an impasse had been reached, when pressure came from an unexpected and unwelcome quarter: the Chief of the General Staff, the Gaullist General Pierre Billotte, urged them not to hold back, in case a Communist-dominated cabinet should result in America cutting off economic and diplomatic assistance. At the same time, Billotte moved troops into the Paris area—largely as a general precaution rather than as a specifically anti-Communist safeguard. At the end of the day, the MRP agreed to enter the cabinet, provided that the Premier was a Socialist—as it happened, Félix Gouin. The result was a Tripartite government of Socialists, Communists, and MRP (26 January–12 June 1946), but with a myriad of warning signals for the future, confirming all too accurately de Gaulle's apprehensions about the direction affairs were taking. When forming his cabinet, Gouin merely said to the Communist and MRP leaders: 'Here are the portfolios I'm giving to your parties. It's up to you to decide who will have them.' It seemed that the old regime of parties was back to stay.

Nor did the prospects seem any better in the Constituent Assembly itself. As expected, the Communists and the left wing of the Socialists under Guy Mollet were pressing for a powerful, unicameral parliament, with the executive securely under its thumb. Léon Blum, however, was insistent on the need for stable government based on a secure parliamentary majority, with the power to resort to referendums if necessary. The MRP likewise wanted firm government, with the additional safeguard of an upper house to act as a brake on the wilder impulses of parliament. The unicameral option won the day, however, both in the drafting commission and in the Constituent Assembly, where the Left pushed it through by 309 votes to 249 on 19 April 1946.

But at this juncture—as de Gaulle predicted—public opinion played a decisive role. When the constitution was put to a referendum on 5 May, it was rejected by 10.6 million to 9.4 million. Opinion polls indicated that at least half of the 'noes' did so for fear of a Communist-dominated Assembly running the country, while the newly acquired female vote was also an important asset to the 'no' lobby and to its principal organizers, the MRP. It was back to the drawing-board once more.

Yet if de Gaulle's faith in referendums had snuffed out the prospect of an all-powerful single chamber, it was no guarantee against a return to the old familiar remedies of the past, or something much like them—despite the over-whelming public demand for new solutions in the previous October. The elections to a second Constituent Assembly on 2 July 1946 reflected the political swing displayed in the May referendum. The MRP picked up eighteen extra seats, while the Socialists lost twenty. But the second Assembly met in very different circumstances from the first. De Gaulle had hitherto avoided making specific recommendations as to what the new constitution should be like. But on 16 June he broke his silence with a wide-ranging speech at Bayeux, which subsequently became the manifesto of the RPF of 1947, and contained a number of proposals which were ultimately to be embodied in the constitution of 1958.

De Gaulle reminded his audience that 'the rivalry of parties in our country is a national characteristic, that of always questioning everything and thus all too often overshadowing the higher interests of the country'; and he called for a semi-presidential system, where the executive would be less at the mercy of the legislature. Although an opinion poll in November 1945 had shown that half the public favoured a President elected by universal suffrage, de Gaulle doubted whether the run-of-the-mill politicians could as yet be persuaded to accept this; and in the meantime he proposed a President chosen by an electoral college. By this he meant the *notables* of the nation—in effect, members of parliament and *départemental* councils, together with representatives of municipal councils from the country at large. In other words, the system of 1958–62 (p. 282).

To muster effective support for this programme, René Capitant launched the Union Gaulliste pour la Quatrième République in July–August 1946. He was at pains to claim that it was not a party, and adherents were encouraged to maintain their existing party allegiances. Within a month, it had half a million supporters, including twenty-two deputies in the Assembly and a further sixty-one close sympathizers. The main effect of the Gaullist campaigns, however, was to encourage the forces of Tripartism to sink their differences and close ranks around a compromise government, for fear that an impatient public, weary with *le régime du provisoire*, should turn a sympathetic ear to the Bayeux proposals.

The compromise that emerged was strongly marked by the MRP, who had played a leading role in the drafting commission. In order to secure Socialist and Communist consent to an upper house and to a reasonable modicum of power for the government, the MRP had regretfully abandoned many of its proposals for a stronger executive. And at the end of the day the final document bore a depressing resemblance to the old constitution of the Third Republic. It was nevertheless approved by the Assembly by 440 votes to 106 (30 September 1946), and nationally accepted in the referendum of 13 October 1946 by 9 million to 7.8 million. With the noes (31.1 per cent) and abstentions (31.2 per cent) outweighing the yeses by nearly two to one, this was scarcely an auspicious beginning for the new regime.

THE ELECTORAL SYSTEM

Paradoxically, the electoral system was not part of the constitution, since experience had shown that discontent with the form of election was likely to arise more frequently than dissatisfaction with the constitution as a whole. It was therefore the subject of a separate law that was easier to alter (27 October 1946). Yet there was considerable irony in the fact that such a fundamental matter, with considerable potential for making or breaking political stability, was never put to the people in the same way as the constitution. It is true that the prime source of French instability was the traditional multi-directional split of the electorate on the deeply felt issues of class, clericalism, and the constitution (p. 35)—to which a fourth 'C', colonialism, was shortly to be added. And no electoral law could alter that.

The problem in the constitutional and electoral systems was how best to strike a balance between representative and workable government. In a country like France, where the divisions were so marked and so numerous, it could be attempted from two directions. The first would be to adopt an electoral system that would bipolarize or at least bridge these differences by obliging the various segments of opinion to minimize their disagreements and line up behind the two most powerful contenders. And the second would be to insulate the government as far as possible from the control and influence of a fragmented parliament. De Gaulle believed that the problem was most effectively resolved by distancing the legislature and executive—making the legislature representative and the executive workable by reducing the opportunities that the legislature had traditionally had for impeding or bringing down the government. Needless to say, this was not a congenial formula in the eyes of a left-wing majority, brought up on horror stories of the Second Empire.

The British system of parliamentary government was regarded with envy by many Frenchmen, since it seemed to combine a vigilant, representative parliament with strong, stable government. What perhaps was not always appreciated in France was that the British system represented not so much the control of government by Parliament, but rather the control of Parliament by government, thanks to an electoral system and a political landscape that guaranteed the government a disciplined majority. Although democracy came into its own at election time, the 'first-past-the-post' system in Britain compounded the rewards, in terms of seats, to the advantage of the leading party—while the fact that British political opinion was broadly split two ways on the issue of 'class', instead of half a dozen or more as in France, reduced the contenders to two main parties. The British Parliament controlled government only in so far as the government emanated from the party with the largest number of seats. But until the next election, the government used the party to ensure that its measures received a smooth passage through Parliament. Moreover, the party was prepared to behave compliantly, since the success of the party depended on the success of the government. Not only would the outcome of the next election hang upon it, but the advancement of individual members within the party was geared to their loyalty to their superiors as well as to their ability.

In France, by contrast, not only did the political landscape make for a multiplicity of parties, none of which enjoyed a majority, but government itself was usually a coalition—thereby depriving the French system of that close link between party and government that characterized Britain. No government was in power long enough to make a deputy's future completely dependent on winning and maintaining its favours; and since parties were much longer-lived than the ephemeral coalition cabinets that ruled France, even ministers tended to put party before government.

While admiring looks were cast across the Channel, it was regretfully admitted that the British electoral system presupposed that the dominant party and its ministers would not abuse their power to bend the system even further in their favour. Both leading British parties accepted the broad structure of the existing parliamentary system, even if they differed on particular aspects of it, such as the powers of the House of Lords. In France, by contrast, deep differences over the constitution were a major feature of political life; and if the electoral system had been changed to enable one party to dominate parliament and form a government on its own, few politicians would have trusted such a party not to exploit its new-found powers. The distrust was all the greater in post-war France, given the electoral strength of the Communists.

At the same time, the Resistance parties, with their ideals of a regenerated France, did not wish to return to what they saw as the parish-pump *esprit de clocher* of the pre-war single-member constituencies. Larger, multi-member constituencies seemed to offer a broader forum for political debate, where the preoccupation would be with national rather than parochial issues. Only the Radicals and other pre-war dab hands at second-ballot compromises at grassroots level mourned the old system—and the MRP were particularly relieved to see it go, since they feared the likelihood of anticlerical alliances between Radicals and Socialists on the second ballot.

The system adopted was very like the one that was used in the Constituent Assembly elections of 1945–6, except that the method of allocating seats between parties in each *départemental* constituency was by 'the highest average' method. This change tended to give more of an advantage to the parties with most votes, thereby in practice providing more seats for the three Resistance parties that currently dominated the Constituent Assembly.

Another factor that had commended proportional representation to the bulk of the parties was the hold it gave them over the candidates. To stand at all, a candidate had to be proposed as a member of his party's list; and his inclusion and position in the list depended entirely on his acceptability to the party and on their assessment of his potential as a vote-catcher. There was no life outside the party, let alone advancement. Bitter experience of the loose discipline of the Third Republic's party politics made this alternative seem long overdue to many politicians—although, as de Gaulle had already vehemently pointed out, it ran the risk of making loyalty to the party the only loyalty in a politician's soul.

Perhaps the most unfortunate consequence of *scrutin de liste* was its tendency to force parties to fight their most savage battles with those who came closest to them in programme and clientele. To some extent, this was true of any one-ballot system, but it was aggravated in the case of *scrutin de liste*, in that the wider *départemental* constituency tempted parties to try their luck, whereas in the narrower confines of a British-type single-member constituency, smaller parties tended to desist unless they stood a reasonable chance of winning. Thus, the Fourth Republic was to witness Communists and Socialists vilifying each other in their competition for the working-class vote, and Gaullists and MRP behaving likewise in their pursuit of the Catholic electorate.

Another disadvantage of *scrutin de liste* was that the methods of apportioning the votes were difficult for the electorate to understand, and either deadened interest or aroused suspicions. It seemed to many that the democratic process had been taken over by the professionals, who could only too easily manipulate it to their purpose, since the general public did not know the rules well enough to be able to exercise an informal watching brief. While it was recognized that every system favoured some parties more than others, it struck observers as unjust that the vagaries of the 1946 system should, for example, give the Communists a seat for every 33,000 votes cast for their candidates, whereas the Radicals had to receive 45,000 votes to obtain a seat. The fact that the 1951 amendments to the law were to reverse this advantage in the other direction merely served to underline the injustices of the system.

THE CONSTITUTION

Despite the *déjà vu* nature of the constitution, the MRP and Socialist members of the drafting committee had insisted that it should afford the government greater protection against the whims of parliament than the old one had done.[1] The government was now constitutionally obliged to resign only in two eventualities. The first would arise if the Prime Minister, with the cabinet's approval, made a forthcoming vote in parliament a matter of confidence—and parliament then rejected it by an absolute majority. The second would occur if parliament passed a motion of censure by an absolute majority. The first eventuality only happened once—to Bidault in 1950—while the second never happened at all. Less than twenty motions of censure were drawn up under the Fourth Republic, of which only five were actually discussed.

At the same time, the government was empowered to dissolve the National Assembly if it brought down two governments within eighteen months— provided that parliament itself had enjoyed at least eighteen months of its allotted five-year span. This weapon of dissolution—though never used before 1955—increasingly tempted Prime Ministers to make crucial votes matters of confidence, hoping that parliament would give the government the majority it wanted rather than run the risk of having to face the electorate prematurely as a result of bringing down two governments within eighteen months. This happened on average about once a month during the 1946–51 parliament. Yet although the threat of dissolution, with all the expense and uncertainty of a new election, might act as a restraining influence on members of parliament, it was always uncertain whether an actual dissolution would bring much benefit to the government. In a multi-party system, the shifts that it might cause in the composition of parliament would not necessarily improve the government's position. Dissolution also had a further hazard. Prior to the amendments of 1954, the constitution designated the President of the National Assembly as the caretaker Prime Minister during a vacancy of parliament. This was not a matter for alarm as long as the President was a respected figure who would not seek to use his brief caretaker powers to push through policies that would be uncongenial to either the outgoing or incoming majority. But uneasiness was to be felt in 1948, when the President of the Assembly, Édouard Herriot, was ill, and the role of caretaker might have fallen on the Communist Senior Vice-President, Jacques Duclos.

Despite the protection that the new constitution sought to afford the government, the Fourth Republic saw a change of ministry virtually every six months. When Prime Ministers were unable to get legislation through parliament, or were anxious to forestall impending defeat, they generally chose to resign before it was necessary rather than risk the humiliation of formal eviction. A strategic withdrawal of this sort increased the chances of bouncing back later, as well as making one's inclusion in the next cabinet more likely. This was reflected in the fact that of the twenty-three Premiers that France had under the new constitution, sixteen had been members of the previous cabinet and twelve were to be members of the cabinet that succeeded them. Moreover, the need for incoming Premiers to gain or keep the support of those parties that had supported their predecessors, resulted in governments retaining three-quarters of the ministers of the previous cabinet, many of them continuing to hold their former portfolios.

It might be tempting to conclude that the instability of government did not matter all that much, since continuity of personnel made for continuity of policy. To some extent this was true, especially of those ministries whose spheres were not central to the main concerns of parliament and which tended to become the province of specialists. Thus, foreign affairs and long-term economic planning, which were the subject of sporadic rather than continued parliamentary debate, enjoyed a certain immunity from the perpetual party sniping that harassed the Ministry of Finance and other spheres that more obviously affected the elector's pocket or his concept of what society should be. The defence of French interests against foreign competitors was a unifying cause rather than a divisive one. Consequently, Bidault and Robert Schuman were able to monopolize the Quai d'Orsay for five years, until Indo-China became too hot a potato for Bidault to handle. In other areas of government, there was admittedly a three-to-one chance of ministers remaining in the new cabinet and an even chance of their retaining their old portfolios. Yet there remained a prevailing air of uncertainty and impermanence that made collective long-term planning extremely difficult—especially since the head of government was, in Philip Williams's phrase, 'the fuse that blows first'. Indeed, the fear of defeat was arguably more disastrous to good government than the actual defeats themselves, in that it condemned France to play-safe, short-term policies.

Given the rapid turnover of governments, it was often assumed that the chances of a deputy or senator achieving ministerial office in France were much greater than in Britain—and that this was a tacit reason why there were only half-hearted attempts to reform the system. This rapid turnover was undoubtedly true of the half-dozen top posts in government, which between 1944 and 1958 were held by 48 individuals in France but by only 27 in Britain; and it was also true of men with full membership of the *conseil des ministres*—122 in France, as against 72 in Britain. Yet because British governmental teams were bigger than their French equivalents, averaging 70 members instead of 30, 295 parliamentarians achieved government rank in Britain in this period, while the French total was only 208.

Until it was modified in 1954, the constitution insisted that the prospective Prime Minister should be invested by an absolute majority of the lower house. It was then his duty to choose a cabinet. Unfortunately, the early Premiers of the post-war period established a well-meant but unnecessary tradition, not stipulated in the constitution, that the Premier should thereupon present the names of his ministers to parliament—which effectively laid him open to the risk of defeat by *interpellation* even before his period of government had properly begun. Although parliament had retained this weapon of *interpellation* that had toppled half of the Third Republic's 109 government, it proved less dangerous in practice than in the past. Only five of the Fourth Republic's twenty-three ministries fell in this way. If the government had a well-disposed coalition supporting it in parliament, then this majority would often refuse to debate hazardous issues; and the chairmen of the standing committees, who determined the Assembly's timetable, would often refuse to allow time for *interpellation* debates—pleading pressure of legislative business—if the government was to their liking. More governments, six in fact, foundered on the rock of the annual finance bill—which affected pockets and governed the ministry's ability to fulfil its policies.

PARLIAMENTARY PROCEDURE

The National Assembly was in session for eight months of the year, but normally met only on three days a week—Wednesday, Thursday, and Friday at 3 p.m. This partly reflected the fact that many deputies were deeply involved in local government and were away for much of the week, conducting municipal and constituents' business as well as their own private professions. Apart from important debates, the average attendance was about forty. Proxy voting was not only allowed but was the norm; and it was a measure of how much deputies had come under party control in the Fourth Republic that it was usual for the party whips to cast the votes of their members whether they were present or not. The placing of a white card in the urn indicated 'yes' and a blue card 'no'. A deputy wishing to vote against the party line had to place two cards of the opposite colour in the urn—the first to negate the card cast on his behalf by the party whip, and the second to record his actual preference. Until the changes of June 1956, discussion of the budget took up half of parliament's time. Thereafter it was restricted to the period 1 November to 15 December. Despite the limited period available for parliamentary debate, parliament dealt with an average of 1,300 bills a year and passed about 230 of them. This was four times the British average; but French bills tended to be less substantial than their British equivalents, since much of the detail was left to outside experts to frame in the form of *règlements administratifs*.

Tactically, the government was at something of a disadvantage, in that the timetable of parliament was decided not by the government but by joint meetings of the chairmen of the standing committees. On the other hand, during those periods when the chairmen were favourably disposed towards the current government, they were in a position to give it considerable covert support. The nineteen committees that they chaired consisted of forty-four deputies each; and the work of each largely corresponded to that of a particular ministry. Since it was the committee's draft of a bill, not the government's draft, that was debated in parliament, its *rapporteur* and chairman played important roles in the legislative procedure of the Assembly. Indeed, the chairman came near to acquiring some of the attributes of a shadow minister, given the absence of a formal or clearly identifiable parliamentary opposition.

With recent memories of the role played by the Senate in the downfall of the Popular Front, it was not surprising that the left-wing parties were determined to reduce its powers in the post-war era. After the failure of the Communists and Socialists to suppress it altogether, they had to make do with a compromise by which its obstructive powers were fundamentally reduced to saying 'not so fast'. And the break with the past was symbolized by renaming it the Council of the Republic.

Its mode of election was also initially changed to ensure that its political complexion would be closer to that of the Assembly. Like the method of allocating seats to the lower house, the complexity of the new system bewildered, and consequently alienated, many ordinary voters—so much so that there was little outcry when it reverted to something closely resembling the pre-war system in 1948 (p. 402 n. 5). Yet this reversion was not so much a recognition of the shortcomings of the 1946 system, as a realization that a mirror image of the lower house was no longer desirable now that the Communists had gone into opposition and the Gaullist RPF was challenging the regime (pp. 155–63). That the Council of the Republic was its old senatorial self again was reflected in the re-emergence of the Radicals as its largest group, profiting handsomely from the fact that two-thirds of the municipal delegates in the *départemental* electoral colleges came from communes of under 3,500 inhabitants, representing less than half of the French population.

The senators—as they were once more called in 1948—were elected for six years instead of the nine under the Third Republic. But half of them were to retire by lot in May 1952, thereby creating a pattern by which there would be a triennial renewal of half of the membership.

Since the initial electoral system of 1946 was intended to secure a close identity of outlook between the two houses, the procedure for passing bills was simplified. Instead of the pre-war *navette*, when bills moved back and forth between the houses until agreement was reached, a bill passed up from the lower house could now only be altered in the upper house by an absolute majority. It was then returned to the lower house which could, if it wished, reverse the Council's amendment—provided that it did so by an absolute majority. But if the Assembly could only muster a simple majority against the amendment, the bill was consigned to limbo, and a new bill had to be worked out. At first the system worked reasonably well, but it rapidly encountered difficulties in 1947, when Communist and Gaullist senators began to adopt obstructive tactics. Despite these contretemps, only about 10 per cent of bills examined in 1948–51 were hamstrung by the system—even though this included some important items, such as the 1950 budget and the 1951 changes in the electoral system of the lower house.

If the Council was gravely referred to as 'the house of second thoughts', its second thoughts were of a depressingly conservative nature. This was scarcely surprising, given that its post-1948 composition had much in common with the pre-war Senate. Sixty per cent of the members were professional men, with a large contingent of lawyers, many of whom had a substantial clientele in the world of agriculture or business.

THE PRESIDENT OF THE REPUBLIC

While the President of the Republic came nowhere near de Gaulle's concept of what the office should be, he was nevertheless far from being a mere ceremonial figure. As before the war, he was elected by a joint sitting of both houses of parliament—and as such he was also dependent on the votes of the colonial representatives, which could play a decisive role. Thus, the first President of the Republic, the veteran Socialist Vincent Auriol, was narrowly elected on 16 January 1947, thanks to the support of eleven members from French Equatorial Africa, hurriedly brought in by plane to tip the balance. As a Radical of the old school bleakly remarked: 'Things have come to a pretty pass when the head of state depends on eleven niggers dropped down from heaven.' In fact, both presidential elections—1947 and 1954—were hard-fought affairs, in which tactical considerations loomed large. As Jacques Duclos commented, explaining the Communist vote in 1947: 'Ce n'est pas pour ses beaux yeux que nous voterons pour Auriol'—momentarily forgetting that Auriol had a glass eye.

Like his predecessors under the Third Republic, the President chaired the *conseil des ministres*—the cabinet in the British sense—but not the larger *conseil de cabinet* which included the junior members of government. The importance of this practice very much depended on the personality and self-appointed role of the particular President. Auriol actively participated from the chair in the discussions—typified in 1951 when he told the *conseil des ministres* concerning its Tunisian policy: 'The policy you want to follow is bad ... But I don't have any political responsibility; I have only a moral responsibility, and I mean to exercise it to the full!'—such discourses usually ending with a dismissive, 'C'est à vous de décider!', delivered in his incomparable Meridional accent. (Would-be imitators could never quite catch the flavour of his favourite phrase, 'Nous ne sommes pas des titangs!') By contrast, his successor, René Coty, restricted himself to the formal conduct of the meeting.

The President could find himself playing a key role in the choice of Prime

Minister. While this function constitutionally belonged to most Heads of State, including the British monarch, it was in practice often a merely formal one, in that the presence in parliament of a party with a strong majority would of itself designate the next Prime Minister. But in France, where no party had a majority, and where the leader of the largest might not be acceptable to the other parties, it was often far from obvious who should be the next Prime Minister. In these circumstances, the role of the President could be crucial; simply the order in which he consulted level-pegging candidates might tip a delicate balance. Auriol went much further, openly arguing in favour of his own predilections. But Coty was far more circumspect. Even so, he was to play a key role in the events that brought de Gaulle to power in 1958.

9

The Pattern of Politics, 1946–1954

THE mid-1940s and the late 1950s were dominated by the six-foot-four figure of General de Gaulle. The interval was filled by the brief life of the Fourth Republic, when it was hard to associate events and developments with specific individuals. The politics of these years seemed like a crowd-scene viewed from an upstairs window, in which one knew that the participants were life-size figures, cumulatively capable of far greater thoughts, actions, and perceptions than any single man; and yet by comparison they seemed Lilliputian and ineffective. Ironically, these were years of great economic achievement, characterized by more adventurous attitudes to growth and a greater readiness to engage in co-operative enterprise; together they laid the foundations of future French prosperity. Yet politically it was a period of disintegration, when the hopes and ideals of the Resistance parties evaporated, and parliamentary government rapidly returned to the *immobilisme* of the inter-war years.

This depressing story had three major turning-points: the collapse of Tripartism in 1947; the break-up of 'the Third Force' in 1951; and the reluctance or inability of conservative cabinets in the early 1950s to extricate government from the stagnant issues in which it seemed becalmed.[1]

THE COLLAPSE OF TRIPARTISM

The pattern of loss and gain in the election to the new Assembly on 10 November 1946 largely confirmed the expectations of the constitution-makers (Table 9.1). Those disparities that did occur partly reflected the high level of abstentions among a vote-weary electorate, but they also stemmed from the burgeoning confidence of the old-style 'moderate' parties. On the other hand, Communist discipline paid even greater dividends than their rivals feared, giving the PCF the largest number of seats in the house. Yet its gains were partly won at the expense of the Socialists, whose losses, together with the parallel gains of the Radicals and conservatives, deprived the Communist–Socialist block of its 1945 overall majority (pp. 139–41). The elections to the Council of the Republic in the following weeks produced a broadly similar pattern of results, but with the Radicals and conservatives proportionally better represented.

Although Tripartism still held a majority in both houses, the electoral

Table 9.1.
National Assembly Election of 10 November 1946 (Results for Metropolitan France only)

	% of vote	Seats		
		Total	%	Total plus overseas deputies
Communists	28.8	165	30.3	183
Socialists	18.1	91	16.7	105
Radicals, USDR, etc.	11.4	54	9.9	70
MRP	26.3	158	29.1	167
Conservatives, various independents, etc.	15.4	76	14.0	93
TOTAL		544		618

Note: The percentage of electorate casting valid votes was 76.6.

campaign had further embittered relations between the Communists and the MRP (pp. 134–5). Since neither Thorez nor Bidault could find sufficient parliamentary support to form a government, Léon Blum undertook to bridge the impasse by forming a totally Socialist government (16 December 1946– 16 January 1947). But following the election of the President of the Republic (p. 149), the ageing and ailing Blum gladly relinquished the premiership to his Socialist colleague, Paul Ramadier. Ramadier was an easygoing old charmer, chiefly known to the general public by the photographs of him walking his cocker spaniel, Soso, or entrenched behind his desk, his papers half-buried beneath the somnolent spread of his giant cat.

The Communists, however, principally remembered him as the only Socialist to vote against their expulsion from parliament in 1940, and so they hopefully asked him for one of the three key ministries. Ramadier accordingly gave François Billoux the Ministry of Defence—diluting the prize by creating three new ministries for the army, the navy, and the air force, and thereby placating the apprehensive MRP. Leaving nothing to chance, the new Army Minister, Paul Coste-Floret (MRP), rushed to occupy the buildings of the Ministry of Defence in the Rue Saint-Dominique before Billoux got there—on the grounds that they had belonged to the Ministry of War in the days before its replacement by the Ministry of National Defence. And the unfortunate Billoux wistfully agreed to occupy an empty house without a telephone in the Rue François 1er. His functions were then distributed between the Prime Minister and the three new ministries (7 February 1947). Symptomatically, a letter he wrote to a French military representative in America was returned to him curtly marked 'sender unknown'.

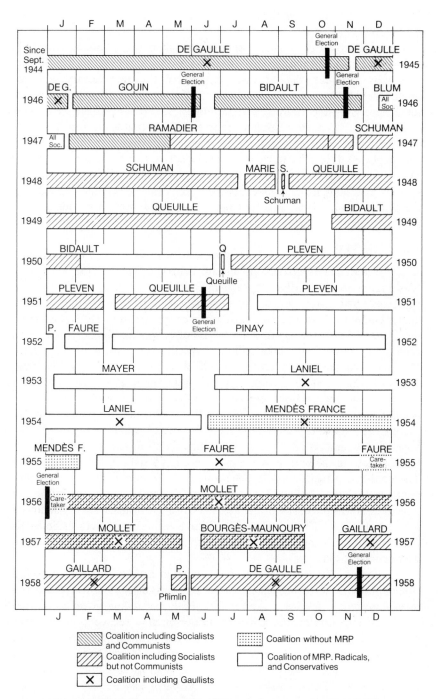

9.1 Duration and Composition of French Governments, 1944–1958

Source Philip Williams, *Crisis and Compromise: Politics in the Fourth Republic* (London, 1972), 33.

A fateful step for the Fourth Republic was taken when Ramadier decided to submit the composition of his cabinet for the approval of the National Assembly. Although Bidault had inaugurated this unnecessary and risky procedure in the pre-constitutional period of government, there was every case for dropping it—especially since it arguably ran counter to the letter of the constitution. Ramadier's cabinet reflected all too faithfully the dangers of choosing ministers mainly as a means of gaining favour with the party chiefs. The expected Tripartite nucleus—nine Socialists, five Communists, and five MRP—was flanked by five members of the Radical RGR and two conservatives, making it the broadest-based cabinet since de Gaulle's first government of September 1944. But, far from being a ministry of all the talents, it was a ministry of all the strings, with each party out to manipulate them as best it could.

The Communist exclusion from government in May 1947 was largely the result of Ramadier's anxiety to placate America. When Blum had been to Washington in May 1946 to secure financial assistance for France, the Americans had gently hinted that there might be more aid if the cabinet no longer contained Communists. Now, in the early months of 1947, the Chief of the General Staff, the Gaullist Pierre Billotte, was unashamedly seeking to exploit American fears about Billoux's recent appointment to the Defence Ministry. Truman's hard-hitting speech of 11 March 1947 against Soviet activity in Europe spurred Ramadier into preparing to expel the Communists, which he had increasingly seen as inevitable, given French dependence on American money. Nor was the risk of losing Russian sympathy as great a sacrifice as it had once seemed. Ramadier's Foreign Minister, Bidault, had lost hope of obtaining Soviet support for French claims to economic control of the Sarre, following the Four Power Conference in Moscow in March–April 1947. The conference had ended in deadlock, with France, Britain, and America rejecting Molotov's proposals for the reunification of Germany. At the same time, Russia was full of resentment over the Franco-British treaty of alliance on 4 March 1947, which ran counter to Blum's earlier assurances to Stalin that French talks with the British government would concern themselves purely with economic matters.

Running parallel with this westward swing in the government's search for support, was its increasing unease at working-class discontent with the government's austerity policies at home. The PCF leadership—loyal to Thorez's continued support for the government's economic recovery programme—maintained a discouraging attitude to workers' demands for strike pressure, despite the conviction of Laurent Casanova, André Marty, and Léon Mauvais that the party was rapidly losing its soul through its continued participation in government. These tensions within the party had already been demonstrated by the refusal of the Communist deputies to follow the lead of the PCF ministers and vote their confidence in the government's Indo-Chinese policies (p. 234). They chose instead to abstain in the Assembly vote on 22 March 1947.

Despite these rumblings from below, the PCF and the CGT dutifully denounced the unofficial strike that had broken out at Renault. (Admittedly, this gesture of loyalty to government was made easier by the fact that the principal strike-organizer was a Trotskyite.) But it was a painful line to hold; and a growing fear that Socialist and MRP newspaper support for the strikers was steadily upstaging the PCF led to a Communist about-turn on 30 April. However congenial to party sentiment, this sudden championing of the workers gave Ramadier precisely the pretext he was looking for to rid himself of his embarrassing bedfellows. He promptly made the government's incomes policy a matter of confidence; and although the government won the Assembly vote easily, by 360 to 186, the Communists, including the ministers, voted against it (4 May). They simply assumed that even if the government resigned, its successor would have to include Communists to be sure of a working majority. Ramadier, however, immediately turned the tables on the party by asking the Communist ministers how they saw their future in the government; when Thorez made it clear that they had no intention of resigning, Ramadier promptly dismissed them.

The significance of what had happened—the effective end of Tripartism—was by no means obvious, least of all to the Communists, who assumed that the eventual fall of Ramadier would see them back in whatever cabinet came to power. Ramadier had anticipated that the dismissal might provoke hostile demonstrations, perhaps even a resort to an armed coup; and, leaving nothing to chance, he had instructed the new army Chief of Staff, General Revers, to keep troops on close alert. Needless to say, no hint of this was allowed to penetrate the unfurnished walls of the lonely Billoux's Ministry of Defence—although Thorez was deliberately allowed to catch wind of it so that it might serve as a salutary warning. If nothing else, the strengthening of troops in the Paris area would provide transport and labour in the event of the Communists calling a general strike.

The Communist ministers were replaced by MRP and Socialist figures, with a certain amount of reshuffling and amalgamation of function. There was also a brisk purge of the broadcasting service, and about a hundred pro-Communist army officers were shunted into less sensitive posts. How far the sacrifice of Tripartism increased American generosity is hard to gauge, since other factors were already inclining America to greater beneficence. General Marshall had returned from the fruitless Four Power meeting convinced that Stalin envisaged the disintegration of Europe and that only massive American aid could prevent it. The welcome result was the liberal-handed Marshall Plan, announced on 5 June 1947, which was to be a major factor in Western Europe's economic recovery. As it turned out, 20 per cent of Marshall Aid for Europe was to go to France—a larger percentage than to any other country except

Britain—while the sacrificial victims of the deal, the PCF, could ruefully lament that after years of outright persecution under Marshal Pétain they were now entering political exile under Marshall Aid.

Quite what was lost with Tripartism is not easy to assess—apart from the obvious blow to hopes of a stable parliamentary majority. The great series of social reforms had already come to an end with the first Constituent Assembly; and there was little sign of significant additions. Neither the Blum nor the Ramadier cabinets had envisaged further advances along the lines of the CNR programme; their main concern was inflation and the rate of exchange. It is true that Jules Moch, Minister of Public Works and Transport in both governments, had put forward a bill to nationalize the principal shipping companies, but the MRP and Radical members of the cabinet emasculated it by a series of fundamental changes.

In the meantime, the Ramadier government resigned (19 November 1947), largely as a result of Mollet's blistering accusations that it had no right to stay in office if it could not implement the SFIO's policies. Yet neither the easygoing Ramadier nor the intransigent Mollet could have guessed that nine years would now elapse before the Socialists would head another government.

THE THIRD FORCE ERA, 1947–1951

The Gaullist Challenge

The exile of the Communists had a dangerous counterweight in the appearance of a new enemy of the Fourth Republic—de Gaulle's Rassemblement du Peuple Français. De Gaulle's disgust with the evolution of post-war France has been outlined in the previous chapter (pp. 139–40 and 141–2). In an oft-quoted phrase, François Mauriac described the RPF as 'the biggest mistake de Gaulle has ever made'. Formally launched on 14 April 1947, it held a series of monster meetings in the summer, where the décor and mass enthusiasm put friends in mind of American political conventions, and led critics to liken them to fascist rallies in the 1930s. Nor was apprehension allayed by de Gaulle's claims that the RPF was not a party but a movement.

The RPF created a parliamentary intergroup at the end of August 1947, open to all deputies except the Communists on the understanding that its members would retain their former party allegiances. Despite this proviso, the MRP and the Socialists prohibited their deputies from joining, with the result that the bulk of the intergroup's membership consisted of conservatives and Radicals, with a handful of disaffected MRP back-benchers. The practice of *double appartenance* was always to cloud attempts to assess the RPF's true parliamentary following in the late 1940s; but pointers to its strength at grass-roots level came with the municipal elections of 19 and 26 October 1947, which saw 38 per cent of the votes go to RPF candidates—most spectacularly in the large cities. Bordeaux, Rennes, and Strasbourg were among those that elected RPF mayors; and the RPF obtained fifty-two of the ninety seats in the town council of Paris, which elected de Gaulle's own brother, Pierre, as its President. Predictably, the main victims of the RPF's success were the MRP and, to a lesser extent, the Radicals and conservatives.

The Gaullists were nevertheless uneasy at the disparate nature of their new-found following, much of it there for negative reasons. Like the MRP in 1945 and 1946, the RPF was somewhat embarrassed by the influx of ex-Vichyites, many of whom recognized in the RPF something much closer to their own concept of politics than the democratic sentiments and promise of mild social reform preached by the MRP. At the same time, the anti-Communist stance of the RPF paid electoral dividends in a period marked by the onset of the Cold War. Added to which, there was mounting disillusion with the record of the democratic parties, who displayed neither vigour nor coherence, let alone success, in the pursuit of their policies. Fear of a third world war also made many electors feel that a man like de Gaulle was more likely to keep a firm hand on the tiller; and by the end of 1947 the RPF had perhaps as many as a million adherents.

Yet if positive support was strong, so too was opposition. To the fury of the Left, de Gaulle spoke on 27 October 1947 of 'an immense power' arising in France that condemned 'the regime of division and confusion'; and when he demanded an immediate election, followed by a new constitution, there was an uproarious response, reminiscent of the 1930s outcry against fascism. De Gaulle's talk of replacing the trade unions with a system of 'capital–labour associations' was regarded as particularly sinister; and, indeed, accusations of neo-Vichy corporatism were to pursue him for much of his career, not least in the final shoot-out in the 1969 referendum (p. 330).

Looking back on 1947, André Malraux later remarked: 'De Gaulle marched us at full speed to the Rubicon—and then told us to get out our fishing rods'. With national elections four years away, and dissolution a dirty word in the Republican lexicon, it would be difficult to sustain the impetus of 1947 until such time as the electorate could make its will felt; and de Gaulle appears to have ruled out any recourse to direct action. On the other hand, he assumed that public exasperation with a succession of ineffective governments would probably increase rather than diminish in the interval, and that a systematic if tacit policy of weakening government through joint hostile votes with the Communists might eventually force a dissolution. Certainly, the next five years were to see the Gaullist deputies in an unholy alliance with the Communists, voting consistently against whatever government was in power.

Such a policy, however, created its own enemies, and the ruling parties were later to win support in unexpected quarters for their defensive measures. This was notably the case with the 1951 electoral law, which effectively skewed the electoral system against the Gaullists and Communists (p. 166). Moreover, the more 'moderate' complexion of government brought about by the creation of the Third Force disposed many former critics of Tripartism to slacken their demands for major changes in the constitution, thereby depriving the RPF of useful auxiliaries. Indeed, many of the RPF's conservative supporters drifted towards the rehabilitated conservative parties in the following years—much as they had drifted to the RPF from the MRP in 1947—but were now in effect belatedly finding their true home. The result was that the RPF, like the MRP in 1948, gradually became more responsive to its left wing.

Its Secretary-General, Jacques Soustelle, displayed energy and imagination, while André Malraux provided the movement with a prestigious literary profile. Yet the RPF's activities were systematically denied adequate coverage by state-controlled radio, and its weekly newspaper, *Le Rassemblement*, was chronically short of funds, despite discreet subsidies from the CIA to the RPF.

The Constitution at Issue

The emergence of the RPF and the exile of the Communists altered the whole tenor of politics. The logic of future government alignments was now inexorably dictated by the arithmetic of the Assembly. The two-legged stool of Tripartism could henceforth muster only 249 votes—65 short of the absolute majority required for investing a new ministry. This left the Radicals and conservatives as the only available substitute for the missing leg.

Not only did the mathematics of the situation throw governments into dependence on coalitions of the Centre rather than the Left, but it resurrected the question of the constitution. The Communists were determined to make government more subservient to its parliamentary critics—as long as they themselves were not in power. Conversely, the Gaullists were aiming at a semi-presidential system, where a strong executive would stem from the electorate rather than parliament. The result was that the regime was under fire from both sides—forcing the other parties to rally to its defence. What Guy Mollet described as 'the Third Force' (October 1947) was gradually taking shape.

The Malaise of the Left

It was in the logic of this new situation that the Prime Minister of the Third Force era should come from the MRP and the Radical party. Although the Socialists participated in government for all but six months of the period before the 1951 election—holding an average of five portfolios in each cabinet—the centre of political gravity had passed to the middle of the spectrum. The Socialists were now faced with the same dilemma that had earlier faced the Communists: should they remain in government in the interests of preserving the system, or should they cut loose from their bourgeois cabinet colleagues and keep their reforming principles intact in the shining purity of opposition?

Most rank-and-file Socialist deputies favoured the middle course of continuing in government—but on conditions. In practice, this meant a lengthy rearguard action to prevent the policies of the governing coalition slipping further to the Right. Six times, the Socialists felt obliged to bring down the government during the current parliament. It was an unhappy role—and one which the MRP was also to find uncongenial when it became the left wing of an increasingly conservative government coalition after 1951. Grass-roots supports fell away from what was already a seriously under-funded party. Indeed, the Socialist newspaper, *Le Populaire*, was shamefacedly in receipt of CIA subsidies via the American Federation of Labor—a well-known fact which the Communists periodically aired in order to counter Socialist accusations that they were subsidized by Moscow.

The PCF was now the millstone around the neck of social progress in France. Committed to a policy of systematic obstruction in parliament, it obliged every government to look to the Centre and the Right for support, thereby ruling out all possibility of further social legislation. It was especially galling for the Socialists that, despite the PCF's destructive role, it would continue to attract the votes of the far Left, thereby depriving the Socialists of the electoral support that would have enabled them to govern in conjunction with parties favourably disposed to reform. As always, it was the Moscow connection that made the millstone so disastrously heavy. Admittedly, it was true that a non-Communist party of the extreme Left, or a militant tendency within the SFIO, would likewise have been an embarrassment to the Socialists. The first would have split the left-wing vote at election time, while the second would have scared off moderate voters. But once the Socialists were in government, extreme-Left elements of this kind would have supported them rather than side with the conservative opposition. The Moscow connection also gave the PCF's policies a capricious character that made left-wing co-operation difficult. And as long as France needed American money, the Russian link ruled out effective partnership in any case. Nor were floating voters likely to forget such disturbing remarks as Thorez's assertion in 1949 that the French working class would side with the Russian army if it were to cross the Frontier 'in pursuit of an aggressor'—a damaging declaration that was subsequently endorsed by the CGT.

The Communists' position was admittedly an unenviable one. Exploratory meetings in September 1947 about setting up the Kominform found the French and Italian Communists under strong attack from their European colleagues for having co-operated with middle-class parties—criticism which brazenly ignored the fact that this co-operation was partly the result of Stalin's own prompting in 1945. At home, Arthur Koestler's *Darkness at Noon* and Kravchenco's *I Chose Freedom* were fast becoming best sellers—as yet unchallenged by Art Buchwald's *I Chose Caviar*. Putting on a brave face, the PCF tried desperately hard to present itself as the party of peace, preserving the world from American imperialism—a portrait in which the Communists depicted the SFIO as being 'l'aile marchande' of the 'parti de Washington'. But such a stance had a decreasing appeal at a time when the bulk of French people thought that the threat of a third world war was gradually receding; 35 per cent considered it likely in July 1947, but only 14 per cent in July 1949. Even so, the late 1940s saw a series of Soviet moves that acutely embarrassed the PCF. The Prague coup of February 1948 was rapidly followed by the condemnation of Yugoslavia's Tito in June 1948. There then came, in devastating succession, the Rajk affair in Hungary in June 1949, the Kostov affair in Bulgaria in December 1949, the attacks on Gomulka in Poland in 1949 and 1951, and the Slansky affair in Czechoslovakia in 1952—all of which received wide publicity in France and seemed to make nonsense of the PCF's claims that Communism was a liberating force.

The wrecking role of the PCF was all the more frustrating, given that much of the Communists' electoral support was an amalgam of various protest votes rather than an expression of genuine pro-Soviet commitment. Figures of over-all party membership are hard to establish (official levels having been inflated for propaganda purposes), but there may have been half a million paid-up members in 1945, dropping to about 300,000 in the mid-1950s, when most parties saw falling rolls as the remnants of Liberation euphoria evaporated. Even so, PCF membership was probably the equivalent of all the other parties put together.

The party organization was characterized by a highly disciplined vertical structure, which ranged down from the Central Committee, through the federations and sections, to the basic cell, containing party members living or working in the same residential block or firm. Only a quarter of the cells were work-based, most members preferring to separate their politics from their promotion prospects. Horizontal contacts between cells were forbidden by the party as detrimental to discipline. Discussion in the cells was encouraged, but not criticism. Discussion was a means of identifying potential trouble-makers, as well as of gauging the reception that new policies were likely to receive.

Communist influence in the CGT was strong. Its Socialist leader, Léon Jouhaux, did not return from Germany until May 1945, and in the meantime the Communist, Benoît Frachon, had emerged as its leading figure, becoming Secretary-General in September 1945. Yet the PCF's periodic attacks on the SFIO created unease both in the CGT and among rank-and-file Communist voters, who saw them as weakening working-class solidarity—a fear shared by Frachon and several of his senior Communist colleagues.

The latter half of 1947 brought all these tensions to a head. Not only was the PCF in formal opposition to the government, but prices rose by 51 per cent and wages by only 19 per cent. The left wing of the union movement believed that only a general strike would force the government to match wage restraint with price controls, whereas Jouhaux's followers disapproved of striking against a Socialist-led government. In a sudden display of financial orthodoxy, the Ramadier government abolished the coal subsidy on 14 November 1947, causing coal prices to rise 40 per cent and substantially increasing the prices of electricity, gas, and transport. At the same time, a CGT call to strike in Marseille on 12 November triggered off widespread strikes elsewhere, which soon involved some three million men as discontent with the current price-rises exacerbated working-class feelings.

While the Communists sought to widen and prolong the strikes, the Socialists were trying hard to bring them to an end. Escalating violence and disagreement about objectives saw a fall-off in support; and, recognizing impending defeat, the CGT hastened to save face by calling off the strike (10 December). A steady return to work ensued, but the damage was done: the Socialists packed their bags and abandoned the CGT nine days later, forming their own union, Force Ouvrière. To add further venom to the break, the expense of the move was partly met by funds provided by the American Federation of Labor. In the meantime, the Assembly granted Robert Schuman's government stern penalties for militant picketing and the right to call up reservists (4 December 1947)—but only after the most violent debate that post-war France had seen, with fist-fights between shouting deputies, and the Garde Républicaine arresting one and forcibly ejecting others. The number of deputies bundled outside would have been higher if the guards' intervention had not been delayed four times by the Communist deputies singing the 'Marseillaise', bringing the guards to attention and transfixing their commander at the salute.

After the secession of Force Ouvrière, the CGT no longer controlled the majority of trade-union members, and it was now far less useful to the PCF as an instrument of militant pressure. Even so, the Communists proceeded to foment strikes for political ends the following year—until resentment, suspicion, and defections among the rank and file of the CGT caused the party to abandon it as counter-productive. An unsuccessful miners' strike (4 October–29 November 1948), organized by the CGT against projected redundancies, saw tanks on the streets and several deaths, with hundreds of miners in gaol and thousands sacked. As a result, the CGT's standing among the miners took a hard knock; and it was not until the miners' strike of 1963 that it eventually recovered something of its former influence.

Despite this hard lesson, the next four years saw a series of similarly counter-productive political strikes, partly directed against American foreign and military policies, but also against specific government measures at home. Best remembered among these was the protest strike against the brief arrest of Jacques Duclos in May 1952, following what the Minister of the Interior insisted on calling 'the carrier-pigeon plot'. The police had found two pigeons in the boot of Duclos's car, intended, as it transpired, for his kitchen rather than the Kremlin. Apart from the unpopularity of these strikes, the PCF's traditional role of champion of the underdog left it vulnerable to the risk of losing touch with the large sections of the working classes who were benefiting from the economic progress and greater social mobility of the post-war decade. As Pierre Hervé later pointed out, it was in danger of becoming 'a Poujadism of the Left', tied to the fortunes of the losers among the working class, just as Poujadism was to be tied to the losers among the lower-middle class. Significantly, the successful strikes of 1953 and 1955, which were for social rather than political ends, owed little or nothing to CGT leadership.

The isolationist strategies of the PCF were equally unpopular at local government level. Communist success in municipal politics was consistently at its best when the party was actively co-operating with other parties. Whereas in 1945 over 40 per cent of towns with 30,000 or more inhabitants elected Communist mayors, control of three-quarters of them was lost in the 1947 municipal elections, when the Communists were doubly disadvantaged by their own break with Tripartism and by the rise of the RPF. Recovery thereafter was slow.

At the top, the PCF lacked attractive leaders. In a party in which discipline was paramount, conversance with orders and an ability to carry them out efficiently and unswervingly counted for more than the persuasive talents that were considered essential to any aspiring politician in the democratic parties. For all his courage and dedication, Thorez was something of an automaton on the rostrum, his dramatic turns of phrase losing their impact through their wooden-faced delivery. And if the avuncular, walrus-moustached Duclos had more of the genial *bonhomie* of the old-style Palais Bourbon maestros, even he found it hard to give conviction to all the twists and turns of the Moscow line. Thorez's ill health, following a stroke in November 1950, resulted in his accepting Stalin's offer of medical care in Moscow, where he stayed until Stalin's death in March 1953, with Duclos keeping house in the interim.

The Liberation era had seen much sympathy among intellectuals for the PCF as the party that seemed most committed to radical social change. The disconcerting dishonesty of its tactics was forgiven in the light of the laudability of its ultimate aims. The Mouvement de la Paix, launched in Poland in August 1948, had many eminent French adherents, including Communists, such as Picasso, Éluard, Joliot-Curie, and Fernand Léger, as well as fellow-travellers such as Vercors, Autant-Lara, and Jean-Louis Barrault. They were prepared to forgive, and even concede some justice in, the Communists' dismissal of Existentialism as escapist; and they made understanding—if regretful—noises when Sartre's books were banned in Czechoslovakia after the coup of February 1948. There was likewise tolerant laughter when the Soviet novelist, Alexander Fadejev, slated Sartre as 'this typing hyena'. However, the mental contortions required by the PCF in following its labyrinthine changes of policy increasingly alienated men of integrity. French intellectuals who had testified against Kravchenko were made to look foolish when his trial (January–April 1949) vindicated his accusations against Stalinism. Similarly, the Lysenko affair of 1949 ended in humiliation. This Soviet scientist's claims that acquired characteristics were inheritable—a welcome hypothesis to Soviet theoreticians of progress—were based on experiments that were subsequently revealed as spurious. Among French Communist intellectuals, only Marcel Prenant, Claude Cahen, and Jacques Monod questioned Lysenko's findings—and were demoted as a consequence. It was only in the mid-1950s that the PCF furtively admitted tó its error.

A further test of loyalty came in November 1949 when David Rousset's denunciation of Soviet labour camps triggered off a wave of PCF denials, intermingled with private Communist comment that even if they did exist, it could not serve the purpose of World Communism to propagate the fact. After much soul-searching, a number of fellow-travellers publicly recognized the justice of Rousset's claims; and the episode served to widen still further the gap between the PCF and independents such as Sartre, Merleau-Ponty, and Vercors, who had hitherto regarded the PCF as a lesser evil in a world in which poverty was the greatest oppression. At the same time, many Catholic Progressistes and a number of Mounier's disciples at *Esprit* broke off contact with the PCF—the whole debate finding its way into Camus's *L'Homme révolté* (1951).

Clericalism Rides Again

If the brief years of Tripartism had seen class and the constitution as the more contentious of the three Cs, the Third Force era was to see the re-emergence of the clerical issue in the shape of the schools question. Despite its being a basic bone of contention with the left-wing parties, the MRP's manifesto of 1944 had stipulated that the State should subsidize private as well as public schools, provided they observed comparable standards. Indeed, the MRP members of the second Constituent Assembly had attempted to get this principle incorporated in the constitution; and, given the support of the right-wing parties, they nearly succeeded—falling short by the narrow margin of 272 votes to 274 (28 August 1946).

Inflation was a very real problem for private schools in the late 1940s, and much of the organizational skill and militancy of private-school parents stemmed from their experience in co-operative schemes for fund-raising. This was quickly recognized by the Gaullists, who, as early as 1947, saw the schools issue as a salutary method of wooing the Catholic vote. The MRP, now ensconced in the Matignon with Robert Schuman as Prime Minister (24 November 1947–19 July 1948), were acutely aware that they were being upstaged by the Gaullists on this as well as on other issues. June 1948 saw a major disagreement between the MRP and their Socialist cabinet colleagues over the Poinso–Chapuis decree of 23 May, which provided funds to help educate the children of needy families, including those at private schools. The

Socialists were already restless with the cabinet's cautious financial policy; and although the government technically fell as a result of Socialist demands for cuts in military spending, the rift over the Poinso–Chapuis decree was the decisive factor—all the more ironic in that the decree was never to be put into practice. Its consequences were nevertheless reflected in the local and senatorial elections, when the Socialists turned to the Radicals rather than to the MRP as their allies.

This division served to highlight the fundamental split in the MRP itself between its socially progressive wing, anxious to remain on good terms with the Socialists, and the remnants of the Liberation influx of homeless conservatives. Although many of these had joined the RPF, and an increasing number were to gravitate towards the rehabilitated conservative and Radical parties, those who were left looked hopefully towards Bidault—Schuman's jealous *alter ego* at the Quai d'Orsay, and rival for the Matignon. Bidault was contemptuous of Schuman, mocking his grave delivery and meticulous habits. Schuman would carefully untie parcels himself, winding up the string for later use and waving away the offer of scissors with scandalized protests of: 'you mustn't do that! That is not the way I was brought up!' Bidault dismissed him as 'an engine that runs on low-grade petrol'—to which Schuman privately responded: 'Not all of us run on alcohol'. While Schuman saw the MRP's future as lying with the Third Force, Bidault had already approached de Gaulle in private to investigate whether there was any common ground between them. He broached the matter with MRP sympathizers in 1949, but they concluded that such a move was unlikely to receive the backing of the party as a whole.

Even so, the MRP was characterized by a marked loyalty of the rank and file towards its leaders; and uncongenial right-wing figures, such as Bidault, Coste-Floret, and Letourneau were dutifully supported, despite the consternation that their opinions and policies often created. The colonial sphere was a particularly contentious area, where the inflexible record of Coste-Floret and Letourneau at the Ministry for Colonies in 1947–50 (p. 231) aroused the misgivings of many MRP back-benchers who personally leaned towards the Socialist preference for conciliating nationalist feeling in North Africa and Indo-China.

The Procession of Ministries

The underlying irony of the Third Force era was that each time the Socialists punished the current government for being insufficiently progressive, its replacement turned out to be even more conservative. The sequence of ministries and their composition is outlined in Figure 9.1 (p. 153). Although they all contained four to six Socialists (with the exception of the final months of Bidault's cabinet), the Socialists were progressively outnumbered by the combined forces of the Radicals and the more modestly represented conservatives. The balance in most ministries was held by the MRP—echoing the role of the Catholic Centre Party under Weimar.

The shape of things to come was already foreshadowed in the Radical André Marie's ministry (26 July–28 August 1948), where the strong representation of conservative financial interests indicated that the state interventionism of the Liberation era was well and truly dead. Brought down by the Socialists, it was eventually succeeded by another coalition under a tried-and-true archetypal Radical of the Third Republic, Henri Queuille (11 September 1948–6 October 1949). Like so many of the stalwarts of the pre-war party, he had started life as a country doctor and had then been a minister no less than nineteen times. Past master of the manipulative skills required to keep a ministry afloat in a multi-party system, he allegedly claimed: 'In politics it is not a question of solving problems, but of silencing those who raise them.' Remembered for the retro-gressive 'reform' of the Conseil de la République (p. 148) and the use of tough tactics against strikes, his cabinet was eventually brought down by the Social-ists, who were now demanding a return to pre-war freedom of collective wage-bargaining, following the recent devaluation of the franc (19 September 1949).

Bidault's coalition (28 October 1949–24 June 1950) was likewise destroyed by the Socialists, who first of all walked out of it on 7 February 1950 and later defeated it on successive wage issues. As it happened, Bidault's fall was the only occasion during the Fourth Republic when the government was brought down by an absolute majority on an issue of confidence, in strict accord with the con-stitutional rules obliging a government to resign.

In the confused weeks that followed, the Socialists under Guy Mollet made a strong but unsuccessful bid for power; and the premiership eventually passed to René Pleven of the UDSR (12 July 1950–28 February 1951). But his broad-based government was deeply divided, not only on economic policy but also on the thorny issue of electoral reform. This was fast becoming a matter of urgent concern, given the proximity of the 1951 elections. Faced with the wrecking tactics of the Communists and Gaullists, it seemed essential to devise a system that would prevent them from achieving a preponderance of seats in the new Assembly. The RPF would be standing in its first election, and the current 24 self-declared Gaullists could conceivably become 150 or more. If, moreover, the Communists succeeded in holding their 177 seats, the two parties between them could bring parliament to a standstill. The problem was, how to reshape the system against the wrecking parties without upsetting the fiefs and grass-roots spadework of their Third Force opponents. Compromise was difficult, and Pleven himself chose to resign rather than accept the proposals that eventually emerged. After another unsuccessful bid by the Socialists to claim the premiership, it was left to the arch-manipulator, Henri Queuille (10 March–10 July 1951), to resolve the matter. This he did by bringing the general election forward to June 1951, which meant that recalcitrant deputies either had to fight the election under the existing rules or under the new proposed system, which, for all its faults, was less fraught with the danger of a Communist–Gaullist predominance. The bill became law on 7 May, despite a desperate attempt by the Senate to substitute the pre-war system of *scrutin d'arrondissement*, the preferred solution of the Radicals and conservatives.

The principal feature of the new law was that it enabled constituency parties to make tactical alliances, thereby reintroducing the opportunity for wheeling and dealing that had disappeared with the pre-war two-ballot system (p. 143). As such, it was favoured by the Radicals and those conservatives and rural Socialists who had effectively exploited 'le vieux bon temps', while putting at a disadvantage the Communists and the Gaullists, whose opposition to the regime made it harder for them to find congenial allies. The importance of the change was reflected in the fact that whereas the 1946 system effectively required the Radicals to gain 45,000 votes to obtain a seat, they needed only 25,000 after 1951. Conversely, the Communists, who had hitherto gained a seat for every 33,000 votes, now had to find 51,000. For its part, the MRP was initially afraid that its indelible clerical tag might prejudice its own chances of making effective partnerships; but it finally voted for the law as a necessary bulwark against the Communists and Gaullists, and its subsequent loss of seats was not as a result of the electoral change.

Under the new system, parties presented their individual lists to the electorate as in the past; but those making a tactical alliance in a particular *département* pooled the votes that their respective lists received. As an alliance they could thereby claim a larger number of seats than if they stood separately. The seats they won were then distributed between the constituent partners of the alliance by the same principle of 'the highest average' that was used to distribute the seats between the rival parties and alliances in the *département* as a whole.

The principal objection to this system was that the same party could make different alliances in different *départements*, depending on its regional tactical requirements. The party might consequently appear to stand for different things in different areas, obscuring the electors' choice and creating conflicts of commitment in the National Assembly—thereby further increasing public cynicism towards party pledges. While this had also been true—up to a point—of the second-ballot deals under pre-war *scrutin d'arrondissement*, these had only pledged individual candidates rather than the parties as such.

A further change to the system stipulated that in the event of an alliance or a party list gaining an absolute majority of the votes cast in a *département*, it would automatically receive all the seats. Since this good fortune was much more likely to befall an alliance than an isolated party, it was yet another attempt to favour the Third Force parties. Indeed, it was alliances that were the beneficiaries of all but one of the forty constituencies in which an absolute majority was reached in 1951. And what the alliance system could produce in practice was exemplified by the notorious case of the *département* of the

Hérault, where, through partnership, the Socialists obtained three seats with a mere 39,028 votes, while the un-allied Communists did not win any seats, despite having the far larger total of 69,433 votes.

An added provision allowed electors to split their votes between two or more lists by replacing names on the list for which they were voting with other names from other party lists. In practice, this so-called *panachage* was to have little effect on elections, since not enough people exercised the right for it to affect the results significantly.

The most controversial feature of the new law, however, was the particular system that was applied to the Paris region. With no attempt to justify itself in equity, this was an undisguised attempt to strengthen the chances of the Centre parties at the expense of the Communists and Gaullists who were strong in the area. The law therefore instituted the mathematical device of allocating seats according to 'the largest remainder' instead of 'the highest average', a piece of quasi-gerrymandering which was made worse by the fact that the majority of electors did not understand the difference and assumed that it was even more dishonest than it was. None of this disposed the public to come to the rescue of the regime when it was threatened with subversion in 1958.

The 1951 Election

Its effect on the elections of 17 June 1951 was much as expected (Table 9.2). Alliances took place in eighty-eight *départements*, with the Radical-dominated RGR and the various conservative groups as the main beneficiaries. Although

Table 9.2.
National Assembly Election of 17 June 1951 (Results for Metropolitan France only)

	% of vote	Seats		
		Total	%	Total plus overseas deputies
Communists	26.0	95	17.5	101
Socialists	15.3	95	17.5	107
Radicals, USDR, etc.	10.1	77	14.1	95
MRP	13.4	84	15.4	96
Conservatives, various independents, etc.	13.5	87	16.0	108
Gaullists	21.7	106	19.5	120
TOTAL		544		627

Note: The percentage of electorate casting valid votes was 77.3.

the RGR's share of the vote had marginally fallen since 1946, it picked up an extra twenty-three seats. Similarly, the conservatives gained a further eleven seats, despite a fall in their share of the poll. Alliances with the Third Force parties also helped the Socialists, who were four seats better off, notwithstanding a drop in vote-share. The MRP, on the other hand, were the principal casualties of the RPF—44 per cent of the Gaullists' votes having come from former MRP supporters, according to current estimates. Yet, despite their misgivings over the alliance system, the MRP actually suffered fewer losses under the new electoral law than they would have done under the old. Not so the Gaullists, whose successes would have been greater but for the change. With de Gaulle personally opposed to alliances, the RPF concluded only thirteen, mainly in the west. With 21.7 per cent of the votes, the Gaullists received 19.5 per cent of the seats—both figures somewhat disappointing when compared to their municipal successes in 1947.

The main intended victims of the new system, however, were the Communists, who were allied with no one. While their oppositional role since 1947 saw their share of the vote fall from 28.8 per cent to 26 per cent, their share of the seats was only 17.5 per cent. Had the election been conducted under the 1946 law, the Gaullists and Communists between them would have commanded a narrow majority of the seats, enabling them to wreck the Republic had they been so minded. The new system had fulfilled the purpose of its architects—whatever the price in public respect.

CONSERVATISM TAKES OVER

The Parties and their Clientele

With the enemies of the regime cut down to substantially less than size, and the economy restored to relative stability, the Third Force parties could now get down to their preferred business of pursuing sectional interests. Table 9.3 gives an approximate notion of the clientele of the main parties, as reflected in an opinion poll of 1952. As one would expect, the PCF had the youngest following and the Radicals and conservatives had the oldest—65 per cent of Radical supporters being over fifty. The Radicals also had the smallest proportion of women—the Radicals' traditional anticlericalism and former opposition to female suffrage acting as disincentives, without the compensating appeal of generous social policies. Conversely, women made up a majority of MRP, conservative, and Gaullist supporters—parties which respected the Church, stood for order, and, in the case of the MRP, were not oblivious to the need for social reform.

When it came to occupations, the industrial working class predictably loomed large among the Communist and Socialist supporters, as did a sig-

Table 9.3.
Social Composition (%) of the Parties' Support among the French Electorate
in the early 1950s

	Total	Com.	Soc.	MRP	RGR	Cons	RPF
Age							
Over 50	37	23	37	34	65	45	37
35–49	29	35	33	35	24	25	25
18–34	34	42	30	31	11	30	38
Sex							
Men	48	61	59	47	64	47	47
Women	52	39	41	53	36	53	53
Industrial workers	19	44	25	15	10	10	18
Farm-workers	7	10	9	3	5	4	4
Civil servants and state service	5	4	9	3	5	6	2
White-collar workers	7	6	4	11	1	3	7
Industrialists and management	5	1	4	3	4	10	7
Commerce and shopkeepers	4	3	3	5	7	7	7
Farmers	16	5	9	20	31	35	18
Liberal professions	1	1	1	2	1	1	1
Rentiers and retired	6	3	7	6	14	8	4
Housewives	30	22	28	30	19	15	32
Size of town							
Over 100,000	24	20	16	19	23	29	22
20–100,000	17	22	18	14	11	11	22
5–20,000	11	10	16	14	6	5	11
Under 5,000	48	48	50	53	60	55	45

Source: Institut Français d'Opinion Publique, Poll of February and March 1952.

nificant section of the landless agricultural labourers. The Socialists also had a fair following of public employees, while, on the other side of the political spectrum, farmers featured largely in the Radical and conservative clientele, and less impressively among the MRP's and Gaullists'. Conservative supporters also included a powerful section of business and management, whose influence, as always, extended much further than their actual numbers.

The appeal of the MRP and Gaullists was the most evenly spread across the various categories of occupation and age, perhaps reflecting the fact that they claimed to stand for certain moral values that were primarily a matter of personal temperament and belief rather than the product of the voter's employment or economic niche. The MRP's catholicism, in both senses of the word, was also mirrored in its relatively even distribution through the various sizes of town and rural community—whereas the Gaullist strongholds were principally in the large towns. But against this, the regional spread of both the MRP and the Gaullists was strongly conditioned by the level of religious observance.

In a famous phrase, written in the 1920s, Albert Thibaudet declared: 'There is no hope for a party that writes on its banner: "interests".' It was still true in the 1950s, in the sense that any party that subordinated itself to a single interest or cause in a multi-party system was likely to suffer heavily from the cumulative electoral strength of counter-interests. And it was usual in French politics for a party to aim to have at least two strings to its bow. Admittedly, by the 1950s the industrial working class was a sufficiently big interest for it to be a safe focus for a party's undivided concern. This had not been the case before the war, when symptomatically the Radicals owed their relative success to the fact that their appeal was spread across several clienteles—peasant, small business, and anticlerical. But the Communists, as self-styled champions of the proletariat, never cornered more than 28.8 per cent of the vote in post-war France, and even that included the protest votes of southern farmers and other disaffected groups that had traditionally voted for whatever party seemed most opposed to the status quo. The Socialists, for their part, had been progressively learning the lesson of pre-war Radicalism—that safety lay in an amalgam of appeals to rural as well as urban discontent, accompanied by vigorous waves of the democratic and anticlerical flags of traditional Republicanism.

The post-war Radicals, increasingly left with the grievances of 'the little man', were vulnerable not only to the erosion of the peasant population, but also to the fact that shopkeepers and small-business men were shortly to find a more vociferous if short-lived spokesman in Pierre Poujade (p. 250). Indeed the very brevity of Poujadism was itself a further demonstration of the force of Thibaudet's dictum.

Although the conservatives were seen as spokesmen of big business, they never expected to be a large force numerically in post-war France. But even they swiftly appreciated the advantages of union on a broader base, as their series of mergers in 1949–51 reflected. When Roger Duchet put the Centre National des Indépendants on a formal footing in January 1949, it created a focal point to which the conservative Parti Républicain de la Liberté was drawn in April 1949. Paul Antier's Paysans likewise gravitated towards them, with the result that the Centre National des Indépendants et Paysans came into existence on 15 February 1951.

The MRP, as spokesman of Catholic interests, ran the obvious risk of being an exemplar of Thibaudet's warning; and although its various members were also enthusiasts of additional causes, these causes often looked to other parties as their prime champions. The vulnerability of the MRP was especially evident when the Gaullists sought to upstage it on the defence of Catholic interests— the Gaullists demonstrating the advantages of a variegated electoral base.

Arguably the least interest-orientated of all the parties, the RPF reflected a cast of mind that could be found in many sections of society—even if its hostility to the constitution and its commitment to the leadership of one man ruled it out for many who might otherwise have been drawn to its pragmatism and cult of self-discipline and national self-respect.

The desire of parties to strengthen their following by taking on board supplementary interest groups often resulted in these groups being able to exercise an influence which was out of all proportion to their wealth and numerical following. In a British or American context, they would have been lost amid the countless other interests competing for the attention of the two major parties, but in a multi-party system their weight still remained useful in the scramble for marginal leads. In an age of universal suffrage and the secret ballot, numbers ultimately counted more than wealth—which was why many parties still competed for the peasant vote, despite its rapid decline, and why most politicians were chary about upsetting the five-million-strong alcohol lobby (p. 245), whose power came from its numbers rather than the undoubted influence of its wealthiest members. The same could be said of the political muscle of the North African lobby, whose weight in the Radical party and elsewhere sprang from the settler equivalent of the party's mainland clientele—the numerous 'little men' of farming and business—rather than from the well-known names of North African big business who spoke on their behalf. On the other hand, house-tenants and war-veterans could always count on a receptive ear, since they combined large numbers with the respectable image of being 'deserving'—an unfortunate consequence of which was a housing shortage, since low rents were a disincentive to landlords to repair and extend their property.

The Continuing Procession

The nature of politics under the new Assembly soon became clear in the early ministries. The Socialists rejected Pleven's invitation to join his cabinet (10 August 1951–7 January 1952), which became in consequence a right-wing team, principally Radical and conservative, flanked with a few MRP perennials. Predictably, it was the Socialists who brought it down, when its budget proposals demanded special powers to streamline the social services and railways in the interests of economy and efficiency. The cabinet was taken over much as it was by the Radical Edgar Faure (20 January–29 February 1952), who vainly tried to bring the Socialists back into the fold by offering various compromises on policy. The conservative elements in his coalition objected, however, and the prospect of tax increases led to his defeat. The lesson was clear: coalitions including the Socialists and the conservative block were no longer possible; and in Paul Reynaud's opinion the time had come to experiment with conservative government—enlisting Gaullist support if need be.

Not that the Pleven–Faure period was devoid of achievement. The Pleven government had steered the principle of index-linked wages through the Assembly (20 September 1951), despite fierce Radical and conservative opposition, while even more momentously, Schuman's Coal and Steel Plan was ratified by all except the Communists and Gaullists (13 December 1951; p. 190. French foreign policy and attitudes towards European union are discussed in Chapter 12 below, pp. 241–2).

Other measures that were arguably less happy in their consequences included the granting of small indirect subsidies to private education. The combination of a conservative cabinet with a strong right wing in the Assembly had encouraged the 'clericals' to push for concessions; indeed, the Association Parlementaire pour la Liberté de l'Enseignement continued to get the nominal support of a majority of the new house. Admittedly, many of these adhesions were little more than an insurance policy in an uncertain world. But a government bill, subsequently named after Pleven's Minister of Education, the Radical André Marie, extended eligibility for secondary-school scholarships to the private sector, while an MRP proposal, drafted by Charles Barangé, envisaged a grant of 3,000 old francs to all parents with children at primary school, including the private sector. Passed by the Assembly on 21 September 1951, the bills divided the Communists, Socialists, and left-wing Radicals on one side from the MRP, RPF, the conservatives, and the right wing of the Radicals on the other. The sums involved were very small and were arguably not worth the deepening of political divisions that they engendered.

While the Left accused the 'clericals' of taking the Republic back to the days of Vichy, similar cries greeted the identity of the new Prime Minister. Antoine Pinay, who had briefly been a member of Vichy's Conseil National, was installed with a conservative government (8 March–23 December 1952) thanks to the votes of twenty-seven Gaullists—something that would have seemed inconceivable in the Liberation era. The left wing of the cabinet was made up of MRP figures; Robert Schuman had been firmly embedded at the Quai d'Orsay since 1948, and Pierre Pflimlin and Jean Letourneau were in charge of colonial affairs, where they were comfortably insulated from party criticism over the government's caution in social matters. There was something of the Orleanist liberal in Pinay's hostility to *dirigisme* and what he called the 'sottises' of the Liberation reform era. Sharing the Orleanists' optimistic belief in the irresistible redemptive powers of good management and technology, he canvassed subscriptions to the famous Pinay loan (26 May) by means of *comités d'honneur* of local worthies, in the best traditions of the July Monarchy. But, equally characteristically, he introduced to France the practice of the Premier buttonholing the nation in friendly radio chats.

Suspicious of the economists and technocrats of the public sector, Pinay preferred market-place gurus such as Jacques Rueff; and he was arguably helped in gaining public support for his policies of cautious financial orthodoxy by the fact that 1952 was a period of relative recession, with industrial production falling by 4 per cent. Ironically, Pinay himself thought that the pause in economic growth was a short-term by-product of his having achieved financial stability—whereas in fact this financial stability was largely the product of the stagnation of the economy. The check in inflation was primarily the outcome of restricted investment in the private sector, rather than the reward of his limiting public investment and his abstention from Banque de France loans. Despite his declared intention to cut public spending by 3 per cent, he was able to do no more than slow down its continuing increase. And even the celebrated Pinay loan, with its very advantageous terms to lenders, only brought in 1.7 per cent of the gold that was thought to be lying in private hands in 1952. Nevertheless, his self-projected image of the good housekeeper found a sympathetic response in many Gallic hearts, and an opinion poll following his eventual defeat found 56 per cent of those questioned regretting his departure, and only 21 per cent approving. Neither de Gaulle in 1946 (47 per cent) nor Blum in 1947 (40 per cent) enjoyed such confidence. Not surprisingly, the Pinay ministry fell as a result of MRP uneasiness at its conservatism—coupled with resentment at Pinay's attempts to keep a guiding rein on the MRP's illustrious tenant of the Quai d'Orsay, Robert Schuman.

Conservatism triumphed once again in the appointment of René Mayer, the leading Radical defender of business interests in North Africa, with a tried-and-true MRP–Radical–conservative team (8 January 1953–21 May 1953). His investiture was especially noteworthy for the support of all but two of the Gaullists.

Gone were the days of 1951 when RPF successes led some observers to predict a Gaullist take-over of a substantial segment of the parliamentary conservatives. Since then, de Gaulle's aloofness from the political scene had steadily disorientated many of his supporters, leaving them vulnerable to the attractions of Pinay's CNI. De Gaulle was both embittered by their growing support for conservative cabinets and disappointed by the RPF's poor showing in the municipal elections of spring 1953, when their share of the seats dropped from 26 per cent to 10 per cent. With iron in the soul, de Gaulle officially dissolved the parliamentary wing of the RPF on 6 May 1953, leaving the disowned herd of Gaullists to seek an eventual identity under the ambiguous title of Républicains Sociaux. By the end of 1954, identifiable Gaullists in the Assembly were down to seventy-two, most of whose votes on legislation fluctuated with the subject-matter—leaving only a hardbitten minority to imitate the Communist strategy of systematically opposing everything.

In fact, the granite-faced front of the PCF concealed similar uncertainty and disagreement in these years. During Thorez's absence in Moscow, his wife, Jeanette Vermeersch, and Auguste Lecœur had brought about the sacking of André Marty and Charles Tillon (16 September 1952). Marty was the forthright voice of the Left within the party, while Charles Tillon was its most distinguished Resistance figure. They also accused Jacques Duclos of wanting to lead the party in unorthodox directions. Duclos personally favoured a more indulgent attitude to the SFIO, based on a 'national front' strategy of 'anti-imperialism' and pro-peace initiatives directed against America. Certainly, the running-out of Marshall Aid in 1951 gave anti-Americanism a wider appeal than before; indeed, de Gaulle was among the louder critics of 'American imperialism' now that CIA subsidies to the RPF had dried up.

However, Thorez's return to France following the death of Stalin on 5 March 1953, helped to dampen down some of these differences; and the change in mood in post-Stalinist Russia gradually disposed Thorez to adopt a strategy more like Duclos's, based on the establishment of closer links with the Socialists. Lecœur was disgraced and Jeanette Vermeersch brought to heel. Even so, these reassuring developments were not enough to persuade Guy Mollet to accept the olive-branch that Thorez eventually offered to the Socialists.

In the meantime, the ministerial merry-go-round continued to turn. The Mayer government was succeeded by a similar conservative government under Joseph Laniel (27 June 1953–12 June 1954), remarkable for the participation of three Gaullist ministers and the adhesion of the bulk of the Gaullist deputies. While this breakthrough was clearly a direct result of de Gaulle's dissolution of the parliamentary wing of the RPF, it was also helped by the fact that Laniel combined impeccable big-business interests with wartime membership of the CNR—unlike so many of his well-heeled fellow runners, whose Occupation years were something of an embarrassment.

Not surprisingly, the principal challenge to the government came from the unions. The summer of 1953 found a quarter of wage-earners earning less than the official minimum wage (SMIG)—while only 40 per cent of the work-force earned more than one and a half times this amount, with the public sector featuring prominently among the low-paid. It was the government's schemes for economies in the postal service, notably through raising the retirement age (4 August 1953), that triggered off the trouble. Within ten days, four million people were on strike, led by Force Ouvrière and shortly followed by the CGT. Laniel resorted to the old *belle époque* ploy of requisitioning labour in the public sector, a manœuvre expressly forbidden by the Statut de la Fonction Publique of 1946. The government adopted a divide-and-rule approach to its opponents, buying off the big battalions of the PTT by exonerating them from the effects of the decree-laws and dealing severally with the rest. Trade-unionism consequently emerged in a state of disarray that was in marked contrast to the major strikes of previous years.

If Laniel's toughness with the unions dismayed the Left in parliament, it assured him of the solid support of the other parties. Indeed, Laniel would have been elected President of the Republic had not the Socialist chairman of the joint session of parliament, André Le Troquer, declared invalid those voting slips that omitted Laniel's Christian name—and might theoretically have been cast for his brother. It consequently took parliament seven days and thirteen ballots to come up with the name of René Coty, a cautious if honest conservative, embodying all the negative virtues (23 December 1953). The farcical nature of the elections was summed up by the bystanders' derisive shouts of 'Don't get lost!' as the new President emerged.

If the dominance of conservative policies at home indicated how profoundly the tenor of political life had changed since the Liberation era, another ominous sign was the growing importance of colonial issues in domestic politics, adding a fourth 'C' to the triumvirate of divisive issues. Indeed, the defeat of the Laniel government over Indo-China on 12 June 1954 was to be the first occasion that a French government had fallen on an imperial matter since 1898—and it was to inaugurate a disastrous pattern that was to lead directly to the death of the Fourth Republic four years later.

10
Economic Growth and Social Change, 1947–1973

THE raw materials of economic success—coalfields, fertile valleys, and the like—rarely coincide in a tidy fashion with the political frontiers of nation-states, even though the frontiers themselves may commemorate endless rivalry for the ownership of these resources. And in the same way, the broad sweeps of economic change seldom conform in a convenient way with the manageable units of political history—even though regimes may rise and fall on the back of economic change. French post-war economic growth was no exception, and its chequered history arguably makes more sense if the 1947–73 period is looked at as a whole rather than parcelled out to match the chronology of other developments. 1947 in many ways represented the return to peacetime conditions, while 1973 marked the end of the post-war era of rapid growth.

As late as the 1950s, it was not unusual to hear English schoolchildren chanting:

> Two froggy Frenchmen,
> One Portuguee,
> One jolly Englishman
> Beat 'em all three!

And it was not just a piece of Victoriana, recited for amusement or insult; it still contained an element of conviction. The instability of French government, its recent history of defeat and occupation, and the continuing inferiority of French economic performance to that of Britain were notorious facts. In 1951, French gross domestic product (GDP) was only two-thirds of Britain's, and her exports less than half that *d'outre-Manche*. But twenty-five years later, French gross domestic product was nearly a quarter higher than Britain's—and in per capita terms the margin was well over a third. As for exports, they too were nearly a quarter higher. What price the jolly Englishman now? It would seem he could not beat even one Frenchman—let alone two, plus a Portuguee (many of whose compatriots were now working in French factories and contributing to France's economic success.)

Nor did the explanation of this turning of the tables merely lie in the decline of Britain. The French GDP was now second only to West Germany's in the list of the ten richest Western European countries, even if in per capita terms, the

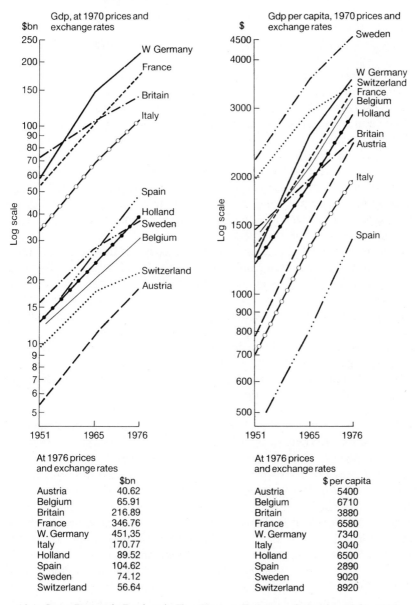

10.1 Gross Domestic Product in Ten Western European Countries, 1951–1976

Source: The Economist, *Europe's Economies: The Structure and Management of Europe's Ten Largest Economies* (London, 1978), 2.

well-scrubbed, commonsensical, war-avoiding Swiss and Swedes retained their position in the top three, leaving France in fourth place (Figure 10.1).

PRODUCTION, POPULATION, AND INDUSTRY

The sector that contributed most to this expansion was industry, accounting for 41 per cent of GDP in 1973, a proportion second only to West Germany's 47 per cent (Figure 10.2). Yet France continued to suffer from her traditional lack of natural resources. Not only was her coal production less than a quarter of West Germany's—and not even a third of her humiliated rival, Britain—but she had no part in the North Sea oil and natural gas discoveries, and her privileged enjoyment of Saharan oil and gas had been as short-lived as the Évian agreements with Algeria (pp. 278). She had, admittedly, struck a large deposit of natural gas at Lacq in the foothills of the Pyrenees in 1951, which for two decades supplied her with a third of her gas requirements, but this welcome find started to tail off in the 1980s. On the credit side, she had continued to develop the hydroelectric potential of her numerous fast-flowing rivers, and in 1966 the Rance Estuary near St Malo became the site of the world's first tidal hydroelectric dam. But although her hydroelectric production was three times greater than Germany's in the early 1970s, her overall electrical production was 40 per cent lower. On the other hand, her other pre-war handicaps—a small domestic market and a restricted industrial labour force—had been significantly offset by an increase in population and a growth in purchasing power, as wages and social security benefits improved.

The Liberation found the French population of 39.84 million scarcely bigger than it had been at the turn of the century (Appendix II, Table 1). Yet the next twenty years saw a greater increase than during the whole of the previous hundred—even if de Gaulle's celebrated plea of 1945, 'en dix ans, douze millions de beaux bébés', was only two-thirds met. As in most belligerent countries, the post-war period saw a baby-boom; but that of France was remarkable in that it tapered off more gradually—with the birth-rate remaining consistently higher than in pre-war decades (Appendix I, Fig. 1). The increase did not reflect a particularly rapid rise in the marriage-rate or in the number of large families; indeed, the *famille nombreuse* of four or more children was becoming a rare sight on Sunday mornings. It was rather the result of more families having a second or third child. Significantly, peasant families were getting larger—averaging three to four children in the mid-1950s—as the traditional constraints of the inheritance laws (pp. 4–5) were increasingly offset by the richer pickings to be had in the towns. It was now the small shopkeeper who had replaced the peasant as the arch-calculator in family matters—one or two children being as many as he wanted to argue over the posthumous takings of his meagre enterprise.

Yet *le petit commerce* was not the only social group to regret additions to the

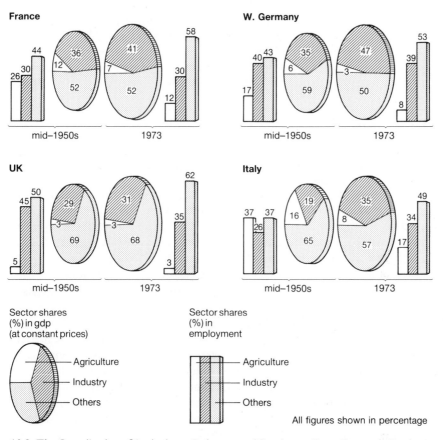

10.2 The Contribution of Agriculture, Industry, and Services to Gross Domestic Product and Employment, mid-1950s and 1973: France, UK, W. Germany, Italy
Source: The Economist, *Europe's Economies*, p. 3.

family. Sample surveys in maternity hospitals in 1959–62 revealed that a third of pregnancies were unwanted, and that without them there would have been no population increase at all. Compared to other developed countries, birth-control remained primitive. Even in the early 1960s, two-thirds of French couples still relied on premature withdrawal and other forms of onanism, with another fifth making use of 'the safe period', a markedly different situation from that in Britain (Table 10.1) and America. Not surprisingly, back-street abortions continued on an enormous scale—over half a million a year—and often in horrendous conditions. An abortion study in the late 1940s showed that half the women surveyed had induced the abortion themselves in lonely isola-tion, while a further quarter had done so with the help of a totally unqualified acquaintance. Two-thirds of these abortions were brought about with some sort

Table 10.1.
Contraceptive Methods in France and the UK, c.1960 (%)

	'Safe period'	Onanism, withdrawal, etc.	Condom	Diaphragm, spermicides, etc.
France (1961–2)	23	67	6	4
UK (1959)	11	31	34	24

of probe—one in twenty with a knitting-needle or pencil—while household disinfectants and other chemicals accounted for nearly one in five of the other. Perhaps as many as 20,000 women a year died as a result of these desperate stratagems. In response to this situation, the Mouvement Français pour le Planning Familial belatedly came into being in 1955; and the first family-planning clinic was set up in 1961. But it needed the legislation of 1967 and 1974 to remove the final restrictions that had been the source of so much misery (p. 342).

France still remained a relatively thinly populated country, despite her post-war growth and her advantages of soil and climate. Even in 1971, she still had only 96 people per square kilometre, compared with 239 in West Germany, 319 in Belgium, and 180 in Italy. Moreover, with the liberated 1960s and the cult of impermanence, the marriage-rate started to decline appreciably—although the effects of this were temporarily offset by the arrival at marriageable age of the post-war baby-boom. The early 1970s also saw a preference for smaller families, even if the number of childless couples did not significantly increase until later. However, these were all tendencies that were slower in coming to France than to her neighbours (Appendix I, Fig. 1). The same was true of the divorce-rate, which although rising from 11.5 per cent to 17 per cent between 1965 and 1978, was only half of West Germany's and two-thirds of that of England and Wales in the early 1970s. As in other countries, it would have been higher but for the parallel phenomenon of 'ongoing relationships' as a trial-substitute for marriage.

Repeating its role of the inter-war years, immigrant labour was a major factor in augmenting the work-force. The one and a half million resident foreigners of the Liberation era became three and a half million by the early 1970s—accounting for over a third of the population increase, and making up nearly 7 per cent of the nation. In addition to the pre-war nationalities—Italians and Spanish, half a million each, and much smaller numbers of Poles and Belgians—there were now nearly 700,000 Portuguese, and, more predictably, some 750,000 Algerians, nearly 200,000 Moroccans, and 100,000 Tunisians. Indeed, the Portuguese immigrants in France were equivalent to a tenth of the population of Portugal and a major contribution to French economic growth. Over half of this influx were working in the motor, engineering, steel, and electrical industries, with heavy concentrations in construction and public works, and another fifth were in services, mainly of the more menial kind. Inevitably, immigrants were to be among the first to suffer when recession hit these industries in the 1970s and 1980s—but they had played an important role in the economic upswing of the 1960s, when French labour demands provoked the heaviest period of immigration. Their arrival had occasioned relatively little tension in the communities in which they settled—and it was essentially the unemployment problems of the 1970s that made immigration a political issue.

Immigration and the post-war growth in family size also helped to counter-balance the growing proportion of the elderly in the population, which had been such a striking feature of pre-war French society. Even as late as the early 1950s the elderly percentage had been higher than that of any other Western European country; yet by the late 1970s, despite having risen from 11.4 per cent to 13.8 per cent, it was lower than that of West Germany (15.4 per cent) or Britain (14.6 per cent).

Welcome though these demographic factors were, their main impact on the French economy did not occur until the 1960s—when immigration was at its highest and when the babies of the 1940s reached working age. As children their main contribution was as consumers. Indeed, the active population—those in employment or self-employed—had increased by a bare 100,000 between 1901 and 1962, and was actually less than it had been in 1929. Moreover, as a percentage of the total population, they were considerably lower than in 1901—42.5 per cent instead of 51 per cent—given the greater life expectancy of the aged, and the raising of the school-leaving age. New sectors of the economy could only be developed by the transfer of labour from others, thereby causing contraction elsewhere and consequent uncertainty and diffidence among private investors, with the result that private investment remained below the Western European average. And of the 1.8 million people who left the land for the towns between 1949 and 1962, the bulk gravitated to the tertiary services, with 550,000 of the rest going into construction, and only 170,000 to manufacturing. So, the expansion of French production in the 1945–52 period is not to be understood primarily in demographic terms.

Productivity, Investment, and Attitudes

Yet if the growth of the labour force was not a major factor before the 1960s, the increase in its productivity certainly was. As Appendix II, Table 5 shows, 1949–69 was an unrivalled period of growth in French labour productivity in all spheres of the economy, ranking next to West Germany's and Italy's in the 1950s (Table 10.2). Growth figures are, of course, only relative terms, whose quantitative significance depends on the threshold from which they start. All

Table 10.2.
Growth Rates of Real Gross Domestic Product, Employed Labour Force, and
Labour Productivity, 1949–1959: France and Selected Countries

	Annual Percentage Rates of Growth		
	Real GDP 1954 Prices	Employed Labour Force	Labour Productivity
France	4.5	0.1	4.3
W. Germany	7.4	1.7	5.7
Italy	5.9	1.1	4.8
Netherlands	4.8	1.2	3.6
UK	2.4	0.6	1.8
USA	3.3	1.2	2.0

Source: Sima Lieberman, *The Growth of European Mixed Economies, 1945–1970* (New York, 1977), 13.

three countries had been theatres of war, with varying degrees of devastation as their baseline; so one would expect a much higher rate of growth in the post-war years than in those countries whose economies had escaped comparable destruction. But French industry had already achieved its pre-war level of production by 1947, and agriculture by the following year; so their continuing rate of growth was all the more remarkable for not being the mere reoccupation of their former positions. Nor was it just a question of the labour force working long hours. It is true that in the interests of national recovery, the unions had not opposed what had become a forty-five- or forty-eight-hour week by 1948; indeed, most workers were anxious to work longer hours in order to make up for the low hourly rates currently offered in an impoverished economy. But even on an hourly calculation, the growth of French labour productivity was well over 5 per cent a year.

While worker motivation was high, what was even more important was the improvement in equipment and organization that enabled him to produce more; and this was fundamentally a matter of investment and changing attitudes in management. Working under pressure to meet German requirements during the Occupation had already brought about rationalization in some industries, and forced a number of mergers and take-overs. Destruction of factories in the course of air raids and the campaigns of 1944 had also afforded the opportunity of rebuilding with more efficient equipment and lay-out. Faced with a heap of rubble, some proprietors had preferred to sell or transfer what was left to younger or more ambitious men, who were anxious to apply modern methods to production and profit. Perhaps most important of all, government loans and financial aid were henceforth dependent on the recipient observing various prescriptions on efficiency and long-term strategy. This was particularly true of Marshall Aid, which, together with earlier American help, provided nearly half of the Fonds de Modernisation et d'Équipement. At the same time, the Commissariat Général du Plan, in recommending specific targets, made producers aware of what the experts saw as their potential if they were prepared to rethink how best to use the resources at their disposal (p. 119). This was especially important, since capital on its own was no guarantee of improved productivity. Capital had been plentiful during much of the economic stagnation of the Third Republic—but most of it had been invested abroad, making the point that it was more a question of the attitude of investors and borrowers rather than lack of capital itself that was the principal problem.

Even so, the reconstruction of the 1940s would have been impossible without the large injections of American funds and the increased investment of French capital in home-based enterprise, which, as a proportion of GNP, was half as high again in the 1950s and 1960s as it had been in the 1930s (Appendix II, Table 6). If this was just part of a general Western European phenomenon—and slightly below the European average—it was none the less a vital increase (Table 10.3).

Table 10.3.
Investment and Productivity in Selected Western European Countries, 1955–1975

	Productive investment as a % of Gross Domestic Product (Average 1955–75 in constant prices)	% change in output per person employed in industry (Average 1956–75)
W. Germany	18.5	5.1
Holland	18.0	6.0
Spain	16.2 (Av. 1958–75)	7.8 (Av. 1955–71)
Sweden (Av. 1961–75)	15.9	4.8
Belgium	15.8	4.2
Italy	14.3	4.8
France	14.2	3.2
UK	13.6	2.8

Increased investment not only enabled the technological rejuvenation of production, but its multiplier effects created demand throughout the economy. And the particular virtue of American aid was that it came in a solidly backed currency, less vulnerable to the self-negating nemesis of inflation that so often attended home-government schemes of finance.

If the amount of increased French investment in industry was still below the Western European average, it must likewise be recognized that technological innovation in France did not play a major part in bringing about the salutary transformation of French industry. Indeed, the direction and speed of techno-logical change itself—as distinct from its adoption and installation—featured remarkably little in the forecasts and calculations of the Commissariat Général du Plan under the Fourth Republic. It was often alleged that the French pen-chant for intellectualizing and abstraction made them better at invention than innovation. British empiricism and German method, on the other hand, supposedly had results of more immediate utility than the far-reaching but not always easily applicable discoveries of French enquiry. Yet the Renault company in effect brought into being a new machine-tool industry in France when it set about supplying its own need for automatic machinery. Among air-craft, the Caravelle air liner (1955) and the Mystère fighter had an inter-national reputation, while the mid-1950s saw a French railway-engine gain the world speed record—a sphere in which France continued to be a leader for the following three decades. In the nineteenth century, French technical expertise had largely been confined to civil engineering and mining, where the École Polytechnique and the École des Mines played a major formative role; but in the 1950s and 1960s, French engineering expertise in other fields was increas-ingly respected throughout the world.

Yet it was more in the willingness of management to install new techno-logy—rather than in the French contribution to the technology itself—that the key to French post-war industrial success was to be found. Even under the Fifth Republic, when French achievement *per se* became both a goal and an instru-ment of presidential policy, the low level of funding for scientific and techno-logical research was something of a national scandal, especially in the private sector of industry. France in the 1960s was spending only 1.5 per cent of GNP on research, compared with 1.8 per cent in West Germany, 2.3 per cent in Britain and more than 3 per cent in America and Russia. France was eighth in the list of countries originating patents—two-thirds of the patents registered in France being foreign—and it was symptomatic of the situation that America sold five times more patents to France than France did to America. Military research was one of the few areas in which expenditure was commensurate with the reputation for excellence and innovation that the government sought to establish. The problem then, as always, was that industrial research, especially when dispensed in private laboratories, involved a lot of duplication, and it was tempting to leave scientific enquiry to those countries with a head start in finance and expertise. It was more often in matters of design and appearance that French products struck the foreign imagination—with the paradoxical result that attractive or unusual models often retained the affec-tion of the buying public and acted as a disincentive to further development. Well-known examples in the car industry were the Citroën DS and 2CV. The parked DS—like a huge, breathing oyster, about to snap off the leg of any pass-ing pedestrian—was esteemed an asset to any house-frontage or prestigious office-block for three decades, while the 2CV enjoyed two separate careers that spanned an even longer period. The 2CV started life in 1948 as the epitome of cost-cutting utility, with no concessions to appearance. Flouting contemporary aesthetics, it combined a curved profile on a rectangular base, like a road-mender's hut—lacking only the glowing brazier. When the occupant got out, it adopted a nosedive position, as though the front tyres had been let down. Yet in the 1980s, it was seen as an expression of period elegance, its original grey, corrugated-iron finish replaced by a maroon livery, emphasizing the shape of the doors, like a vintage Hispana-Suiza or a horse-drawn carriage.

At the end of the day, French industry probably owed most to the new men and attitudes that took control in the post-war decades. This is not an area of economic history that is easily amenable to quantitative research; but it remains a matter of great regret that the personnel of industry have not received anything approaching the degree of scholarly attention that has been lavished on the backgrounds and attitudes of politicians and civil servants. Apart from the changes in personnel brought about by the upheaval and demands of the Occupation and Liberation, France's relatively modest loss of life in the Second World War—compared with the First—saved her from the unhealthy reliance on the older generation of management that had occurred in the 1920s. Indeed, it was a measure of the premium now put on ideas and energy that the effective maximum age for appointing middle-rank executives was brought forward from about fifty in 1945 to forty in 1951. All this had its counterpart at a governmental level—but much more markedly, in that ministers and the senior civil service were much more directly affected by changes of regime and official outlook. The new École Nationale d'Administra-tion—the dream-child of Michel Debré, founded in 1945—became increasingly imbued with what commentators have called a 'Keynesian consensus', which grew stronger in the 1950s and 1960s. Its products, *énarques*, glowed with a belief in the necessity of state intervention, with themselves as its prime instrument.

Planning

But in the meantime, a strong lead had been given by the Commissariat Général du Plan—as earlier pages have outlined (pp. 117–19). The key figures in French planning liked to claim that in many respects it was a misnomer. They insisted that their planning was 'souple'. In Charles Kindleberger's words:

Knowledge of income and industry projections and faith in the inevitability of expansion are communicated to firms at intra- and inter-industry meetings. This ... has a faint resemblance to a revivalist prayer meeting ... In addition ... the Planning Commission uses a series of controls—powers to fix prices, adjust taxes, control credit, lend govern-mental capital, and authorise construction—to encourage firms to expand ... The total polity is bent on expansion ... expressed, so far as possible, in numbers. Given the under-lying faith in expansion, the numbers tend to confirm themselves, within limits.[1]

It is often claimed that the West German 'economic miracle' was carried out without *planification*, and that the contribution of planning to French growth has therefore been greatly exaggerated. This is to ignore Germany's traditional advantages of rich mineral and demographic resources, and above all the well-established strength of German entrepreneurial attitudes. French *planification* was intended precisely to supply French industry with the cast of mind that already existed in Germany and which did not require the leadership of Bonn to bring it into being.

The First ('Monnet') Plan of 1947–50/52 was designed to lay the foundations of post-war growth, and it put its emphasis on coal, electricity, steel, cement, tractors, and transport. Originally conceived as a three-year plan, it was extended to 1952 so that its span would conform with that of Marshall Aid. Like most success stories under the Fourth Republic, it was not submitted to parliament; and like its successor, the Second ('Hirsch') Plan of 1954–7, part of its objectiveness lay in its insulation from the sniping and interparty rivalries of the Palais Bourbon and Luxembourg. As Table 10.4 demonstrates, production came near to most of the plan's targets—which helped to create widespread

Table 10.4.
French Industrial Output, 1929–1952

	Actual Production					Planned Output Target for 1952
	1929	1938	1946	1950	1952	
Coal (million tons)	55.0	47.6	49.3	52.5	57.4	60.0
Crude Steel (million tons)	9.7	6.2	4.4	8.7	10.9	12.5
Cement (million tons)	6.2	3.6	3.4	7.2	8.6	8.5
Petroleum (million tons)	0.0	7.0	2.8	14.5	21.5	18.7
Electricity (billion kWh)	15.6	20.8	23.0	33.1	40.8	43.0
Tractors (1,000 units)	1.0	1.7	1.9	14.2	25.3	40.0
Nitrogenous Fertilizer (1,000 tons)	73.0	177.0	127.0	236.0	285.0	300.0

Source: Sima Lieberman, *The Growth of European Mixed Economies, 1945–1970* (New York, 1977), 11.

confidence and an increased willingness on the part of producers to take the Commissariat Général's forecasts and recommendations to heart. It also helped gently to introduce traditionally protectionist minds to the principle of foreign competition—a prospect that was to become a reality much sooner than most of them anticipated.

Informed criticism of the plan mainly centred on its relative neglect of other sectors of the economy which found self-investment difficult in the aftermath of devastation and upheaval. It was true that overall industrial growth between 1948 and 1954 compared badly with that of the former Axis powers—France's average of 2.8 per cent a year looking very sorry beside West Germany's 14.1 per cent, Austria's 11 per cent, and Italy's 8.2 per cent. But these countries had suffered greater devastation than France, and their initial recovery rates were bound to seem impressive in percentage terms. More reassuringly, the French rate was rather better than Britain's (2.5 per cent), Belgium's (2.5 per cent), and unscarred Sweden's (0.4 per cent). French steel production had markedly improved on pre-war levels—averaging 10 million metric tons in the early 1950s, as against 7 million in the late 1930s—but this still compared badly with West Germany's 17.4 million and Britain's 17.2 million, and Germany's lead was to lengthen in the years to come.

The Second Plan (1954–7) made some attempt to address itself to living standards by including specific targets for food production and housing—which for the most part were comfortably exceeded (Table 10.5). If its targets for the traditional bases of industry were less evenly achieved, this partly reflected the growth in consumer industry in response to the rising standard of living. Less predictably, the Suez crisis of 1956 was to thwart the realization of the petroleum programme—even if the government itself was largely to blame through its participation in the ill-judged venture (pp. 256–7).

While the nationalized industries undoubtedly profited from government favour in the period of post-war investment, it would be hard to assess whether the change of ownership made them more profitable. Obviously, the proximity of nationalized industry to government at a time when it had large sums of American money to disburse was an important factor in its modernization, just as the prominent place given to the nationalized industries in the First Plan likewise guaranteed them close government support. But this stemmed primarily from the fact that fuel and transport were the *sine qua non* of economic recovery, rather than that they were nationalized. Certainly, the nationalized industries between them absorbed half the total investment made during the Fourth Republic. But their levels of productivity were not always commensurate with what they received; and how they would have performed under private ownership can only be a matter for conjecture. A look at the car industry alone reveals an impressive range of conflicting evidence, providing ammunition for both sides of the argument. Peugeot and Citroën (owned by Michelin) were family firms that found distinction in radically different directions, the first in

Table 10.5.
The Second Plan: Production Targets

	Actual Production		Second Plan Target	Fulfilment as % of Target
	1952	1957		
Coal (million tons)	57.4	59.1	61	97
Electricity (million kWh)	40.8	57.5	55	105
Petroleum (million tons)	21.5	25	30	83
Steel (million tons)	10.9	14.1	14	100
Cement (million tons)	8.6	12.5	10.8	115
Meat (1,000 tons)	2,065	2,500	2,500	100
Wheat (million *quintals*)	84.2	110	95	116
Housing (units completed)	74,920	270,000	240,000	111

Source: Sima Lieberman, *The Growth of European Mixed Economies, 1945–1970* (New York, 1977), 18.

efficient production and the latter in innovation, both outpacing by far the limited liability company, Simca. As against this, the nationalized Renault company excelled in production and foreign sales.

What mattered much more than the fact of state or private ownership were the particular circumstances of the firm. The rapid increase in productivity of French nationalized industry, compared with its British counterpart, stemmed not from state ownership *per se* but from the speed with which it replaced war-damaged or ageing equipment with modern, efficient machinery and methods. The calibre of management within each enterprise was likewise of far-reaching importance. Outstanding public-sector managers included Pierre Lefaucheux of Renault, Pierre Massé of Électricité, and Louis Armand of French Railways. The importance of these considerations was compounded by the fact that the nationalized industries were run very much like private enterprises. Thus, Électricité de France offered its obligations on the stock market in the same way as a private company—but with the important difference that it could boast a state guarantee.

It was hoped by many planners that the nationalized industries would act as pace-makers for the private sector, and that their size and resources would encourage rationalization and concentration among their free-enterprise rivals. While the period undoubtedly saw a reduction in the proportion of people who worked in establishments employing ten men or less, it was essentially the middle-range enterprises, employing between a hundred and five hundred men, that grew more numerous (Appendix II, Table 7). The position of the large units, with over five hundred employees each, remained relatively stable—until the late 1960s, when the stimulus of the EEC led to a rapid growth in the size of enterprise, with explosive social results in 1968.

Although the destruction and pressures of the war period and Liberation era had brought about a number of mergers and take-overs, the subsequent rate disappointed expectation. Both management and workers tended to be hostile to take-overs. Apart from the loss of family prestige which it was thought to entail, there lurked the suspicion that an amalgamated growing enterprise was more vulnerable to the possibility of nationalization. And the workers, for their part, feared that streamlined concentration might involve the shedding of labour. Once again, it was the EEC and attendant government policies that brought about the main increase, with mergers rising from an annual increase of sixty-one in the 1950s to two hundred in the late 1960s.

However, the post-war decade saw considerable concentration of a less absolute kind, taking the shape of group-forming and links between firms that otherwise retained their original identity. Thus the late 1940s saw much of the iron and steel industry gravitate towards three major groups—Sidelor, Usinor, and Lorraine-Escaut—while chemicals showed a parallel tendency to concentration. In the case of the motor industry, it was more a question of the less powerful firms going out of business—with the big four controlling nearly 90 per cent of production by the early 1950s.

The European Challenge

The main catalyst of concentration, the EEC, had its roots in the Coal and Steel Plan of 1950. Proposed by Robert Schuman on 9 May 1950, it was largely the brain-child of Jean Monnet and his team. It was loudly welcomed by the Germans, who saw it as a symptom of their rehabilitation, while the Americans hoped that it would ease the way to closer military co-operation against Russia. Britain, on the other hand, regarded the Franco-German combination with deep suspicion, foreseeing stiff competition to her own newly nationalized mining and steel industries. Opinion within France was likewise divided. Communists predictably distrusted it as an instrument of Western solidarity against Russia, while the Gaullists denounced it as a derogation from national sovereignty. The Socialists, on the other hand, viewed it with cautious interest, their uneasiness at the clerical clique of Schuman, Adenauer, and ultimately other Christian democrats such as Gasperi, outweighed by the opportunity it seemed to offer of Western co-operation and French industrial propulsion. The most embarrassing objections, however, came from the French steel-masters themselves and from their allies in the Conseil National du Patronat Français, who feared German competition and saw the proposal as effectively torpedoing their own schemes for cartelizing the market. Even so, the government could count on the support of those nationalized industries which wanted a plentiful supply of inexpensive steel—notably French Railways, Renault, and the mines.

Although the plan initially pivoted on Franco-German co-operation, it specifically offered participation to other interested countries; and by June 1950 the Benelux countries and Italy had signified their desire to join. The result was the Coal and Steel Community which formally came into existence on 18 April of the following year. Despite the opposition of the French steel bosses, conservative politicians such as Paul Reynaud and a large section of their following eventually decided to back the plan; and their support was suffi-cient to get it through parliament (p. 172). Few of those voting, however, could have foreseen the speed of the community's evolution into the EEC, taking in the whole sphere of economic production with the Treaties of Rome (25 March 1957). And the fact that the apprehensive steel magnates of 1951 came to support this evolution was not a sour piece of vengeance on perfidious colleagues in other industries; it reflected their genuine satisfaction with the way in which the community had developed.

The most striking effect of the EEC on French industry was its encourage-ment to larger firms to break out of their traditional alliances with their smaller and less efficient brethren, and start cutting prices competitively. Indeed, it was a condition of survival in the new market. But it also led to the spate of take-overs and mergers that were such a feature of the 1960s. The changing nature of the game, however, was only slowly appreciated in France—even in the Commissariat Général du Plan, where the game had originated. This was evident in the quasi-protectionist preconceptions of the Third Plan (1958–61), in which imports were still regarded as a constraint rather than a stimulus. There was still little hint of the EEC vision of a thriving market, where competi-tion would improve standards and lower prices, and where each member would discover his particular forte and develop it to the advantage of all. Indeed, the Third Plan's intentions were to try to hold imports to the 1956 level, while raising exports by a third. As it happened, this sector of the plan's intentions were exceeded. With the help of two devaluations, exports increased by 71 per cent between 1956 and 1961, even if imports from outside the franc zone rose by 25 per cent. In other respects, however, the plan was only partially success-ful. The financial crisis of 1958 necessitated a downward revision of its GDP growth target to 23 per cent, which was unevenly achieved—agriculture falling substantially short of expectation.

Given the growing experience and increasingly sophisticated information acquired by the Commissariat Général du Plan, the third and subsequent plans were more broadly conceived than their predecessors, in that they envisaged social as well as economic targets. Thus the Third Plan envisaged the building of 300,000 homes, which in fact became 316,000. Even so, the new maestro, Jacques Rueff, believed that austerity in consumption and welfare spending was essential if France was to transcend her current financial difficulties. And the result was a reduction in government subsidies, with consequent price-rises in food and in the products of the nationalized industries. At the same time, family allowances, war veterans' pensions, and other benefits were allowed to lag behind the cost of living. Rueff was at heart a liberal economist, who disliked the complicated mechanism of cogs and ratchets that linked prices and wages across the whole spectrum of the French economy. He consequently abolished the indexing of agricultural prices to those of industry; and his encouragement to industry largely took the form of generous depreciation allowances in the 1959 budget, enabling factories to modernize their equipment without burdening the state and the consumer with unconditional industrial subsidies or artificial price supports.

But membership of the EEC, with its implicit acceptance of the evolving pattern of the community's trade, invalidated the Planning Commissariat's customary policy of setting specific production targets, and accordingly none were included in the Fourth and Fifth Plans. But if they eschewed targets *per se*, they permitted themselves to engage in 'projections', which were widely publicized in much the same way as targets—and, in the case of the Fourth Plan (1962–5), were handsomely realized (Table 10.6). This was as much a tribute to the skill and realism of the plan's forecasters, as to the achievements of producers.

Table 10.6.
The Fourth Plan: Projections for 1965 and Achievement Levels

	1965 Index (1961 = 100)		Annual Growth Rate (%)	
	Projected	Achieved	Projected	Achieved
Gross domestic output	124.0	124.1	5.5	5.5
Consumption	122.5	124.6	5.2	5.7
Gross capital formation	130.0	136.6	6.8	8.1
Productive investment	128.0	125.1	6.4	5.8
Housing	125.4	159.2	5.8	12.4
Government	150.0	151.5	10.7	11.0
Imports	123.0	151.1	5.3	10.9
Exports	120.0	130.3	4.7	6.8

Source: Sima Lieberman, *The Growth of European Mixed Economies, 1945–1970* (New York, 1977), 171.

The Fifth Plan (1966–70), while retaining the broad socio-economic concerns of the Fourth, put particular emphasis on the future growth of industry, and broke new ground in paying much greater attention to regional development. Indeed, it was the Fifth Plan which saw closer co-operation with the regional economic development committees (CODERs) created in 1964(–8). By now, the breadth of the plan's remit involved a network of committees and subordinate working parties that totalled well over 5,000 members; and unkind critics likened it to the proliferation of offices under the *ancien régime*. This, however, was to pay insufficient tribute to the effectiveness of their achievements and to ignore the fact that many of them combined these functions with other duties, or were on temporary secondment.

Membership of the EEC and the rapid decolonization of the early 1960s saw major changes in the pattern of French foreign trade. It had been a long-established feature of the French economy that it was less geared to exports than her more urbanized neighbours, such as Germany and Britain. Since France was less dependent than they were on imports of food, she did not need to export as much to keep body and soul together. Even so, her growing urban-ization and improved economic performance saw her exports rise from 8.2 per cent of GDP in 1951 to 12.8 per cent in 1970. And if this still remained well below the ratio for Britain (15.9 per cent) and West Germany (18.2 per cent)— Table 10.7—it was, in straightforward quantitative terms, a close challenge to Britain which was to be successfully driven home in 1971.

Even more striking, however, was the reorientation of trade. As Table 10.8 shows, the shift of French international trade from the franc area towards the EEC countries was already apparent in the 1950s, well in advance of the main impact of decolonization and the Treaties of Rome. And from that point of view, France's transfer of institutional commitments from the Third World to Europe reflected the logic of her commercial evolution. Even so, the institu-tional transfer vastly accelerated the rate of commercial reorientation, as Table 10.8 demonstrates. The significance of the franc area as a market for French exports fell from just over a quarter in the late 1950s to a mere 5 per cent in the late 1970s, while as a source of French imports it dropped from just under a fifth to less than 3 per cent. In contrast, French exports to the EEC countries were to swell from under a third of total exports in the late 1950s to well over half in the late 1970s, while as a supplier of France the EEC's share grew in similar proportions.

At the same time, the strongly industrialized character of most of her EEC partners gave a welcome fillip to French agricultural exports. Despite the expansion of French industrial sales abroad, agriculture's share of national exports actually rose from just under 13 per cent to over 16 per cent in the course of the 1960s. Indeed, by the end of the decade, French farmers were providing 42 per cent of the EEC's total exports of food and raw materials —compared with French industry's 17 per cent share of EEC exports of

Table 10.7.
Volume of Exports as Percentage of Gross Domestic Product, 1951–1976:
France and Selected Western European Countries

	1951	1960	1970	1976
France	8.2	9.8	12.8	15.9
Austria	7.4	12.4	20.0	23.6
Belgium	24.1	29.5	48.4	59.8
UK	14.1	13.2	15.9	19.6
W. Germany	8.3	12.6	18.2	24.1
Italy	4.4	7.3	14.2	18.2
Holland	16.2	23.6	37.2	47.8
Spain	3.2	4.8	6.5	8.7
Sweden	12.7	15.2	20.6	23.1
Switzerland	13.3	16.9	24.3	30.4

Source: The Economist, *Europe's Economies: The Structure and Management of Europe's Ten Largest Economies* (London, 1978), 3.

Table 10.8.
French Foreign Trade Structure, 1952–1979 (% Shares in Total)

	1952–54	1958–60	1968–70	1977–79
Imports				
From the Franc Area	24.7	18.2	9.8	2.8
From the EEC	21.2	30.4	54.6	50.4
From other countries	54.1	51.4	35.6	46.8
Exports				
To the Franc Area	38.4	26.2	12.0	5.1
To the EEC	25.3	32.6	52.5	52.1
To other countries	36.3	41.2	35.5	42.8
Primary products	31.4	26.9	26.9	23.8
of which: Agricultural products	12.9	12.7	16.3	14.5
Manufactures	68.6	73.1	73.0	76.2
of which: Chemicals	7.7	8.5	10.4	11.6
Semi-manufactures	33.6	31.1	22.1	19.1
Machinery and transport equipment	17.5	24.0	31.3	36.3
Export performance[a]	8.9	8.9	8.4	10.0

[a] Share in the exports of manufactures of the 12 major exporting countries.

Source: Andrea Boltho (ed.), *The European Economy: Growth and Crisis* (Oxford, 1982), 454.

manufactures. Given the strength of West Germany and the rise of cheap Italian manufactures, the French industrial share was unlikely to improve dramatically; but in quantitative terms its record was good, especially in the field of machinery and transport equipment—these products rising from a quarter to a third of the nation's total exports in the course of the 1960s.

Growth rates, like shares of production, are admittedly purely relative terms, their significance and impressiveness depending as much on the performance of one's predecessors as on one's own current efforts. Growth rates transfer into vertical, chronological terms the ambiguities that shares of the production nurture in the horizontal, spatial dimension. It is true that France's weak economic performance in the past would always show her later periods of growth to statistical advantage. Yet it remained an impressive fact that her first fifteen years of EEC membership saw her with the highest GNP growth rate in the community—5.5 per cent per annum, as against Italy's 5.3 per cent, West Germany's 5 per cent, Belgium's 4.9 per cent, and Holland's 4.2 per cent.

Like other advanced industrial countries, her growth rate fell in the recession of the 1970s—for the reasons to be outlined in Chapter 18 below—but at 2.8 per cent it remained above the Western European average of 2.3 per cent. Admittedly, those of some other countries were better; but this partly reflected the fact that they were still enjoying the statistical advantage of a sizeable switch of population from agriculture to industry, with its higher productivity rates. This stage in economic development had largely been accomplished in France during the previous years, when she too had enjoyed the statistical face-lift that it gave.

It was France's good fortune—and a tribute to the courage and acumen of her economic advisers, and, indeed, to the ministers who headed them—that she launched the European idea at a time when West Germany was still not fully recovered from the devastation of war. The competition and opportunities within the Coal and Steel Community therefore acted as a beneficial stimulant to her industry, and enabled it to sustain and respond to the greater challenge and opportunities of the EEC. Britain's major misfortune—and unquestionably her greatest mistake in the post-war era—was to postpone interest in joining until it was too late; de Gaulle's excluding tactics in the 1960s (pp. 312–13) were merely an episode in the much longer history of Britain's suspicion and vacillation towards the European idea. The outcome was that when Britain eventually joined the EEC in 1973, its other industrial members had already developed such a head-start that its competition was a traumatic rather than a stimulating experience, with imports from the Continent greatly outweighing British reciprocal exports. The respective fortunes of Britain and France were often likened to two men running to catch a moving bus. The one who attempts to mount during the early stages of move-off may well be stimulated into running speeds he would not normally attempt—but his efforts get him safely aboard. The one who leaves it until the bus has greatly accelerated, will probably miss his footing and be dragged along the ground, his clothes and knees in tatters.

Admittedly, France had specific national advantages in the heavy EEC demand for her agricultural produce—but this was only one factor in her success. And indeed, had Britain accepted founder-membership in the 1950s, she and West Germany might well have brought about a Common Agricultural Policy that would have been less damaging to them, and much less wasteful in general. Apologists for Britain's dilatoriness also liked to point to the advantages that France had derived from the need to replace damaged plant and communications after the war: 'If only more of our railways and factories had been destroyed in the blitz, they would now be as good as the French—and we would have competed much more effectively in the Common Market.' All of which underlines the psychological basis of so much of the pattern of success and failure in the post-war European economy. It should not have required the destruction of war to bring about the modernization of plant—indeed, modernization is physically much easier if supply and back-up systems remain undamaged. But psychologically some such imperative may be the decisive factor.

The imperatives of EEC competition likewise gave a sharp spur to investment and rationalization. Productive investment as a proportion of GDP grew from 12 per cent in the mid-1950s to nearly 15 per cent in the late 1960s, but this ratio was only three-quarters of West Germany's and was not enough to provide German industry with a serious challenge. Self-finance had been slow in developing, representing just over 60 per cent of overall investment in the early 1960s—compared with just under 80 per cent in West Germany and the Netherlands. And although the second half of the decade saw a substantial increase in government loans to industry, over half of them went to the nationalized sector. The result was that private firms remained heavily reliant on bank credit, which at interest rates of 10 per cent or more substantially increased industry's costs.

While EEC competition was the principal goad to concentration, these salutary developments were also given close encouragement by the government's Fourth and Fifth Plans (1962–70). Medium- and long-term credits were offered to facilitate mergers, and tax exemption was granted to the capital gains of those firms that were in the process of merging. As already indicated, mergers in the 1960s affected some of the largest enterprises in the country. The Sidelor and de Wendel steel companies gave birth to a subsidiary, Sacilor, in 1964, and then merged themselves four years later. Citroën's earlier acquisition of a quarter of Panhard's shares in 1955 turned out to be merely the prelude to a complete take-over in 1965, while Sud-Aviation represented a logical coming-together of the aircraft manufacturers, Sud-Est and Sud-Ouest. At the same time, the average factory size, in terms of work-force, grew by nearly 15 per cent in the brief span of 1962–8—discounting those with less than fifty employees.

AGRICULTURE

Industry's gain was superficially agriculture's loss. Yet the EEC provided some-thing of a golden age—or at least an Indian summer—for France's dwindling farmers. These were arguably their happiest years since transatlantic food imports first shattered their world in the late nineteenth century. The post-war decade was a depressing period for French agriculture. Its total output in 1952 was only 3 per cent higher than in 1938, and although it still employed just under a third of the work-force, it provided less than a fifth of the nation's wealth. Table 10.9 indicates how poor its achievements were compared with

Table 10.9.
Agriculture in 1954: France and Selected Western European Countries

	Production per active male	Production per hectare of agricultural area	Yields per milking cow per year	Use of Fertilizer per hectare of agricultural area
France	100	100	100	100
Denmark	203	200	169	292
W. Germany	120	170	135	341
UK	207	117	136	177
Italy	47	110	85	76

Source: Sima Lieberman, *The Growth of European Mixed Economies, 1945–1970* (New York, 1977), 28.

some of its northern neighbours; and it was no consolation to see that Italy's were worse—Italy having few of France's intrinsic advantages. Despite the growing move to the towns, the average French farm in the early 1950s was still under fifteen hectares of land (37 acres), scattered in over forty *parcelles*. Not only was it impossible to use modern, labour-saving techniques on these *parcelles*, but the owner lost up to two hours a day moving from one to the other. Those on sloping land took little account of the contours, and gave rise to continual soil erosion. The Vichy consolidation law of March 1941 (pp. 94–5) was adopted by the de Gaulle government in 1945, but it was only implemented very slowly. By October 1950, twenty-five *départements* were still untouched by the law, while the process had only just begun in thirty-nine others.

There were, however, wide disparities of scale and efficiency. The large, relatively modernized farms of northern France were often better than those in West Germany. And although agricultural methods in Alsace, Lorraine, and the Mediterranean coast were less advanced than in the north, hard work and attention to detail often gave yields that were comparable to those in Germany.

Elsewhere, however, levels were markedly lower. Paradoxically, the stimulus of the EEC in the 1960s tended to broaden these disparities; and already by 1968 the top 10 per cent of farmers were producing 60 per cent of the total output.

It had been the complaint of many farmers in the 1950s that when the more venturesome of them tried to adopt progressive methods, their only reward was a poorer price for their product: modernization, like a good harvest, was a mixed blessing. And the only answer to this type of problem was an unpopular one: a reduction in acreage of those crops that were periodically liable to glut, and, more especially, the elimination of large areas of vineyard in the south. No one then imagined that salvation was shortly at hand in the shape of the monstrous wastage of the Common Agricultural Policy. It did not occur to even the most nostalgia-ridden farmer that Europe would take on the mantle of the Third Republic and pay him to grow crops no one wanted—no one, that is, except the starving and undernourished of the Third World, who unhappily did not fit easily into Brussels's way of thinking.

The EEC, however, was also to offer opportunities of a more positive kind in the shape of lucrative urban markets across the Rhine and elsewhere. The fact that French farmers rose to the challenge so effectively reflected changes that were already beginning to affect them well before these opportunities occurred. The post-war expansion of services and industry and the attraction of urban wages had reduced the farming population from about a third of the country in 1946 to about a quarter when the EEC was created. Indeed, it was the exodus of rural labour to the towns that obliged the remaining farmers to give serious thought to rationalization and modern methods. Thus, the number of tractors increased fivefold in the 1950s—substantially in advance of the real impact of EEC opportunities.

In quantitative terms, this move from the land was at its most marked between 1954 and 1962; and by 1968 the total post-war migration had reached twelve million, including the small-town retailers and artisans whose livelihood melted with their peasant clientele. Rural France and its dependants had shrunk from 46 per cent of the population to 34 per cent in a matter of two decades. Nor did the bulk of the migrants seem to regret the move. A sample enquiry among farmers in the 1960s found just over half declaring that they would leave the land if they could; and of those who had left, only 2 per cent had serious thoughts of returning.

With the number of farms dropping from well over two million in the 1950s to one and a half million by 1970, there was a corresponding increase in their size, as Appendix II, Table 2 demonstrates. This also encouraged the renting-out of land, with the result that in the mid-1950s rented farms had already risen to nearly 40 per cent of the total, while owner-occupancy was down to 55 per cent and still sinking. As for share-cropping, barely one in twenty farms were now run on this basis. The government itself had made renting a more attractive proposition in its farming statute of 17 October 1945—which took several leaves from Vichy's law of 15 July 1942. Contracts, which were to hold good for nine years, entitled the outgoing tenant to an indemnity for the improvements he had effected. Moreover, he had first option on the land should the owner decide to sell it—and tribunals were instituted specifically to guarantee a fair price. Nor could an owner evict a satisfactory tenant unless he or his son intended to live on the farm and cultivate it for himself.

The export opportunities of the EEC, together with the CAP incentives, encouraged an increasing switch to monoculture—and *remembrement* became much more feasible as farmers abandoned mixed farming and opted for consolidation of their acreage. The process was fortuitously aided by the increasing use of bottled gas in peasant households. While undoubtedly an advantage, in that it liberated peasants from the chore of chopping wood and tending fires, it also resulted in the grubbing-up of the hedges that had hitherto been preserved as a source of fuel. Marking the boundaries of traditional holdings, these hedges had often been a practical obstacle to the *remembrement* and rationalization of peasant farms. Yet their disappearance was a loss to the environment that the ecological movements of later decades were loudly to deplore.

The 1960s unquestionably found a more optimistic attitude among peasant families than had been the case a decade earlier. Foreign observers, such as Laurence Wylie, had been conscious of a debilitating nostalgia in the post-Liberation years for the mythical peasant life of pre-1914, with its faith in the family as a self-sufficient unit and its confidence in the power of the State to preserve it against the threat of internal and foreign pressures and military attack.[2] The wretched condition of the pre-1914 peasantry scarcely warranted such nostalgia; but two invasions in the last thirty years, inflation, the fall in the value of gold, and the growing threat to income represented by the competition of more efficient farmers elsewhere, all challenged the old certainties of traditional peasant thinking. The fact that inflation had solved the deep indebtedness of many of them was scarcely acknowledged by the older generation.

With the growing prosperity and opportunities of the late 1950s and 1960s, however, there was a greater readiness to resort to credit to buy more efficient equipment—especially among the younger generation of farmers—as long as incomes kept pace with inflation. Even so, many peasants who had bought tractors on credit in the 1950s subsequently claimed that their profits were too low to justify the outlay. Changing attitudes were mirrored in the village councils, where the dominant older generation in the early 1950s saw their role as the traditional one of minimizing taxes and resisting government interference. As long as the school was reasonably heated and the worst pot-holes in the roads filled in, that was enough. But the 1960s found younger councillors in charge, who saw government in positive terms as an instrument of local improvement, such as better hygiene and recreational facilities.

These attitudes had their counterpart at national level. Relics of pre-war agrarian agitation, and older figures such as Henri Dorgères and Paul Antier, were more interested in pressurizing government into guaranteeing attractive prices for farmers, rather than the modernization of French agriculture. When government support for farm prices was overtaken by inflation (which this support had itself helped to worsen), they were joined by younger militants in the physical demonstrations that soon became a familiar pattern of rural protest in the early 1960s—the dumping of produce in front of administrative offices and the blocking of roads with farm vehicles. These were a particular feature of the depressed small-scale agriculture of the west and south. Both the Confédération Générale d'Agriculture and its more successful rival, the Fédé-ration Nationale des Syndicats d'Exploitants Agricoles, were torn between conservative and progressive elements, with the conservatives normally win-ning the day on union policy. FNSEA continued to demand guaranteed prices and a minimum of government interference, except when in their interest; but it found itself progressively upstaged by the Assemblée Permanente des Présidents de Chambres d'Agriculture (APPCA), which by 1960 was claiming an exclusive right to negotiate with the government on behalf of farmers. More-over, 1957 saw FNSEA's more progressive Catholic wing break off to form the Centre National des Jeunes Agriculteurs—a union with firm roots in the Jeunesse Agricole Chrétienne which had been active in developing more radical thinking during the Vichy period.

Under the Fifth Republic, the CNJA'a attitudes were viewed with increasing favour by Premier Michel Debré, who brought its leading figure, Michel Debatisse, into the government's Conseil Économique et Social. Moreover, the *loi d'orientation agricole* of July 1960 and Edgard Pisani's *loi consultative* of July 1962 borrowed heavily from CNJA proposals, even if the government's modifications did not always please the CNJA. The two laws instituted a network of regional Sociétés d'Aménagement Foncier et d'Établissement Rural (SAFERs), with powers of intervention to buy land that came on the market. The SAFERs would then sell or rent it with a view to creating a better overall balance in the locality. In most cases, it was a question of adding land to existing farms to improve their viability—although a quarter of the land acquired by the SAFERs in their early years was used to create new farms. The twenty-nine SAFERs that came into existence were eventually responsible for transferring well over 7 per cent of the arable land in this shaping exercise; and, more importantly, their continual watching brief affected the market as a whole, since sellers often preferred to accept an offer that was in clear conformity with government policy rather than risk SAFER intervention. In some areas of the deep southern heartland of smaller owner-occupancy, the SAFERs negotiated up to a third of the land transactions.

An equally imaginative but less successful provision of the 1962 reforms, was the institution of Groupements Agricoles d'Exploitation en Commun (GAECs), corporate groups of neighbouring farmers who would manage their land jointly while retaining their existing ownership. By the early 1970s, only 0.3 per cent of the land was under such management, although the scheme helped to encourage similar initiatives on an informal basis among neighbours and related families, thus reinforcing a tradition that already existed on a number of farms. Perhaps the most appreciated feature of the Pisani law was the establishment of Indemnités Viagères de Départ, which enabled old farmers to retire and make way for younger. With 40 per cent of French farmers over the age of fifty-five—including 400,000 over sixty-five—this facility found ready takers, the annual figure rising from some 13,000 in 1965 to well over 100,000 four years later.

The outcome of all these changes was that by the mid-1970s, French productivity per person was higher than in most Western European countries (Table 10.10).

Table 10.10.
Agricultural Productivity, 1964–1975: France and Selected
Western Countries

	Output per person employed 1975 ($)[a]	Annual increase 1964–75 (%)
France	4,020	5.0[b]
Austria	2,900	7.3
Belgium	6,650	5.1
UK	4,870	5.8
W. Germany	3,590	5.8
Holland	7,720	7.9
Italy	2,870	6.4
Spain[c]	1,550	6.6
Sweden	5,140	5.5

[a] 1970 prices and exchange rates.   [b] 1970–5.   [c] 1964–74.

Source: The Economist, *Europe's Economies: The Structure and Management of Europe's Ten Largest Economies* (London, 1978), 28.

Parallel to these developments in farming itself, rural family life also underwent substantial changes.[3] The traditional peasant household of three generations was increasingly under pressure in the 1950s. Grandparents were moved into separate living-quarters—a partitioned section of the farmhouse if it was large enough to permit this—or otherwise a cottage or apartment in the village. At the other end of the age-scale, the children's place in the home was challenged by the counter-summons of educational and employment opportunities. The decline in the traditional assumption that children would follow in their parents' footsteps called into question the sufficiency of the village primary school as a preparation for employment. The post-war period saw increased rural interest in the *cours complémentaires* (pp. 16–17), which, although established in 1886, had been less frequented by children in the more remote villages, who had often tended instead to stay on in the limbo of the top reaches of the village *école primaire* until they were legally of an age to leave school. Between 1945 and 1960, the enrolment in *cours complémentaires* increased nearly two and a half times, whereas the normal pressure of population on primary-school growth was still only of the order of 50 per cent or so. Situated mainly in towns, the *cours complémentaires* were brought increasingly within the orbit of the villages by the chartering of school buses. Improvements in rural buses likewise enabled a larger proportion of young people to commute daily to office- or shop-work in the larger towns, until such time as their wages or marriage permitted them to rent a room or an apartment there. This, too, was a factor in eroding the hermetic attitudes of village life.

For the older residents, however, the main emissary of the outside world was television. The 1950s and early 1960s saw the gradual spread of television transmitters into the more remote corners of upland France, their gigantic masts rivalling the traditional hilltop landmarks of cairn or cross. Generally, it was one of the more prosperous village cafés that took the initiative in buying the village's first set; but sometimes it was the parish council that took the plunge, installing a set in the parish meeting-room and limiting the viewing to what the *curé* or parish council considered suitable. As more villagers acquired sets, however, what had been a custom-drawer to the cafés became a rival. Families saw more of their fathers in the evening, even if communication with them was scarcely greater than before. A particular casualty was weekend village sport, where *boules* and other afternoon activities found it hard to compete with top-rank sport on television.

The French Desert

Arguably the basic biblical text of the economic historian should be: 'To him who hath, it shall be given.' It was a commonplace of economic history that progressive farming and progressive industry were often found in close proximity—both depending on a prosperous market, as well as on their own particular forms of input, be it rich soil or a ready source of motive power. On the other side of the medal, rural depopulation was inevitably most marked in those areas in which the farmers were least favoured by climate and soil; and the urban employment to which they gravitated was often situated in regions of established agricultural prosperity. The 1960s consequently saw a growing tension between the have and have-not regions—a confrontation that roughly corresponded to a diagonal line running from Le Havre to Grenoble. Two-thirds of the population now lived to the north of this line, and the imbalance would have been worse but for the influx of immigrant labour in the south. Whereas at the beginning of the century, only one in eleven of the nation lived in the Paris region, it was now more like one in six. The stagnant regions felt themselves to be deprived of aid and resources, which were increasingly being channelled towards the favoured areas of economic growth on which the nation's international trade depended. A particular bone of contention was that the north was given prior claim on the natural gas resources of the Pyrenees, despite their geographical situation. By the early 1960s, a third of the houses in many southern villages were unoccupied—and to many of the surviving southerners, it seemed that insult was being added to injury when they were put on the market as cheap holiday cottages for northerners or foreigners, with some still unsold twenty years later.

The government admittedly encouraged the foundation and development of some remarkable industrial centres—notably Fos, west of Marseille—but these were islands of achievement that served only to emphasize the depressed nature of other parts. It was mainly in the 1970s that governments seriously addressed themselves to this problem (p. 343); but by then the prosperity of the 1960s was on the wane, and the resources and incentives for substantial development were much more limited.

SERVICES

A dominant feature of post-war industrial societies was the growing proportion of the population engaged in services (Appendix I, Fig. 2, and Appendix II, Table 8). The development of international trade, improved communications, and the complexity of modern technology all demanded armies of people to run and repair them, as well as to instruct others in their use and potential. At the same time, the increasing sophistication of factory machinery was steadily reducing the ratio of labour to output in industry itself. The mounting diversity of services defies even the most facile of generalizations. Apart from the new world of high-tech operators, there existed the traditional spectrum of activities. These ranged from the circumscribed world of shop assistants and café waiters, a few of whom still slept over the premises in attic rooms, to the itinerant bargemen, whose floating homes were redolent of distant places, as they passed beneath city bridges, their cabin windows bright with potted geraniums— happy dogs and waving children on the decks, and billowing washing slung between the masts. And then there were the legions of public servants. Ranging from the vertiginous realm of the *sapeurs-pompiers* to the subterranean world of the *égouttiers*, they embraced a multitude of conditions, each of which was periodically commended to the attention of the public, in its place and season. The *sapeurs-pompiers* enjoyed a favoured place in Bastille Day processions and other celebrations of national valour and achievement, while the hazardous life of the *égouttier* was graven on children's memories by dramatic municipal warnings: 'Vous pourriez tuer un égouttier' if you flushed noxious chemicals down the lavatory. Thus, people who had never knowingly seen an *égouttier* in their lives, bore his fragile existence in mind as they went about their daily business—thereby exemplifying that Fraternité meant as much to a good citizen as Liberté and Égalité.

UNIONS

The only comprehensive economic link between these multifarious people was that they were consumers—victims and beneficiaries of inflation and government social spending. A small percentage of them sought to escape their diversity and find a common feeling and a greater force in one or other of the three main unions. Given the limitations of parliament as an instrument of working-class pressure in a multi-party State, the unions were superficially an attractive alternative weapon. But the unions, too, were divided, and even in 1970 covered little more than 20 per cent of the work-force (as against well over 40 per cent in Britain and 35 per cent in West Germany). The specific year-to-year activities of the unions are discussed in other chapters, as are the sparring and jockeying for position of the three main rivals for working-class allegiance. The CGT, with its five million members, offered amalgamation to the 700,000 members of the CFTC in September 1945. Not only was this turned down, but the Socialist wing of the CGT broke off in December 1947, to form CGT-Force Ouvrière, taking with them three-quarters of a million members (pp. 160–1). Although collective contracts were re-established by the law of 23 December 1946, negotiations remained under the control of the Ministry of Labour; and it was not until February 1950 that free collective bargaining was restored, after a sharp tussle with Bidault. In the meantime, however, union membership was falling rapidly, until by 1955 it was only a third of its high tide in 1945–8.

CIVIL SERVANTS AND MANAGEMENT

The rights of civil servants to enjoy full union protection and resort to collective action had been a long-standing matter of dispute in France. But a series of strikes resulted in their receiving a Statut Général de la Fonction Publique in October 1946, largely based on a draft by Maurice Thorez and broadly corresponding to the demands of the civil servants' section of the CGT. A grey area that the statute did not attempt to resolve was the traditional change in administrative appointments following a change in the government's political complexion. Affecting mainly advisory and responsible posts, it did not amount to an American-style spoils system, in that the great majority of displaced personnel were given appointments elsewhere, sometimes returning to their previous posts or to some assignment that was politically less sensitive. Eventually, a decree of 21 March 1959 specifically listed the six hundred senior civil service posts that an incoming government was at liberty to dispose of as it saw fit. It was in any case usual, but not invariable, for the *directeurs* of ministries to lose their posts with the advent of a government of different persuasion. The British concept of the same man dutifully covering up for a succession of ministers was alien to a political culture in which party differences entailed a continuing guerrilla war, where the generals preferred to rely on their own followers. At the same time, this essentially political shuffling of personnel needs to be distinguished from the general, continuing movement of state functionaries that characterized French bureaucratic life. In a highly centralized country, where the government was traditionally at pains to prevent its provincial representatives from being drawn into the comfortable complicities of local life, three years was as long as men in responsible positions of authority expected to stay in one locality; and since promotion and general advancement were contingent on there being a fairly high level of mobility, senior civil servants accepted this as part of their lot.

On the other side of the fence from the unions, stood the employers' organizations. The Liberation had found the *patronat* anxious to keep a low profile, especially the *gros patrons* with wartime profits to explain. But in a period when many important economic and social decisions were being taken at a national level, they swiftly felt the need for some defensive organization to protect their interests. 1945 consequently saw the emergence of three important groups. The more militant small-scale employers gravitated towards Léon Gingembre's Confédération Générale des Petites et Moyennes Entreprises (January 1945), while the Jeunes Patrons came into existence shortly afterwards. But with France entering into a series of important elections, the *gros patrons* recognized the dangers of being without proper representation, and duly formed the Conseil National du Patronat Français in December 1945–June 1946.

LIVING STANDARDS

Wages

In a period of inflation and post-war shortages, the unions had a hard fight to keep wages even within shouting distance of the continual rise in the cost of living—as demonstrated in Chapter 9. And in the immediate post-war period, the unions had respected the calls by all parties, including the Communists, to subordinate wage demands to economic recovery. Indeed, industrial real wages per hour did not regain their pre-war value until 1955; and even then it was only the longer working day and the welfare legislation of the Liberation era that put the worker in a position of clear financial advantage *vis à vis* his pre-war counterpart. Moreover, the early years of the Fifth Republic saw a distinct shift of income from labour to capital with the implementation of the Rueff plan (pp. 190–1). Real wages stagnated in 1958–9, as a result of the government allowing the cost of living to rise substantially without compensating wage increases; and the fraught nature of the political situation discouraged the unions from taking firm industrial action. Nor did the standard of living of the mass of the population keep pace with the rapid advances in economic growth in the 1960s, even if it grew steadily in an unspectacular fashion. One forgotten benefit of the period was that 1963 saw the widespread institution of a fourth week of paid holidays in many firms, without there being a statutory obligation to provide it. But French wages compared badly with those of all her EEC partners, except Italy, and also with Britain's (Table 10.11); and the disparities

Table 10.11.
Average Hourly Earnings ($) in Manufacture, 1965 and 1975:
France and Selected Western European Countries
(Men and Women)

	1965	1975
Sweden	1.6	5.2
Switzerland	1.1	4.4
W. Germany	1.0	3.9
UK	1.0	2.8
Holland	0.8	3.8
Belgium	0.8	3.8
France	0.7	2.7
Italy	0.7	2.4
Austria	0.6	2.5
Spain	0.3	1.5

had increased by 1975, despite the more generous wage settlements following the upheaval of 1968. On the other hand, the Giscard presidency (1974–81) saw significant improvements; and by 1978 real wages were already over two and a half times higher than they had been in 1938 (Appendix I, Fig. 3). Indeed, if social benefits are added, the improvement was almost threefold.

Yet these increases were unevenly spread across the spectrum of occupations. As Table 10.12 shows, the distribution of wealth between the classes in the mid-1950s was not significantly different from what it had been twenty years earlier. And subsequent pay-rises tended to perpetuate or even accentuate these inequalities. Thus, while management salaries went up by 40 per cent in real terms between 1956 and 1964, skilled workers' real wages increased by only 25 per cent, while the legal minimum wage (SMIG) rose by less than 4 per cent. The

Table 10.12.
Distribution of Income in France among Occupations in the mid-1950s

	% of active population (1954)	% of income (1956)
Farmers	20.7	10.3
Farm-workers	6.0	1.8
Industrial and commercial employers	12.0	15.7
Senior management and liberal professions	2.9	7.0
Middle management	5.8	5.9
White-collar workers	10.8	5.5
Blue-collar workers	33.8	20.8
Service workers	5.3	1.3
Other categories	2.7	2.4
Not part of active population	—	12.1

legal minimum wage, instituted in 1950, was geared to the current prices of some 213 staple articles, some of which were already outmoded and unlikely to figure in a working-class budget, while more recent everyday necessities were ignored by the official index. As for the other major underprivileged class, a decree of 30 July 1946 had theoretically established equality of wages for women; but, as always, a large range of the better-paid jobs were virtually closed to them. And even where there was no male monopoly, employers were often reluctant to offer an appointment involving extensive in-service training to a woman, for fear that marriage or child-bearing would deprive the management of the fruits of their investment.

Social Benefits and Housing

Wages, however, were only part of the family lifeline. By 1958, the social welfare reforms of the Liberation era provided perhaps 20 to 25 per cent of working-class family income, compared with less than 3 per cent in 1938. Even for management and the liberal professions, the various forms of welfare were worth perhaps 7 or 8 per cent of what came into the household.

As in most advanced industrial countries, old-age pensions represented the most expensive single item (Table 10.13)—and grew increasingly so with the improvement in life expectancy. For men, the prospect of death receded markedly from 62 years to 67 during the course of the Fourth Republic, and then more slowly to 69 by 1973. For women, with their greater domestic competence and resilience to retirement traumas, it retreated from 67 to 73, and then to 76 in the same periods. Women were also less addicted to alcohol and tobacco—a major factor in a country where the average alcohol consumption was 39 litres a year, compared with 14 in Italy. Other factors affecting the partiality of death were climate and occupation. Those who resisted the northward lure of urban wages,

Table 10.13.
Structure of Social Security, 1975: France and Selected Western European Countries

	% breakdown of social benefits by function 1975					
	France	W. Germany	UK	Italy	Holland	Belgium
Sickness benefit	26.5	29.6	25.8	27.1	30.4	23.1
Old age pensions	40.8	41.7	45.5	34.2	36.4	38.4
Invalidity (incl. disablement of all types)	6.2	10.8	11.4	20.9	15.8	11.7
Unemployment benefit	2.7	3.6	5.8	2.8	6.0	7.1
Family benefits	19.6	10.2	10.6	11.8	11.0	14.9
Total, including other benefits	100.0	100.0	100.0	100.0	100.0	100.0

Source: The Economist, *Europe's Economies: The Structure and Management of Europe's Ten Largest Economies* (London, 1978), 26.

could expect an extra five years of life in the dry if impoverished latitudes of the south. And an even longer bonus—seven years—was added to those who served out their working life in a white-collar rather than a blue-blouse job. All of which still left the worker comparing his brief enjoyment of a couple of years' pensioned retirement with the decades of 6 per cent weekly deductions that had slowly built up his entitlement. His only consolation was that it had cost his employer twice as much or more in contributing towards it (between 10 and 15 per cent)—which might add a mild tinge of bitterness to the *patron*'s fireside thoughts as he sat out his extra seven years of slippered retirement.

The original Liberation ordinance of 19 October 1945 had theoretically envisaged a 20 per cent pension at the age of sixty (pp. 131–2). But in practice, sixty-five was the normal retirement age; and with the accumulation of five years' extra contributions, the real pension was 40 per cent of the notional basic salary. A quarter of a century later, a law of 31 December 1971 raised it to 50 per cent. An unusual feature of the French scheme was that the beneficiary could draw his pension even if he decided to continue working beyond retirement age. This latitude perhaps stemmed from the fact that the pension was entirely funded from workers' and employers' contributions, with the State having few grounds for the initiative-killing type of prohibition that made state aid such a two-handed friend to the needy in a number of neighbouring countries.

The State likewise contributed nothing to sickness and maternity benefits, which were also covered by the workers' and employers' contributions to the pension funds. By contrast, however, the State alone was responsible for keeping afloat unemployment benefit, thereby enabling it to keep a tight control over what, in other hands, might become a covert system of strike pay. Unemployment benefit itself was very niggardly—about six francs a day in the late 1950s, depending on the region and on a means test. And although its Gradgrind character was softened by a discretionary series of grants to dependants—with a ceiling equivalent to two-thirds of the household's customary income—workers and management were very conscious of the scheme's inadequacies. They consequently supplemented it by an independent series of collective agreements, formalized on 31 December 1958, which provided a benefit of 35 per cent of wages for a nine- to twelve-month period—the system to be financed by a 0.25 per cent payroll contribution by the employer and a 0.05 per cent wage deduction from the worker. Eventually, the two parallel schemes were united on 1 April 1979. Happily, unemployment was not a major problem in France before the 1970s, rarely rising above 2 per cent of the work-force.

As indicated in Table 10.14, the employer's contributions to the overall social security system was high compared with most countries. In addition to the contributions already listed, the *patron* financed virtually the whole of the family allowance system—involving him in contributions equivalent to 13.5 per cent of his payroll—and, more conventionally, he was entirely responsible for accident insurance on the premises, which added on average a further 3 per cent to his payroll levy. The effect of this burden on management undoubtedly tended to depress wages in France, and caused foreign observers to ask why the State did not take on a larger share. To have done so, would have obliged the State to have instituted an entire restructuring of the taxation system—which arguably was long overdue, but which no French Republican government could have got through parliament. The existing system was heavily dependent on indirect taxation, which proportionately hit the poor much harder than the rich. This is particularly evident from the comparative figures in Tables 10.16 and 10.17, and the Table of household income, 10.15. The only way the State could have

Table 10.14.
Financing Social Security, 1964: France and EEC Countries (% shares)

	State	Employer	Employee
France	7	69	24
Germany	18	40	42
Italy	7	77	16
Holland	7	40	53
Belgium	26	40	34
Luxemburg	20	40	40

Note: Family allowances are not included in these calculations.
Source: Yves Trotignan, *La France au xx siècle*, i (Paris, 1976), 341.

Table 10.15.
Indices of Disparity between Household Incomes in France, 1975
(100 = National Average for Each Column)

Head of household	Gross income	Gross income and family allowances	Gross income less tax	Net income
Farmers	45.8	48.9	47.2	50.6
Agricultural workers	61.0	67.3	65.5	72.1
Artisans, small shopkeepers	117.7	115.3	109.1	107.1
Industrialists, business men	272.8	263.1	225.2	216.9
Liberal professions	262.1	253.1	214.6	207.1
Higher management	218.1	211.6	207.3	200.6
Middle management	127.6	126.1	129.8	128.1
White-collar workers	95.5	96.1	98.9	99.4
Skilled labour	83.3	86.2	87.7	90.7
Unskilled labour	71.2	76.2	75.7	81.1
Not working	68.8	66.4	70.4	67.6

Source: D. L. Hanley, A. P. Kerr, and N. H. Waites, *Contemporary France: Politics and Society since 1945*, 2nd edn. (London, 1984), 95.

taken on a greater share of the social security burden without overloading the masses with higher indirect taxation, would have been to bring income tax into line with what countries like Britain were charging. But, as Chapter 2 explained (pp. 37–8 above), this would have required a revolution in attitudes which still remained unaccomplished three decades later.

After pensions, family allowances were among the most costly of the social provisions, and remained generous compared with those of most countries

Table 10.16.
Sources of Tax Revenue, 1965 and 1975: France, UK, W. Germany, Italy (%)

	Taxes on incomes and profits		Social security contributions		Taxes on goods and services		Other taxes	
	1965	1975	1965	1975	1965	1975	1965	1975
France	16	18	34	40	38	34	12	8
UK	37	44	16	18	33	25	14	13
W. Germany	34	35	27	34	33	27	6	4
Italy	18	22	34	46	39	29	9	3

Table 10.17.
Revenue and Social Spending in the Early 1970s: France and Selected
Western European Countries

	% of total personal income in 1975 taken by		Public expenditure as % of GDP in 1974		
	taxes on incomes	social security contributions	social security	health	education
Sweden	23	10	13	8	8
Holland	13	18	16	6[a]	8
Belgium	13	13	15	5	6[b]
W. Germany	12	14	7	5	4
Austria	13	11	10	4	5
UK	13	10	10	5	7
Switzerland	12	9	7	4	5
France	5	16	11	6	5[c]
Spain[d]	2	7	6	4	2[e]

[a] 1972.   [b] 1973.   [c] 1973.   [d] 1970.   [e] Central gov. only.

(Table 10.13). In the 1960s, they amounted to 22 per cent of the notional basic wage for each of the first two children, rising to 33 per cent for the third. In the more poorly paid categories of occupation, a mother at home might receive more than her husband earned, which would certainly be the case with the wife of an agricultural day-labourer with four children. Net family allowances did not keep pace with the cost of living. In the 1950s, the real income of a childless worker increased by about three-fifths, whereas that of a worker with three children only went up by about a third.

Next in order of financial weight was sickness benefit, equivalent to half the beneficiary's notional salary—or two-thirds if he had three or more children. Here, French expenditure was roughly on a par with that of comparative countries (Table 10.13). Housing, however, fared less well. Well over 10 per cent of French housing had been destroyed in the war, and although the government commissioned the building of a large quantity of cheap apartment blocks for rent (HLMs), the supply came nowhere near satisfying the demand. It had initially been assumed that the post-war intention to build large quantities of public housing would rapidly make existing rent controls unnecessary, since an abundance of accommodation would automatically keep down prices. But the failure of the public house-building programme to meet current needs led to rising rents and the consequent retention of rent controls. And, as in other countries, these controls discouraged private contractors from building rented accommodation. Moreover, the obstacles that the law put in the way of terminating unprofitable lets decided many landlords to leave their property empty.

The HLMs undoubtedly helped to relieve some of the pressure on the private sector. Yet in many towns, they became synonymous with dreary grey rectangles, compounding the desolation of the *banlieues*, like the undemolished relics of an unprofitable trade fair. And the 1960s brought the oppressive dimension of high-rise architecture, when, as elsewhere, there appeared mush-rooming alphavilles, without the green spaces and trees that redeemed many of their British equivalents. They admittedly brought an end to the oil-drum and tarpaulin encampments that had characterized the worst years of the housing shortage. But with more imagination—and not that much more money—these problems might have been tackled with less of a blighting impact on the life and appearance of the urban perimeter. And despite the building boom of the 1960s, more than half of French houses were still without a bath or shower, while nearly half had no inside lavatory.

Social Mobility and Education

Contrary to the hopes of many contemporaries, the economic changes of post-war decades saw no particular upswing in social mobility (Table 10. 18 and Appendix Table 8). Those changes that did occur in the social structure were primarily the result of sideways movement, arising out of the growth of the services and industrial sectors, and the corresponding contraction of the peasantry. Although urban wages generally represented an increase in income for the rural migrant, this had likewise been true of similar movements in the past. Taking the active population as a whole, 58 per cent in 1953 had branched out from their father's occupation or level; and by 1964 the proportion had risen to 66 per cent. But among the industrial working class, the movement was on balance the other way: 47 per cent in 1953 had moved away from their father's occupation or rank, but by 1964 the percentage had dropped to 32 per cent, rising slowly to 39 per cent in 1970.

The social inequalities of post-war France seemed destined to further pro-longation by the inequalities of the educational system. In 1956, Guy Mollet's Education Minister, René Billères, estimated that only 21 per cent of working-class children went on to secondary education, while in the case of farm-workers' families, it was only 13 per cent. For management and the professions, by contrast, the proportion was well over 80 per cent. Mollet's government fell before Billères's proposed remedies could become a reality; but they served as a base for the changes brought about by Jean Berthoin and Christian Fouchet in the early years of the Fifth Republic. Berthoin decreed the raising of the school-leaving age to sixteen (January 1959)—but only with effect from 1967. The existing *cours complémentaires* were replaced with *collèges d'enseignement*

Table 10.18.
Post-tax Household Income, c.1970: France and Selected Countries (Decile Shares)

	Lowest-earning households							Top-earning households		
	1	2	3	4	5	6	7	8	9	10
France (1970)	1.4	2.8	4.2	5.5	7.4	8.8	9.7	13.1	16.6	30.5
W. Germany (1973)	2.8	3.7	4.6	5.7	6.7	8.2	9.8	12.1	15.7	30.6
UK (1973)	2.4	3.7	5.3	6.9	8.5	9.9	11.1	12.9	15.4	23.9
Netherlands (1967)	3.2	5.9	6.8	7.7	8.3	9.2	10.4	12.1	14.5	21.8
Spain (1971)	1.5	2.7	4.4	5.8	7.8	9.0	11.0	13.0	16.5	28.5
Sweden (1972)	2.6	4.7	6.3	7.8	9.0	10.0	11.6	13.1	16.4	18.6
Norway (1970)	2.4	4.2	5.7	7.3	8.7	10.2	11.7	13.0	15.0	21.9
USA (1972)	1.7	3.2	4.6	6.3	7.9	9.6	11.4	13.2	16.0	26.1
Canada (1972)	1.6	3.6	5.2	6.8	8.3	9.7	11.2	13.0	15.8	24.7
Australia (1966–7)	1.6	3.2	5.3	6.9	8.3	9.5	11.1	13.0	15.7	25.2
Japan (1967)	2.7	4.4	5.7	6.7	7.8	9.0	10.1	11.6	14.1	27.8
Average	2.2	3.8	5.3	6.7	8.1	9.4	10.8	12.7	15.6	25.4

Note: Each column corresponds to 10% of the national total of households; each figure is the % share of national income entering these households.

Source: OECD.

général (CEGs); and the first two years of both the new CEGs and the *lycées* were to consist of a *cycle d'observation*, which was theoretically intended to allow restreaming of pupils in the light of their progress. In practice, however, the *cycle d'observation* was itself largely streamed, thereby reinforcing the assumptions made about pupils at primary level. And although the entrance examination to the *lycées* was abolished, the decision on whether a pupil should follow the *cycle d'observation* in a *lycée* or a CEG was also based on assessments made before leaving primary school. In order to reduce the dangers of premature type-casting, Pompidou's Education Minister, Christian Fouchet, transferred the *cycle d'observation* to a new middle school in 1963, the *collège d'enseignement secondaire* (CES), and extended its length from two to four years. It took time, however, for the system to come into operation, and the first 1,500 CESs were not finished until 1968, although building continued rapidly thereafter.

But, once again, streaming continued and partly defeated the object of the exercise. Although the *cycle d'observation* now took place under one roof, the two upper streams broadly corresponded to the old *lycée* and CEG programmes, with a third stream for those who were not up to either. Transfers from one stream to another were theoretically possible, but tended to take place only in the more obvious cases of misplacement. In the early 1970s, Joseph Fontanet proposed a more flexible system, involving a conscious mix of streamed and unstreamed classes; but ministerial changes put an end to the scheme, and nothing significant was done before the Haby reforms of 1975, which are examined in a later chapter (p. 348 below).

(p. 348 below)

L'ÈRE DES TRANSFORMATIONS

The changes that came in the wake of the economic prosperity of the 1960s were not to everyone's liking. The high-rise menace of the *banlieue* waste land had its counterpart in the up-market world of city offices and prestige premises—which burgeoned during Pompidou's presidency, and was then restrained by the joint action of his successor and the economic recession. The west end of the Champs Élysées had long been seen as a warning of what could happen to Paris if insensitive modernity was allowed its head; and its American-style bars and aggressive company offices had provoked uneasy thoughts in the 1930s—some less savourily expressed than others. Georges Bernanos had notoriously asked whether the future of the world would lie with 'some little Yankee shoe-shine boy, a kid with a rat's face, half Saxon, half Jew, with a trace of Negro ancestry in his maddened marrow, the future King of Oil, Rubber, Steel, creator of the Trust of Trusts, future master of a standardised planet, this god that the universe awaits, god of a godless universe'.[4] At least the transatlantic end of the Champs Élysées observed the roof-line of the existing buildings—which was more than the modernity of the 1960s was to do. And those who regretted the desecration of well-loved vistas by the monstrous creations of the 1960s could take some consolation in the knowledge that appalling though they were, they were as nothing compared with what the architects wished to do, had they been allowed. The visionary Swiss, Charles Jeanneret ('Le Corbusier'), whose genius created the chapel of Notre-Dame-du-Haut at Ronchamp (1950–5), entertained for much of his life a chilling plan for the rebuilding of Paris, beside which Fritz Lang's *Metropolis* would have resembled an inoffensive garden suburb.

Not surprisingly, the post-war emphasis on the necessity for economic and social planning bred its own counter-fear of an ultra-disciplined, technological society—with René Clair and Jacques Tati among its better-known critics, inimitably aware of its tragi-comic potential. Nostalgic Francophiles, such as Richard Cobb, wistfully looked for relics of the old France beside 'abandoned railway lines, or little-used canals, behind breakers' yards and . . . gas-works', in the shape of 'two-decker trams converted into houses and covered in greenery', or 'a tiny garden decorated with scallop shells arranged in rings stuck in earth filling old petrol cans . . . a small temple of individuality, an artisanal folly, by a Sunday architect'.[5]

Not that the big clean-up of the 1960s was necessarily all loss; on balance it perhaps brought more benefit than destruction. Only the most devoted *belle époquard* could regret the arrival of modern sanitation—even if the removal of some of the finer wrought-iron *pissoirs* was an aesthetic as well as a physiological loss. Even in the 1950s, many hotels advertising 'Confort moderne' had still been equipped with 'Turkish'-style lavatories that required of the user a highly insensitive nose and the leg muscles of a gymnast. As for the 1970s, the best-known architectural creation, the Centre Pompidou (1977), was ironically the product of an Anglo-Italian partnership, Richard Rogers and Renzo Piano.

FLESH AND SPIRIT

As in the past, what drew foreigners to post-war France were the familiar ingredients of landscape, architecture, culture, and cooking. To those from across the Channel or across the Rhine, it seemed that the pleasures of life were still cared about with an intensity and calculation that had few rivals elsewhere—even if evidence of enjoyment was often obscured by the *désabusé* cast of countenance and speech that made 'gay' the last word one would readily apply to Paris. Indeed, the night-club singers that drew the largest audiences dealt almost exclusively in a weary mix of sentiment and cynicism. Not that the familiar formulas of Édith Piaf, Juliette Gréco, and Charles Trénet had it all their own way; George Brassens and the Frères Jacques brought humour too— though there was no one to equal the great performers of the *belle époque*, such as Yvette Guilbert (1865–1944), whose reissued recordings were a continuing reminder of how a sharp ear for inflection could invest the simplest of statements with a multitude of meanings.

Admittedly, 'to live like God in France' was a more expensive business than pre-war generations of visitors had found. Even after the initial post-war food shortages, the rise in labour costs affected all classes of restaurant; and customers carefully scrutinized the price-list before entering. The number of meals bought in Paris in the 1950s was only a quarter of those consumed before the war—and still remained less than half the pre-war figure in the affluent years that followed. At the same time, the proliferation of works and office canteens was another major cause of this contraction. And then, on top of these material problems, there came an assault from a further quarter—the dire warnings of the medical profession. But French chefs were nothing if not resilient. They listened attentively to the mounting condemnation of the high-cholesterol ingredients that were the essence of so many of their best-loved dishes; and from the late 1960s culinary high priests such as Paul Bocuse developed what came to be known as 'la nouvelle cuisine' for the small but growing minority of diners who chose to heed the doctors' warnings. But, as elsewhere, the down-market end of catering became increasingly dominated in the 1960s and 1970s by fast-food shops and cheap Chinese restaurants.

The traditional lead of Paris in the world of women's fashions was rapidly re-established after the war. Christian Dior outraged an austerity-minded Europe in February 1947 by introducing 'the New Look' calf-length hem-line; but its evocation of the cool elegance of more leisured days quickly assured its international success. The late 1950s and 1960s, however, saw Italy, Scandinavia, and London making a significant impact on the French fashion market, especially among the young. And with magazines such as *Elle* and *Marie-Claire* spreading clothes-consciousness to the less affluent sections of society, designers were increasingly aware that it was here that turnover and potential profit were greatest.

Thought and Culture

As in the 1930s, the ability and propensity of the French to discuss intellectual and artistic matters in an engagingly articulate, confident, and concise manner always guaranteed their writers and publicists a ready audience elsewhere as well as in the enclosed, incestuous world of the Paris reviews and literary-prize committees. And France still continued to provide a congenial home for expatriate talent—even if the affluence and climate of America increasingly tempted many of the more noted refugees to move on to California and other lotus-lands. As before the war, critics lamented the absence of great figures comparable to the giants of the *belle époque*. And while it remained true that prophets are not recognized in their own age, let alone in their own village, it was a matter of concern that comparisons with neighbouring countries were not always flattering. People constantly asked—without receiving an answer— whether there was anyone comparable to Picasso, Stravinsky, or Thomas Mann—all three admittedly now long in the tooth, exiles from their own countries, and with few obvious successors among their compatriots.

Even when it came to 'isms' and the nurturing of new currents of thought and artistic creation, America and Germany were formidable rivals. New York was now much more important than Paris as an art centre; and although pre-war figures such as Matisse, Braque, and Léger continued painting, members of the new generation enjoyed only fluctuating favour. While the abstracts of Pierre Soulages (1919–) were held in steady if modest esteem, the mayfly popularity of Bernard Buffet (1928–) was virtually confined to the late 1950s. In music, Messiaen continued to impress audiences and inspire students; but among those who profited from his teaching, Pierre Boulez (1925–) was based in the musical world of south-west Germany rather than in France. It was not until the opening of the Institut de Recherche et de Coordination Acoustique-Musique in the bowels of the Beaubourg in 1977 that his dynamic energies were once more pivoted on Paris. As for literature, critics claimed that it was in double peril. In the 1950s, the irascible Herbert Lüthy complained that:

The main stream of French literature flows, not from lyricism but from the social art of conversation, and the gravest threat to it today is that conversation is gradually using up and pulverising all the elements of action. French writing consists of conversation drama, conversation novels, aphorisms and scraps of conversation out of diaries, in which every-thing is talked about and talked to pieces without its ever leading to anything; and this is also the gravest threat to France herself, for in this respect she faithfully reflects her literature.[6]

The other alleged danger was in the isms themselves, which perhaps more than ever thrived on half- or at least three-quarter truths, uttered as aphorisms or asserted by characters drawn from history or legend, whose lives were redrawn to exemplify these assertions.

Existentialism was a case in point. And, once again, much of its inspiration, if not all its conclusions, came from abroad. Lüthy's exasperation was not an isolated reaction:

Perhaps the most obvious symptom of the crisis of the French intellect is that for ten years it has been expressing itself in a quasi-profound jargon borrowed from Hegel, Marx, Heidegger, and Nietzsche . . . It accepts Marx's prophetic utterances of 1848 as the last word about the human situation, Zarathustra's 'God is dead!' as the last word in enlightenment, and *Beyond Good and Evil* as the last word in ethics; and it proclaims these things to a nation in every hole and corner of which they have long been discovered and discussed. As a bold and revolutionary counter to the petty bourgeois nostalgia for the French *fin-de-siècle* it has discovered the German *fin-de-siècle*; and the most original post-war achievement in the literary life of Saint-Germain-des-Prés has been imitation, even in details of pose and accent, of the 'great nausea' of the previous post-war period in

Germany . . . Jean-Paul Sartre . . . in his simultaneous roles as philosopher, pamphleteer, playwright, literary eschatologist, and political 'activist', has continued to perform his balancing act with existentialism and popular Marxism, and . . . has refused to answer the question of how he reconciles the two; whether he regards as true and real the 'self choosing' man of his philosophy as a total product of his own self, or the man postulated by his sociology as a total product of society . . . [For to do so] would be the old-fashioned morality of all the western moralists since Hesiod and Montaigne, who have reconciled in the human state freedom and determinism, individualism and responsibility, mastery of fate and submission to it, man's grandeur and his insignificance.[7]

Despite the inherent tension and contradictions between Existentialism and Marxism, both attracted a large segment of educated opinion in the post-war decade, especially among the young. Gossip columnists and fashion writers were quick to take up the cult of what they saw as Existentialist life-styles; and Left Bank cafés were thronged with tight black sweaters and trousers, their wearers displaying a studied indifference to conventional forms of social behaviour, lest they betray the individual's duty to create his own world in a world without meaning. Like all bohemianism, the stereotypes it engendered rapidly palled. People returned to sitting on chairs rather than on the floor or the window-sill, and the practice of greeting newcomers with cold silent stares gave way to the old slack habits of affability and informative conversation. To sample something of its flavour today, one has to return to the café scenes in such films as Jean Cocteau's *Orphée* (1950). If Existentialism was the opium of the intellectuals and would-be intellectuals in the late 1940s, its literary consequences were undoubtedly impressive. The case for commitment contained in Sartre's astringent post-war plays and novels, drew equally distinguished caveats in Albert Camus's writings of the same years—tragically terminated by his death in a car accident in 1960. Camus's exploration of the dilemma of the individual in a world without meaning ranged from hope in *La Peste* (1947) to the apparent despair of *La Chute* (1956), and was arguably in the great tradition of the French 'conte moral' that had its roots in the *ancien régime*.

Very different from this, were the concerns of the mainstream of the so-called Nouveau Roman, which from the 1950s sought to move away from subjective questions of motivation and duty, and concentrate on the objective delineation of man as part of his surroundings and the material world in general. Thus, the *chosisme* of Alain Robbe-Grillet and Michel Butor presented the individual's sensations as on a par with his physical environment—and often at odds with it, since the pattern and workings of the material world ultimately transcended and eluded him. Similarly, Nathalie Sarraute's sharp depiction of dialogue and sensation sought to explore their intrinsic shape and thrust in their own right, rather than as the traditional expression of the thoughts and motives of the speakers themselves. Indeed, the divorce between utterance and inner feeling was also part of her concern, and a major source of the humour of her novels.

This reduction of man to being no more than an item and a causal link in the overall network of material existence had its counterpart in other forms of intellectual activity. Structuralism became the dominant ism of the 1960s and 1970s. Sociologists, social anthropologists, historians, philosophers, and literary critics sought to discern the basic structures that underpinned human activity and the world that mankind inhabited—structures that lay deeper than the conscious levels of thought and behaviour. Claude Lévi-Strauss, Jacques Lacan, Roland Barthes, Michel Foucault, and Jacques Derrida were prominent exponents of Structuralism in their respective disciplines—their suggestions and hypotheses encouraging specialists in disparate fields to look to each other for supportive evidence and mutual enlightenment. Some of their books sold widely and became fashionable topics for party conversation—notably Foucault's *Les Mots et les choses* (1966), in which the public avidly seized on such comments as: 'Man with a capital M is an invention: if we study thought as an archeologist studies buried cities, we can see that Man was born yesterday, and perhaps that he will soon die.'[8]

In contrast to this broad-fronted shift in ideas, the French contribution to post-war developments in poetry and the theatre was much less evident. The major poets, René Char (1907–), and Henri Michaux (1899–), and Francis Ponge (1899–), had already acquired firm reputations in the 1930s, while Louis Aragon was arguably past his best. In the theatre, the dominant indigenous names were likewise established pre-war figures—notably the ever-prolific Jean Anouilh and the earnest, overcast Henry de Montherlant

Table 10.19.
Leading Daily French Newspapers of the mid-1970s

	Founded	Circulation
France-Soir	1944	980,000
Le Parisien Libéré	1944	980,000[a]
Ouest-France (Rennes)	1944	695,000
Le Monde	1944	548,000
Le Figaro	1826	492,000
Le Progrès (Lyon)	1859	490,000
Sud-Ouest (Bordeaux)	1944	420,000
La Voix du Nord (Lille)	1944	415,000
Le Dauphiné Libéré (Grenoble)	1944	408,000
L'Aurore	1944	384,000
L'Humanité	1904	194,000
La Croix	1880	137,000

[a] Circulation figure includes provincial editions.

(1896–1972). From the late 1950s, however, the bright stars in the firmament were foreign exiles—Eugene Ionesco, Samuel Beckett, Arthur Adomov, and Fernando Arrabel, who, in their individual ways, were the main representatives of the Theatre of the Absurd in France. As for performers, the two outstanding figures of the period were Jean-Louis Barrault and Edwige Feuillère, who, unlike the leading talents of the 1930s, were arguably at their best in tragedy, both having a larger-than-life intensity that was reminiscent of the great actors of the nineteenth century.

As in other countries, the written word was increasingly giving way to the visual media as a source of instruction and entertainment. If the 1960s and 1970s saw many cinemas succumb to the competition of television, film directors found their work reaching a wider audience than ever before, as TV channels filled out their programmes with films that had completed their initial cinema-showing. Directors who had enjoyed little more than a cult following among the *cognoscenti*, were drawn to the attention of millions of habitual tele-viewers, even if they were not necessarily liked or remembered by the many who saw their work. Among the younger directors of the post-war decade, the sensibility and restraint of René Clément and Robert Bresson took the so-called 'realist' tradition of the late 1930s to new levels of achievement, with Clément's *Les Jeux interdits* (1952) and *Gervaise* (1956) remaining perhaps the finest examples of the genre.

The late 1950s, however, brought a sharp break with this tradition in the so-called Nouvelle Vague, represented by the widely differing aims and styles of such directors as Alain Resnais, Louis Malle, François Truffaut, and Jean-Luc Godard, who had little in common other than their desire to explore new ground. *Les Cahiers du cinéma* and Henri Langlois's entourage at the Cinéma-thèque had nurtured a growing impatience in the 1950s with the immobility of current cinematic conventions. Resnais's *Hiroshima mon amour* (1959) was the first of a series of films that made extensive and inventive use of kaleidoscopic flashbacks and juxtapositions to create a visual stream of consciousness, reminiscent of the experimental cinema of the 1920s or Cocteau in the 1930s and 1940s—and which led to accusations of obscurity and intellectual élitism from large sections of the cinema-going public. The most innovative figures of the Nouvelle Vague became increasingly divorced from the mainstream of commercial film-making, even if individuals such as Truffaut successfully maintained their appeal in both camps. Some directors popularly lumped with the Nouvelle Vague owed more to their 'realist' predecessors; this was notably the case with Claude Chabrol, whose depiction of small-town life in the early sections of *Le Boucher* (1970) displayed the qualities associated with the rural classics of the 1930s. By contast, the most controversial director of the 1960s and 1970s was Godard, whose bleak comments on the consumer society were mirrored in his systematic paring away of the customary craftsmanship and packaging of what was conventionally regarded as a well-made film. The consumer society also came in for criticism from a very different direction when the leading humorist of the post-war cinema, Jacques Tati, moved from the straightforward comedy of *Jour de Fête* (1949) and *Les Vacances de Monsieur Hulot* (1953) to the wistful, if no less comic, indictment of modern gadgetry and life-styles in *Mon Oncle* (1958) and *Playtime* (1967).

At the end of the day, the directors who made the most lasting impression on their audiences were arguably those whose assertions on human imperfection were leavened with some affection for it, and who could portray it with the perception and humour that stems from a broadly based and commonly shared sense of proportion. This was markedly true of Éric Rohmer's series of 'contes moraux'—notably *Ma Nuit chez Maud* (1968) and *L'Amour, l'après-midi* (1972). These were later followed by his no less subtly observed 'comédies et proverbes', the quality of their dialogue and his eye for revealing visual detail giving them an unobtrusive strength and discipline that had few rivals among his contemporaries.

The economic recession of the 1970s was to see a notable deceleration in innovatory activity in the arts. Unhelpfully termed the Post-Moderne, this preference for the tried and true was often accompanied by a nostalgic mix of old styles, verging on pastiche. It was perhaps inevitable that the expansion of the 1960s, with the aggressive modernity of its huge monolithic buildings, should provoke countermovements favouring smaller-scale urban landscapes, and incorporating a variety of fashions with familiar human faces. Yet it was arguably no accident that the growing strength of these countermovements in the arts should coincide with new economic and social precepts preaching that small was beautiful and that a mixture of economic strategies, including past successes, was a safer way of ensuring steady if unspectacular progress. A pre-Keynesian monetarism was likewise to gain ground in many Western countries in the late 1970s, which in France was also to be accompanied by a short-lived current in intellectual circles questioning democratic assumptions—the brief phenomenon of 'la Nouvelle Droite'. And if its wider appeal in fashionable magazines failed to survive the Socialist victory of 1981, it nevertheless left influential offspring.

Religion

While the most popular and respected writers and creative artists were those whose skilled dissection of human motivation was performed with a wry sympathy, fellow-feeling was likewise an issue in the Churches. The experience of French priests in German prison camps and factories made many aware that it was only in the sharing of a common experience that the Church could establish contact with a working class that in many areas had grown up outside its influence. The rehabilitation of Catholicism as a political force in post-war France had found many middle-class Catholics more open in the expression of their religious allegiance, especially in those branches of the public service where it had been something of a liability under the Third Republic. Indeed, the level of church-going among the professions and civil service was probably higher in the 1950s than it had been since the nineteenth century. This was not the case, however, with the working classes. It was true that the official attitude of benevolent neutrality to the Church saw a modest increase in the number of men going to Mass in those dioceses in which official anticlericalism had kept the level artificially low in the early years of the century; and the example of a church-going professional class encouraged imitation in some other working-class quarters. But in general the overall trend continued to be downward, and regular Mass-going among adults dropped from well over 20 per cent to less than 15 per cent in the course of the 1960s (pp. 296–8).

Drawing on wartime experience, a hundred or so priests were authorized in the post-war decade to take full-time jobs in factories and other secular work-places, with the object of winning the workers' confidence and bringing them into closer contact with religion. In pursuit of these aims, a number of these worker-priests became heavily involved in trade-union activities, including committee work for the Communist CGT. Others were involved in the fellow-travelling Mouvement de la Paix; and two were arrested during the anti-General Ridgway demonstrations of 1952 (pp. 161–3). Although the worker-priest movement had been inaugurated by the Archbishop of Paris in January 1944 as a response to alarming evidence of the low level of religious observance, the experiment had always been regarded with deep suspicion by conservative elements in the Vatican. The involvement of priests in Communist activities decided Rome to bring the enterprise to an abrupt end, and the worker-priests were ordered to withdraw in January 1954. They could theoretically continue working on a part-time basis—for a maximum of three hours a day—provided that they stayed clear of union and political activity. But limited part-time work was not an available option in many factories; and, faced with the prospect of abandoning the fruits of years of patient activity, two-thirds of the priests refused.

This was inevitably seen as confirming the worst suspicions of Rome; and there followed a firm assertion of ecclesiastical discipline right across the Church in France. The various laymen's Catholic organizations were instructed to restrict their activities to religious matters, and were expressly ordered to distance themselves from the pursuit of social goals. Leaders who remonstrated were purged; the Association Catholique de la Jeunesse Française was a notorious example. At the same time, over-speculative theologians were silenced; and the three provincials of the Dominican Order in France were all revoked in January 1954 for failing to keep a tight rein on their subordinates. It was all depressingly reminiscent of the witch-hunt years of Pius X—who, symptomatically, was declared a saint in May 1954.

In these dispiriting days of the mid-1950s, few could have guessed the adventures that lay round the corner. The death of Pius XII in October 1958 was to see the papal tiara pass to the former nuncio in Paris, Angelo Roncalli, who, as John XXIII, was to bring into being the Second Vatican Council—and all that that entailed (pp. 297–8).

11
France Overseas, 1944–1954

EMPIRE in the 1950s had lost something of its pre-war glamour. Admittedly, it took a dyspeptic, land-locked Swiss to describe its development in these terms: 'The empire was something with which the French people had nothing whatever to do, and its story was that of machinations of high finance, the Church, and the military caste, which tirelessly re-erected overseas the Bastilles which had been overthrown in France.' And then, turning specifically to the European settlers in Tunisia, Lüthy referred to them as 'people who have settled like rats in the cheese of the protectorate and have populated every office building, down to the last post office, bank, and tax-office counter, with members of their clan. After the second or third generation nothing European is left about them except their trousers.'[1]

Outwardly the French post-war empire was much as it was in the 1930s. Only the mandated territories of Syria and Lebanon had become independent, and imperial trading links in the late 1940s were as important a part of the French economy as in the pre-war years—a quarter of French imports coming from the colonies and 35 to 40 per cent of her exports going there. It was to take the European movement of the 1950s and the loss of Indo-China to challenge this; and even in 1959 these proportions were little altered, the only significant change being the swing of French exports towards her EEC partners (Table 11.1).

The new constitution of 1946 sought to strengthen the feeling of community by extending representation in the French parliament to all overseas *départements* and territories—the only unrepresented areas being the 'associated states' of Morocco, Tunisia, and the Federation of Indo-China. But the colonial deputies, including the Algerians, made up only 12 per cent of the National Assembly's membership; and although their votes very occasionally played an important role on close-fought issues, their diversity of interests prevented them from acting as a coherent political weight in parliament, able to obtain colonial concessions by promising support to mainland parties on other issues. All overseas territories, including the associated states, were formally linked in the newly created French Union—a term intended, like the British Commonwealth, to lessen the sense of colonial subservience. Its representative institutions, however, were largely of a symbolic nature, the real business of government taking place elsewhere.

This outward display of imperial solidarity was still unable to conceal the

Table 11.1.
Distribution of French Trade, 1929–1962 (%)

	Imports				Exports			
	From EEC	From EEC and UK	From EEC, UK, Spain, and Switzerland	From the franc zone	To EEC	To EEC and UK	To EEC, UK, Spain, and Switzerland	To the franc zone
1929	23.52	34.27	37.90	12.0	30.78	45.78	55.90	18.5
1938	17.53	24.56	27.13	27.1	25.70	37.33	44.9	27.4
1947	10.8	13.95	16.24	24.1	19.15	26.0	32.15	41.63
1950	16.8	20.5	24.0	26.1	20.3	29.4	35.5	35.95
1953	17.8	22.35	25.9	25.0	18.6	24.0	34.0	37.0
1956	21.1	26.3	32.47	23.38	25.3	30.9	39.2	32.15
1959	21.9	27.3	30.2	24.0	27.3	31.75	36.95	31.7
1962	33.58	38.77	42.46	20.76	36.84	41.57	49.6	20.10

Source: Fernand Braudel and Ernest Labrousse (eds.), *Histoire économique et sociale de la France*, iv, pt. 3 (Paris, 1982), 1,383.

ambitions unleashed in the colonies by the Second World War. The spectacular victories of Japan in the Pacific had demonstrated that Asian peoples could challenge Western nations with their own weapons—and win. Despite Japan's ultimate defeat, the impact of her initial successes on the Third World's imagination was enormous, and many lessons were drawn. The speed of Japan's post-war recovery was likewise impressive. The wartime trauma of the old colonial powers was particularly acute in the case of France, given the humiliations of 1940–4. Less had been expected of the military performance of small colonial countries like Belgium and Holland, and their governments had departed into respectable exile in London. The French empire, by contrast, had seen Frenchmen fighting Frenchmen—with Gaullist expeditions contesting the control of Vichy-held territories. This civil-war aspect of French rule encouraged both sides to vie for support from the native population—and raised the issue of whether substantial concessions should be offered in response to their demands for more control over their own affairs. On the other hand, the indignities of 1940–4 had created hypersensitivity in post-war France on matters of national prestige. While there was readiness to widen the electoral basis of the colonial assemblies and to give the colonies greater representation in the French parliament, there would be widespread opposition to anything suggesting sovereign independence.

BLACK AFRICA

De Gaulle's Brazzaville conference of French African governors and administrators in January 1944 had specifically declared that 'the object of the civilising work accomplished by France in the colonies excludes any idea of evolution out of the French empire. There is no question of instituting self-government, even in the distant future.' Reforms were to be limited to improving the lot of the native population.

In the traditional and, in many ways, artificial debate between assimilation and association (p. 31), the issues were now much more blurred. On the side of assimilation, the colonies were given stronger representation in the French parliament; and at the same time the vote was extended to a much wider section of the native population: in the case of Black Africa, to anyone who had held a job for at least two years, or was literate in French or Arabic, or had served in the army. Further extensions were to take place in 1951 and 1956. On the other hand, the associationist argument was recognized in so far as the distinction was more firmly drawn between voting rights and full French civil status. The latter continued to carry the obligation to observe French social customs such as monogamy, an obligation which many traditionally polygamous peoples were reluctant to adopt. The extension of the franchise was no longer conditional on the voter foregoing these traditional customs, while the abolition of the *indigénat* (p. 31) reduced the significant advantages that full French civil status entailed. Each of the Black African territories now instituted a general council, which was a development of the pre-war *conseil d'administration*. A third of the membership was elected by voters who had opted for full French civil status, while the other two-thirds represented the rest of the voting population. The council's main function was the apportioning of taxation among the various sections of the territory's population. Otherwise it was a sounding-board, not a legislature.

1947 saw France set up the Fonds d'Investissement et de Développement Économique et Social des Territoires d'Outre-Mer (FIDES), with a tacit mandate to develop commodities that would otherwise cost dollars to buy elsewhere. By 1957, it had invested $542 million in West African territories—with the result that West African politics became increasingly preoccupied with the distribution of these funds. Although West African deputies were initially disposed to look to the Communists in France—given the PCF's commitment in principle to colonial independence—this ceased when the PCF went into opposition and no longer had a finger on the purse-strings.

The Rassemblement Démocratique Africain, launched by the Ivory Coast leader, Félix Houphouët-Boigny, to hasten self-government in the French Black colonies, grew increasingly conscious of the dependence of their economic development on French aid, and consequently adopted a moderate attitude towards Paris. Yet, of the various Ministers for the Colonies between 1947 and 1951, only Mitterrand (July 1950–July 1951) made serious attempts to build on this situation and establish some sort of dialogue with the RDA. He widened the franchise in Black Africa, but in return he expected the RDA to take a firm anti-Communist line and support French policy in Indo-China.

ALGERIA

Turning to North Africa, post-war Algeria now had a population of nine and a half million, a million of whom were Europeans, of predominantly Italian and Spanish descent rather than French. The nineteenth-century French practice of confiscating lands for white settlers had put a third of the arable area of Algeria in European hands—a factor which helps to explain why Algerian agriculture was able to draw so heavily on French government aid. Wheat and wine subsidies alone amounted to £18 million a year in the 1950s. Modern farming methods, however, led to severe unemployment among the Muslim population, some 10 to 20 per cent of whom were permanently without work, with a further 20 per cent only temporarily employed. The average Muslim farm-holding in 1954 was only 11.6 hectares, providing an annual income of about £100, while the average European equivalent in Algeria was 123.7 hectares, worth some £2,800 a year. Taking all occupations into account, Muslims were earning little more than 3 per cent of what Europeans were getting. This partly reflected the fact that even in 1954 only about 20 per cent of Muslim boys received schooling, while among girls it was little more than 6 per cent. Symptomatically, the half-million Algerian Muslims working in mainland France were sending home a sum of money equivalent to a third of the total income of Muslims working in Algerian agriculture.

A major cause for concern was that the Muslim population was increasing much faster than the European. Not only did this threaten future living standards—as the history of post-independence Algeria was to demonstrate—but the resident European population feared its political consequences, not only in Algeria but in France. If France continued the fiction that Algeria and mainland France were one, and fulfilled its long-term promises of a progressive widening of the franchise, then the dominant voice in French politics could eventually become Algerian.

Various concessions had been made to the Muslim population by de Gaulle's wartime government in order to secure their loyalty. An ordinance of 7 March 1944 had theoretically opened government jobs to Muslim and Frenchman alike in Algeria, and had extended the French voting-roll to some 70,000 Muslims who qualified by virtue of various services to the State. The Muslim voting-roll was now to include all men over twenty-one; and the local assemblies which they elected were henceforth to be 40 per cent Muslim in membership. But the immediate Muslim response was to create an alliance of nationalist groups behind Ferhat Abbas, l'Association des Amis du Manifeste et de la Liberté, which pledged itself to 'une République algérienne autonome fédérée à une République française rénovée'—a commitment strongly urged by Messali Hadj's Parti du Peuple Algérien. Working-class national feeling tended to look to Hadj's PPA, while the educated Muslim middle class, many of whom had previously felt interest in the national advantages of French assimilation, became increasingly drawn to the moderate nationalism of Ferhat Abbas. Running parallel with those movements was the specifically Islamic programme of the *ulemas*.

As it happened, it was a VE day demonstration of Hadj's party which led indirectly to the Sétif massacre of May 1945, which did so much to poison Muslim–European relations in the post-war period. Exchanges of shots with the police were followed by five days of Muslim attacks on Europeans, in which over a hundred were killed and a number of women were raped. The sight of mutilated corpses, women with their breasts cut off, and men with their severed genitals in their mouths, provoked savage European reprisals, in which whole villages were bombed and 6,000 or so Muslims were killed, many of them the victims of indiscriminate white lynching. The result was that Algerian nationalism won large numbers of new adherents and sympathizers. Prominent among these were Muslim soldiers returning to Algeria from the European battlefields, including Ben Bella and other future leaders of the independence movement.

In an attempt to assuage Muslim aspirations, the Ramadier government legislated for an Algerian Assembly of 120 members to deal with local matters. Sixty of its members were to be elected by the 464,000 holders of French citizenship and by 58,000 'meritorious' Muslims—civil servants, graduates, holders of civil and military awards, etc.—while the remaining sixty were to be elected by the other 1,400,000 entitled to vote. The first elections of April 1948 were subjected to widespread rigging, with the result that critics of French policy were represented by only seventeen of the sixty Muslim members—nine from Messali Hadj's Mouvement pour le Triomphe des Libertés Démocratiques, and eight from Ferhat Abbas's Union Démocratique du Manifeste Algérien. The extent and blatancy of French manipulation alienated moderate Muslim opinion and aroused widespread criticism in the mainland press: thirty nationalist candidates were arrested during the campaign, while unwelcome results were diluted with fictitious pro-government votes, or in some cases not sent forward at all. Such patent malpractice was largely a response to the municipal elections of October 1947, which had seen considerable successes for Hadj's MTLD with its threatening slogan of 'the suitcase or the coffin' as the alternatives open to the settler population. As for the European members of the Algerian Assembly, the majority belonged to the die-hard Union Algérienne, who, together with the Gaullist RPF, occupied all but six of the sixty seats. The rigging and the uneven weighting of the electoral system ensured that liberal measures passed by the Paris parliament would be effectively blocked by the Algerian Assembly when those pertaining to Algeria came up for ratification by the necessary two-thirds majority—a provision which the *colon* deputies in Paris had succeeded in tacking on to the statute of 1947. This was the fate of several of the most important clauses of the statute itself, including the extension of full local rights to the so-called *communes mixtes*, the introduction of female suffrage, and the implementation of extensive educational and agrarian reforms. And to ensure that the filter would continue to be effective, the 1951 elections to both the Algerian Assembly and the French National Assembly were, if anything, more systematically stage-managed than those of 1948. The discovery of Saharan oil in the mid-1950s—after a decade of prospecting—was to give yet a further turn of the screw to French determination to remain the masters.

TUNISIA AND MOROCCO

French policy towards the protectorates of Tunisia and Morocco was closely conditioned by what were seen as the defence requirements of Algeria. It was firmly held in Paris that any move towards independence would have to be safeguarded by guarantees of close military and diplomatic co-operation with France. The character of the French senior administrative staff in the post-war

Maghreb varied markedly between the three main territories: whereas in Algeria it was predominantly Radical, in Tunisia it was largely Gaullist, and in Morocco it was pronouncedly Vichyite.

The French stake in Tunisia was a limited one, with settlers of French stock numbering only 150,000 in a population of over three million. It was largely recognized that neither Vichy nor the rival Algiers government had much to be proud of in their handling of Tunisian affairs during the war years. Algiers had violated the status of the protectorate by deposing the Bey, Moncef, in May 1943 for what it regarded as his over-independent political behaviour. Although the new Bey was more docile, the release from prison of the Néo-Destour leader, Habib Bourguiba, provided an important focus for nationalist sentiment. Rejecting the modest reforms offered by the French in February 1945, Bourguiba presented the Tunisian case vigorously before the Arab League and the United Nations. Unlike so many nationalist movements that were content to base their support on the middle classes and student activists, Néo-Destour was at pains to mobilize the Tunisian working class through its newly created and highly effective trade union, the Union Générale du Travail Tunisien. 1949–50 saw an optimistic period of negotiation with France, in which Robert Schuman, Bourguiba, and the new Bey, Lamine, seemed fair set, in Schuman's words, 'de conduire la Tunisie vers le plein développement de ses ressources et vers l'indépendance qui est l'ultime objectif pour tous les territoires de l'Union française' (10 June 1950). But these initiatives provoked a bitter response from the French population in Tunisia, supported by right-wing and military opinion in France: so much so that Pleven's government felt obliged to take a different tack in the autumn, angering Néo-Destour and the UGTT. Riots followed in November 1950; and by 1952 Tunisia had become a prey to rival terror campaigns by the *fellaghas* and by the settler Main Rouge. The following two years were characterized by an extraordinarily insensitive handling of the situation by the new Resident-General, Jean de Hauteclocque, who arrested Bourguiba and deported a number of the Bey's most trusted ministers. Successive governments in Paris were either too brief or too pre-occupied to assert their authority, and by the summer of 1954 it seemed that there was no middle way between full-scale war and a settlement on Bourguiba's terms.

As in Tunisia, the French presence in Morocco was relatively small, 266,000 in a population of eight and a half million; but the immediate post-war period had seen considerable investment by French business interests, alarmed at the leftward swing of politics in France and its possible implications for surplus assets in France. The French colonial lobby could always claim that the nationalist party, Istiqlal, would be swamped by feudal war-lords, who would use independence to turn back the clock. Istiqlal nevertheless launched a campaign in January 1944 for the termination of the French protectorate—

which the arrest of its leaders in the following month did little to abate. Like the Tunisian Néo-Destour, it loudly rejected the modest reforms produced by the Paris government. French unease grew in April 1947, when the Sultan, Sidi Mohammed ben Youssef, publicly spoke in praise of the Arab League. The result was the recall of the liberal Resident-General, Éric Labonne, who had initiated a promising programme of economic regeneration, and his replacement by General Alphonse Juin (15 May), who repeatedly threatened the Sultan with deposition.

Juin's successor, the equally inflexible General Guillaume (August 1951), turned a blind eye to the schemes of the Pasha of Marrakesh, El Glaoui, to replace the Sultan with a pliable puppet of the war-lords. Although the Foreign Minister, Georges Bidault, went through the motions of ordering Guillaume to restrain El Glaoui, Guillaume knew Bidault favoured the deposition of the Sultan, and acted accordingly. On 20 August 1953, the Sultan was bundled off into exile, and a motley assembly of El Glaoui's colleagues elected their puppet, Arafa, as Sultan. Laniel's government made no attempt to assert its authority: as Bidault allegedly remarked: 'How can we punish people who have succeeded so brilliantly?' Only François Mitterrand saw fit to register his disgust by resigning; and the Finance Minister, Edgar Faure, protested, but that was all.

Even so, there was increasing disquiet in France at the character of French government in the Maghreb—and in the overseas territories in general. While liberals disliked it for its repressive nature, there were conservatives who were uneasy at the spectacle it provided of weak cabinets in Paris being confronted with successive *faits accomplis* by their supposed representatives on the spot, speciously justifying their initiatives on the grounds of the need for swift action and of the difficulty of referring to Paris for instructions. The fact that ministers were as often as not only too grateful for these initiatives, which relieved them of the anxiety of making bold decisions, made the situation worse, bringing the whole authority of government into contempt. Indeed, the regime was shortly to die of it.

Unease at the colonial situation was a stimulus to the spread of a number of new liberal periodicals, notably *L'Observateur* of 1950—soon to become *France-Observateur* (15 April 1954)—and *L'Express* (May 1953). Among Catholic publications, *Esprit*, *Témoignage Chrétien*, and *Réforme* likewise shared this disquiet, differing sharply from the conservative *France Catholique*, which still saw the flag as the principal ally of the Cross in penetrating the Third World and edging out the Crescent. A significant section of the Catholic episcopate felt similar misgivings, and were aware that the Vatican was becoming increasingly attentive to Third World demands, given the changing balance and distribution of Catholic demographic strength. Prominent among Catholic critics was François Mauriac, who conducted a verbal duel with General Juin which embittered the proceedings of the Académie Française. The MRP was particularly exercised over these issues, its left wing sharing the liberals' concern over current colonial developments, while others were held back by a sense of loyalty to those of their party leaders who carried so much responsibility for the inept inflexibility of recent French conduct in the overseas empire.

The Socialists, for their part, tended to a more empirical approach, distinguishing between the feasibility of independence in Tunisia, with its 'modern' nationalism and close trade-union links with Force Ouvrière, and the difficulties in Morocco, with its war-lord ethos. The Radical party, by contrast, was split between the liberalism of a small minority of members, such as Mendès France and Edgar Faure, and the representatives of *petit blanc* colonial opinion, who had a powerful spokesman in Martinaud-Déplat, Minister of Justice—and then the Interior—in the critical 1952–4 period. With the rightward slip of French politics, and the consequent need for a Radical presence in successive cabinets, colonial interests were assured of powerful advocates at court.

On the other hand, the Maghreb was becoming an increasing liability to France's foreign policy. While the anti-imperial denunciation of the Eastern block was too familiar a part of the international landscape to count for much in French calculations, America was a different matter. Unlike Indo-China, where Communism was a potent presence, the Maghreb was not an area in which the French could count on sympathy from the Americans. Indeed, both the Tunisian and Moroccan trade-union movements had close links with the American Federation of Labor; and the Néo-Destour looked to the United States as a potential friend, an expectation which received encouragement when the United States joined in the criticism of French policy in Tunisia expressed at the United Nations following the unpunished assassination of a Tunisian trade-union leader by settler interests (5 December 1952). Fatefully, the issue of American involvement in North-West African affairs was to be a link in the chain of events that finally brought down the Fourth Republic six years later.

MADAGASCAR

France could nevertheless derive some comfort from the fact that other colonial powers were still deeply entrenched in Africa, providing some element of consolidation in complicity. This was no longer the case in Asia and the Indian Ocean. With British India on the road to independence, expectations were aroused in Madagascar of a similar future. The Mouvement Démocratique de la Rénovation Malgache, founded in February 1946, was making fast progress in local elections and among trade unions. Looked on with a benign eye by many members of the French administration, it seemed to offer the possibility of a gradual and non-violent evolution towards self-government within the framework of the French Union. However, the return to Madagascar of native soldiers who had fought in the French campaigns of the Second World War, introduced into political life a hardened element who had witnessed the defeat and humiliation of French armies, and who were both expert in using weapons and had little desire to resume the subservient existence of pre-war colonialism.

Hoping a little naïvely that America would favour their enterprise, an insurrection broke out on 29 March 1947—from which the MDRM distanced itself, and which had only limited support from the mass of the population. Some five hundred Europeans and 1,900 Madagascans were killed by the insurgents—numbers soon to be dwarfed by the war of subjugation, in which 18,000 troops accounted for 89,000 inhabitants, with all the attendant horrors of such wars. Despite its reluctance to be involved in the uprising, the MDRM was dissolved; and two of the five Madagascan members of the French National Assembly had their parliamentary immunity lifted and were imprisoned—with only the Communists and a section of the overseas deputies protesting. The repression was followed by an unofficial campaign of terror organized by the Ligue des Intérêts Franco-Malgaches—which successive High Commissioners tried to moderate but with only limited success. Dialogue was at a virtual end, even with the leaders of moderate national opinion.

INDO-CHINA

The fact that Madagascar was an island, with no immediate powerful neighbours nearby to fish in troubled waters, enabled France to preserve her position there until the collapse of the Fourth Republic. The reverse was the case in Indo-China.

During the Second World War, Indo-China had presented a complex alignment of forces. The French administration under Admiral Decoux had remained loyal to Vichy throughout its period of effective existence, steering a dexterous course between the Japanese on one hand and Indo-Chinese self-esteem on the other. Decoux had combined a modest reform programme with visible assertions of French imperial will, with the result that the governor-generalship perhaps enjoyed greater prestige among the mass of the ordinary people than it had in the pre-war era.

The various native independence movements differed in the strategy they envisaged to achieve their goal. The oldest group, the centrist republican Phuc Quoc, had looked to Japan for assistance—launching a rebellion in 1940 in the hope of Japan's support in return for the use of Indo-China as a springboard for attacks against China. But the Japanese had come to terms with the Vichyite administration, and the rebellion was suppressed while the Japanese looked on.

By contrast, the left-wing Viet Minh coalition under its Communist leader, Ho Chi Minh, pinned its hopes on China in its schemes to get rid of both the French and the Japanese. Chiang Kai-shek was glad enough to profit from such expectations, but disliked the left-wing complexion of the leadership. Arresting Ho Chi Minh in 1942, he expanded the Viet Minh into a much wider coalition, the Dong Minh Hoi, and gave it a more conservative leadership. He soon discovered, however, that the new leadership lacked Ho Chi Minh's dynamism and he was accordingly obliged to release him. Moreover, the success of Ho's followers in supplying the Americans with detailed information on Japanese troop movements won them Roosevelt's sympathy. Roosevelt was in any case increasingly convinced that Indo-China should be taken from France and put under international trusteeship at the end of the war. He was reported to have said: 'France has had the country ... for nearly one hundred years, and the people are worse off than at the beginning'. Chiang Kai-shek likewise favoured Indo-Chinese independence, since he hoped that the country would then rapidly become a Chinese satellite.

By March 1945, the Japanese feared that the declining fortunes of both Japan and Vichy would persuade the French administration in Indo-China to transfer their allegiance to de Gaulle. To forestall this, the Japanese took over government on 9 March, and encouraged the Emperor Bao-Dai of Vietnam and Prince Sihanouk of Cambodia to declare their countries independent, on the understanding that they would govern in Japan's interests. Faced with this challenge, de Gaulle's government responded with modest Brazzaville-type concessions (24 March), but these were firmly within the context of the projected French Union, and in certain respects they were less progressive than those that the Vichy administration had instituted during Decoux's governor-generalship.

Events thereafter seemed superficially to move increasingly in Ho Chi Minh's favour. The defeat of Japan in August 1945 left Bao Dai precariously exposed to the Viet Minh forces that were present in strength in the hinterland; and, recognizing where the future probably lay, he abdicated on 25 August in favour of the Republic that Ho's followers had declared in Hanoi a few days previously. But the death of Roosevelt in April 1945 had deprived Ho Chi Minh of a powerful sympathizer; and, with France a signatory of the Japanese capitulation, the prospects for the reassertion of French control over Indo-China were less bleak than they had seemed a few months earlier. Britain was prepared to support a fellow imperial power, while China, which had hitherto favoured Indo-Chinese independence, was bought off by the French surrender of her surviving toe-holds and privileges in China.

De Gaulle sent out a High Commissioner—Admiral Thierry d'Argenlieu, a close collaborator of his London days and a former naval officer turned Carmelite, who had reverted to naval service at the beginning of the war. His austere intransigence had made him a loyal and courageous colleague in wartime, but he was to show himself singularly unsuited to the delicate diplomacy that was required in post-war Indo-China.

The situation in the north was much more fraught, in that Bao Dai's abdication in August 1945 had been followed by a whole series of social reforms, instituted by Ho Chi Minh's revolutionary government which included a large number of non-Communists. Moreover, an election in North Vietnam in January 1946 demonstrated the breadth of popular support for Ho. The logic of the situation was not lost on perceptive minds in Paris, and a skilful Paris emissary, Jean Sainteny, negotiated a working agreement with Ho, envisaging virtual autonomy for Vietnam within the French Union (6 March 1946). But Thierry d'Argenlieu subsequently claimed that Cochin-China in the south, with its rich rice-growing lands, could not be included in the agreement—a claim that was quickly taken up by French commercial interests and supported by the Gaullists and MRP. D'Argenlieu rapidly set up an autonomous Republic of Cochin-China (1 June 1946), under a government that could be relied on to be subservient to French wishes; and the French Premier, the ever-*en avant* Georges Bidault, was happy enough to ratify the initiative.

The next stage in the Indo-Chinese tragedy was also the result of initiatives by men on the spot. The Viet Minh complained that French customs officials in Haiphong were carrying out functions that were in contravention of the Sainteny agreement of 6 March. Ho Chi Minh sent a telegram to the new French Premier, Léon Blum, on 15 December, proposing a settlement. The appeal might conceivably have borne fruit, given the pre-war acquaintance of the two men, when Ho Chi Minh had been an assiduous participant in the discussions of the left-wing parties in France. (He had even been present at the celebrated Congress of Tours in December 1920, which had witnessed the Socialist–Communist split (p. 46).) Blum was sympathetic to the idea of Vietnamese independence—although the balance of expressed opinion in France was not. The MRP and Radicals strongly opposed it, while most Socialists and Communists kept an ambiguous silence. But whatever potential for peace the telegram might have had was nullified by its being deliberately held up in Saigon by d'Argenlieu's officials, who were determined that their plans for a show-down should not be jeopardized by negotiations of this sort. By the time the telegram arrived on 26 December, the more intransigent wing of the Viet Minh, under Giap, had already launched an attack on Hanoi on 19 December, and Blum found himself in the familiar but disastrous position of claiming that negotiations would have to await the restoration of order. This, as the history of decolonization has repeatedly demonstrated, was a certain recipe for prolonged bloodshed.

The first part of 1947 saw a concentrated French effort to clear the towns and main highways of Viet Minh forces; but its relative success blinded officialdom to the power of the Viet Minh in the countryside. As a result, it confidently rejected Ho Chi Minh's repeated attempts to obtain a truce, assuming them to be signs of weakness. Ho had even offered to get rid of Giap, following the welcome resignation of Thierry d'Argenlieu.

Although Bao Dai had abdicated, he still had a following; and the French increasingly regarded him as a possible 'solution' for Vietnam. Bao Dai's indigenous supporters were a disparate collection of anti-Viet Minh groups, incongruously entitled the National Union Front, which principally looked to China's Kuomintang and to America. Bao Dai himself was a far more astute figure than the Farouk-like playboy of press legend. Conducting a circumspect game, he recognized that his own supporters were far less representative of popular feeling than the Viet Minh, whose acumen he secretly admired; and he was aware that his survival in Indo-China would depend on his being able to extract better terms from the French than the Viet Minh. To the fury of both the Viet Minh and the colonial die-hards, Bao Dai obtained from France 'solemn recognition of the independence of Vietnam' in return for Indo-Chinese adherence to the French Union and an undertaking 'to give primacy ... to French counsellors and technicians' (5 June 1948). Moreover, the formal agreement of 8 March 1949 guaranteed that Cochin-China would be part of an independent Indo-China, the quid pro quo being French freedom to move troops and to make use of military bases throughout Vietnam. Yet French attempts to strengthen Bao Dai's credibility as the head of an independent Vietnam were continually undermined by the bullying attitude of the latest High Commissioner, Léon Pignon—another disastrous MRP protégé, whose treatment of Bao enabled the Viet Minh to pose as the only truly independent expression of Indo-Chinese nationalism. In the meantime, neighbouring Laos and Cambodia had become nominally independent states within the French Union on 20 July and 8 November 1949.

Vietnam in the late 1940s and early 1950s rapidly became the hunting-ground of French racketeers who had ceased to find post-Liberation France a profitable terrain. Currency manipulation, trading in import and export licences, and the more traditional opium traffic created wealth for the un-scrupulous—including, it was claimed, close associates of the MRP, the Social-ists, and the RPF. The situation was complicated by the rivalry of the MRP and Socialists for administrative posts in Indo-China. Each accused the other of involvement in the widespread corruption of the time; and the campaign of mutual acrimony reached its nadir in the so-called 'Affaire des Généraux' of 1949–50, a highly complicated and inflated scandal which gave the MRP the opportunity of accusing the Socialists of passing military information to the nationalists under cover of peace-seeking moves. It led to the disgrace of Generals Revers and Mast, guilty perhaps of naïve misjudgement rather than of corruption.

The inflexible Pignon was eventually replaced by General de Lattre de Tassigny, a man who combined military ability with political realism. While reassuring Vietnamese national aspirations, he also worked to convince the

Americans that large subsidies to the French forces were the best guarantee against a Communist take-over. American concern over Ho Chi Minh's popularity had grown steadily since January 1950, when Communist China and Russia had formally recognized the Viet Minh as Vietnam's legitimate government. Indeed, by the time of Tassigny's death in January 1952, America had already provided a third of the 830,000 million francs that France had spent in fighting the Viet Minh. With the Vietnamese war now taking 40 per cent of the French defence budget, American help had become indispensable; the final four years of the French presence in Indo-China were to see the United States bearing 80 per cent of military and development costs—a sum equivalent to more than a year's total French revenue.

But even this was not enough. Government counter-propaganda was difficult in a country where literacy was low and radios scarce, while the redistribution of abandoned land, set in motion in June 1953, came too late to be an effective pro-French factor in competing for native allegiance. As Mendès France pointed out in October 1950, the only alternatives were either to negotiate with Ho Chi Minh or to triple military and economic investment. But conservative governments, committed to financial caution on the one hand and to their right-wing supporters on the other, felt unable to do either. And for the left-wing parties, the Indo-Chinese war was 'la sale guerre', destroying villages and killing thousands of peasants in defence of the rich businesses and shady speculators of the cities. With France committed to Vietnamese independence, albeit within the tenuous links of the French Union, a growing number of voices from both halves of the political spectrum questioned the purpose of continuing the war at such expense. It was argued that America was in a far stronger financial position than France to worry about whether an independent Vietnam was ruled by Bao Dai or the Viet Minh, and that the logic of the situation was to shuffle on to America as much of the responsibility for Indo-China as possible.

In the context of these growing if unwritten assumptions, French military action in Vietnam was no longer an attempt to reconquer the kingdom, but rather a holding operation, designed to ensure that the situation handed over to America should reflect some credit on French military capability. A rapid deterioration was feared as a result of Viet Minh gains in the north-west frontier regions, which greatly facilitated the transport of supplies from neighbouring countries and threatened the security of Laos. And it was to check this development that Generals Navarre and Cogny established by airlift an advanced post at Dien Bien Phu (20 November 1953), despite the scepticism of some of their seniors. The result was disastrous. Encircling the French positions at Dien Bien Phu, Giap launched an all-out attack that brought about a French surrender on 7 May, with the loss of nearly a tenth of their total forces in Indo-China. The trauma was all the greater for Dien Bien Phu being the only set battle in the whole process of French decolonization; it became the equivalent of the Italian defeat of Adowa in the collective memory of the French army, generating similar neuroses and refusals to face reality.

During the weeks preceding the fall, Georges Bidault had approached the Americans for air support—a discussion which even included the hypothesis of using nuclear weapons. But Churchill warned against the consequences of internationalizing the Indo-Chinese struggle, and advised America to leave the matter for the forthcoming Genevan conference to resolve. Needless to say, Dien Bien Phu was the death-knell of the Laniel government; and Georges Bidault's handling of Asian affairs was the subject of prolonged and bitter parliamentary criticism. Prominent in the attack was Pierre Mendès France: so much so that when the government was eventually brought down on 12 June, it was to Mendès France that the nettle was passed.

12

Janus and Cassandra
The Mendès France Ministry and
What Followed, 1954–1955

THE eight brief months of the Mendès France ministry were among the most productive of the Fourth Republic. They were a tantalizing glimpse of what the founders of the regime had hoped it might become. And, unusually for a ministry that had just fallen, the public's epitaph on it in February 1955 was that it had a better general record than any of its predecessors since 1946.

The Janus-like image of Mendès France sprang from the energetic individualism of his style—which looked forward to the Fifth Republic, when the executive would boldly take initiatives over the heads of the party leaders—but his attitudes also embraced much of the conventional outlook of the Republican tradition, as embodied in his own party, the Radicals. Many politicians found him disconcerting. His quick, bird-like movements and inquisitively cocked head put one in mind of a blue tit about to tackle a tough milk-bottle top. The party chiefs resented his attempts to choose ministers on an individual basis without recourse to their parliamentary masters—an approach reminiscent of Gaullism. He incurred the traditional animus that the parliaments of the Third and Fourth Republics harboured against strong men who might challenge their domination. They were suspicious of his radio broadcasts to the nation, which, although inaugurated by the safe and respectable Antoine Pinay, suggested an attempt to appeal to the people over the heads of the politicians. They were likewise suspicious of his economic think-tank, which included future luminaries such as Giscard d'Estaing, and seemed yet another example of the government preferring the advice of non-mandated experts to that of parliament and the party chiefs. Of a piece with this image, was his practice of rapid personal visits to the scenes of controversy or crisis, and making them the occasion for bold statements of policy. It was likewise characteristic of him that he should personally take the current hot-potato portfolios—first of all foreign affairs, and then the economy. His above-party stance was also reflected in his continuing to press through legislation with alternative majorities—the 'for', 'against', and 'abstention' columns in the parliamentary record of his administration show remarkable changes in party voting from issue to issue.

None were more disconcerted than his own party. A Radical who rose above party and local ties was at marked variance with their traditions—the more aggravating in that he paradoxically combined it with an insistence on greater discipline within the party. His image as an energetic reformer more than doubled the working-class support for the Radicals—which increased from 5 per cent to 11 per cent by the time of the 1956 election. Yet it was precisely his role as maestro of a state-planned, mixed economy that most alarmed the individualist, small-business ethos of many of his party members. Although firmly committed to the *laïque* inheritance of the Third Republic, his pro-gressive policy won the Radicals new adherents among those left-wing Catholics who would have felt out of place in the SFIO, with its Mollet-style neo-anticlericalism.

Mendès France was invested as Premier on 17 June 1954 by an astonishing majority of 419 votes to 47, the opposition and abstentions coming mainly from the conservatives and the MRP. The Communists were giving their positive support to an investiture for the first time since 1947, following their decision at the Drancy conference of 1953 to support progressive ministries. Their support was all the more remarkable, in that Mendès France had announced in advance that he would not count their votes as contributing to the requisite absolute majority—a rejection which caused Jacques Duclos to refer to him in the lobby as a 'gutless and cold-footed little Jew'. Mendès France's motive in discarding them was to reassure the United States and secure close American support in the forthcoming Geneva discussions, as well as to win the votes of the anti-Communist section of the National Assembly.

Not surprisingly, the Radicals provided the largest contingent of ministers in his government, notably Edgar Faure at Finance; but he had enemies in the colonial lobby, particularly Martinaud-Déplat and René Mayer. On the other side of the political spectrum, Mendès's above-party style was privately appreciated by de Gaulle; and the government initially contained four Gaullists, including Jacques Chaban-Delmas (Public Works). But Mendès's policies on the EDC and on German rearmament were major stumbling-blocks to sustained Gaullist co-operation. Among his enemies, the MRP were alienated by his traditional anticlericalism and his hounding of Bidault over Indo-China; but it was essentially the EDC issue that was to embitter their relations. Yet the rank and file of the MRP were broadly sympathetic to his economic and colonial policies; and it was symptomatic of this attitude that the Overseas Ministry should be held by the MRP deputy, Robert Buron.

INDO-CHINA

At his investiture, Mendès France promised to resign if he did not succeed in securing a Vietnamese peace settlement within the month. The undertaking impressed both the Assembly and the general public, who realized that northern

Indo-China could only be held by massive injections of money and men, including conscripts, who had hitherto been left out of 'la sale guerre'. This was a price that few politicians were prepared to envisage and which public opinion would certainly oppose. The war had already cost the French empire 92,000 dead, 114,000 casualties, and 28,000 captured. Whereas an opinion poll in July 1947 had revealed 52 per cent of the public in favour of fighting to keep Indo-China, the proportion had dropped to a mere 7 per cent by February 1954. The Viet Minh, for their part, were likewise disposed towards a negotiated settlement by the escalating expense of the war, while their Russian and Chinese mentors were seeking more stable relations with Washington, following the Korean confrontation (p. 241). The Communist powers were also conscious that a prolonged Vietnamese conflict might further strengthen the prestige of the Viet Minh and make it less amenable as a future satellite.

There was no problem about the formal recognition of the independence of Vietnam, Cambodia, and Laos, but the immediate future of Vietnam was a major sticking-point. The Viet Minh wanted an election throughout Vietnam in six months time, fully confident that they would gain 80 per cent of the votes. Fearing that their prediction was all too likely, Mendès France was anxious to postpone elections as long as possible. The Americans, for their part, wanted to go further and postpone an election indefinitely—their aim being to create in the meantime an independent South Vietnam which could eventually become the launching-pad for a reconquest of the north. While Mendès France felt no personal enthusiasm for reopening hostilities with the north, his prime concern was French withdrawal from her costly entanglement with Indo-China—which in practice meant leaving the subsequent responsibility for what might happen there to America. Faced with pressures for a rapid settlement, the Viet Minh reluctantly agreed that a two-year interval should elapse before the all-Vietnam election, and that in the meantime the country should be temporarily divided along the seventeenth parallel, with the north in the hands of the Viet Minh, and the south entrusted to Ngo Dinh Diem, who had been Bao Dai's chief minister. The dividing line of the seventeenth parallel represented a major concession on the Viet Minh's part, since it was not only less logical than the thirteenth, which they proposed, but it gave South Vietnam the strategic air-base of Tourane—the main objective of French haggling.

With the Geneva proposals secured within the month, and French conscripts no longer threatened with service in Vietnam, the National Assembly expressed its approval by 462 votes to 13 (23 July 1954). Thereafter, the French disengagement proceeded very rapidly. Following Franco-American agreements on 29 September, the United States formally joined France on 13 December in the campaign against Viet Minh guerrilla activity in South Vietnam. Mendès France was well aware of the unpopularity of Diem, whose position stemmed from the previous Bao Dai arrangement; and he was also well aware that the Geneva settlement resolved nothing, other than enabling France to step out of the back door. In this unedifying game of passing the parcel-bomb, it was now up to the Americans; and all the evidence suggested that their intentions were a recipe for yet greater misery in Vietnam. The tragedy was that a more flexible French attitude to Ho Chi Minh in the 1940s might have saved the Indo-Chinese at least part of this *via dolorosa*.

WESTERN EUROPE—PAST AND PRESENT

It was one of the many ironies of Mendès France's brief ministry that, like de Gaulle, he should be principally remembered, and with most gratitude, for what opponents called 'une politique d'abandon'. It was all the more ironic in Mendès France's case, in that the withdrawal from Indo-China was a much simpler and less hazardous undertaking than de Gaulle's disentanglement from Algeria. Western European defence, by contrast, was bound to burn the fingers of whoever handled it; and yet Mendès France's conscientious courage in see-ing through a scheme he did not particularly like got scant praise either at the time or subsequently.

Like his predecessors, he inherited a difficult package of unresolved issues from the past, which had their origins in the Brussels Treaty of 17 March 1948. Despite the modest understatement of its self-declared aims, the Brussels pact was in effect the nucleus of NATO, in that its signatories, France, Britain, and the Benelux countries, were to link themselves only a year later with the United States, Canada, Norway, Denmark, Iceland, Portugal, and Italy in the North Atlantic Treaty of 4 April 1949. The most disquieting aspect of this military alliance, from France's point of view, was its patent numerical inadequacy, and the growing probability that other members would see German participation as the only solution to its shortcomings. The nine divisions of France, Britain, and Belgium put at the disposal of the Brussels pact of 1948 had become twelve under NATO, plus a thousand aircraft—but these faced twenty-seven Soviet divisions and six thousand aircraft. And, embarrassingly for France, there was now in existence the Federal Republic of Germany, able and willing to play a role—the fruit of the Washington agreement of 8 April 1949. Admittedly, France could take some sour satisfaction from the fact that the new German state did not include the Russian zone, which was following its own independ-ent course as the German Democratic Republic (7 October 1949). It was also true that the Sarre had remained a separate political entity since 1947, with all of its coal production put at the disposal of the French with effect from April 1949. But the outbreak of the Korean war in June 1950 deepened pessimism over the future of East–West relations; and America made it clear to Britain and France in September 1950 that she wanted a German contingent within an integrated NATO force within a year.

Faced with American determination—supported by the British—and the deeply felt anxiety of other NATO members, Premier René Pleven felt that he had no option but to put forward a proposal of his own; one which would maintain the fragile unity of his government. So, Robert Schuman and the ever-resourceful Jean Monnet set about devising a military equivalent of the Coal and Steel Plan, with the aim of controlling the German military beast if its birth was unavoidable. And the bold scheme they sent to parliament on 24 October 1950 envisaged a European army, responsible to a European Assembly and a joint Minister of Defence. To the relief of its architects, it was accepted by 343 votes to 225, with the predictable opposition of the Communists and Gaullists. But although France saw a European army, with a stiff supranational safety-catch, as preferable to the American proposal of September 1950, there was only a lukewarm welcome from the other NATO states. From the outset, the British refused to participate. After much muttering and amendment, the version of the plan which eventually made its way into the Paris Treaty of 27 May 1952 was a very loose-limbed counter-scheme to the carefully controlled apparatus of Monnet and Schuman. Gone was the joint Minister of Defence and several of the other safeguards of supranational control. And if its twelve German divisions were a modest total compared with the 136 which had swept into France in 1940, the French contingent of fourteen divisions was also modest. Shorn of its superstructure, the scheme was no longer the concept that Pleven's team had so painfully and skilfully sold to parliament. Consequently, and understandably, successive governments were reluctant to bring the treaty to the Assembly for ratification; and in the country at large, opinion hardened against it. Whereas in May 1953, 30 per cent of those polled were in favour of the EDC and only 21 per cent were against it, public discussion in the summer of 1954 saw hostile opinion predominate.

Mendès France himself felt no particular enthusiasm for the EDC. As Herbert Lüthy observed: 'during the two years in which the EDC had been the centre of controversy, he had never taken up a definitive position towards it. It was a subject which did not interest him; the idea and ideology behind it were alien to his realistic mind . . .'[1]—just as later he was sceptical of France's advantages in joining the EEC, and later still was suspicious of a President of the Republic elected by universal suffrage. His realism was by no means infallible. Had Britain been a member of the EDC, his scepticism might have been less marked. Yet he recognized that the matter could no longer be shirked; and he tried hard but unsuccessfully to obtain changes in the EDC proposals that might facilitate their passage through the National Assembly. Among other modifications, he wanted to delay the supranational features that most offended French national sensibilities; and he wanted to prevent the possibility of German troops being stationed on French soil. But his slender chances of obtaining these changes were further narrowed by the MRP, when the party confidently assured other nations that the EDC plan would get through parliament as it stood. In fact, the National Assembly rejected it by 319 votes to 264 (30 August 1954).

The MRP were furious that Mendès France had not made the vote a matter of confidence backed by the threat of dissolution—a constitutional weapon that was theoretically at his disposal until November 1954, when the eighteen-month period of permitted usage would elapse (p. 145). But Mendès France was insufficiently attached to the EDC to risk his credit in this fashion. There was no guarantee that the threat would work, and he had already lost his Gaullist ministers by acting as midwife to the ill-fated project (14 August 1954). Even so, the MRP were never to forgive his refusal to gamble on the matter.

Apart from Italy, France was now the only country not to have ratified the EDC. It was widely and loudly hinted in international circles that Germany would have to be rearmed whether France liked it or not; and, sensing the inevitable, Mendès France wearily gave his support to Anthony Eden's plan to breathe new life into the old Western Union, established by the 1948 Brussels Treaty. The admission of Germany and Italy into the union—and Germany into NATO—would at least have the virtue of ensuring some element of international control over a future German army, however slight; and it would perhaps make the prospect of German rearmament less unpalatable to the French Assembly. The scheme had the further advantage of involving Britain—which the six-power EDC had not—and thereby guaranteed the continued presence of British forces on the Continent until the expiry of the Brussels Treaty in 1998.

It was signed in Paris on 23 October 1954, and then gingerly put to the French National Assembly on Christmas Eve. Once more, suspicion and intransigence won the day; parliament rejected it by 280 votes to 259, with the embittered MRP joining the ranks of obstruction. Conscious of France's increased isolation, Mendès France bitterly turned the MRP's former advice to him against them, and made the issue a matter of confidence. To add to the MRP's gall, the manœuvre worked: parliament passed it by 287 votes to 260 six days later, thanks to the conversion of a sizeable minority of Gaullists and a number of hitherto hesitant Radicals and conservatives. But the narrowness of the victory disturbed France's allies. Churchill warned Mendès of the danger of 'the empty chair' should exasperated neighbours feel obliged to cut France out from the deliberations of the NATO countries. Once again, France was made to realize the limits of the options open to her in the international field.

DOMESTIC POLICIES

It was characteristic of Mendès France's bifocal vision that he should seek to strengthen the executive while wishing to restore some features of the Third Republic. Remembering his own narrow failure to become Prime Minister in 1953, he amended the constitution on 7 December 1954 to make a simple majority sufficient to invest an incoming Premier—instead of an absolute majority as before. But the prospective Premier had to make known in advance the composition of his cabinet. The same law also empowered the Prime Minister to continue as caretaker in the event of a dissolution of parliament—instead of the President of the National Assembly as hitherto. Although this change did not significantly increase the likelihood of governments in France resorting to dissolution, it did at least remove one of the inhibiting factors from its path: the fear that an uncongenial President of the Assembly might suddenly become head of government.

The law's other major changes were more controversial—and scraped through by only two votes. Mendès France wished on the one hand to restore to the Council of the Republic some of the functions of the previous Senate, while limiting its obstructive powers on the other. The Council regained its right to initiate legislation that did not affect government finance, and the old *navette* system was reintroduced into the transfer of bills from one house to the other. Bills could now shuttle backwards and forwards between the two houses, without the need for an absolute majority in one house to counter the amendments of the other. A time-limit of a hundred days ensured the system against abuse by filibusters, and in the case of unresolved disagreement, the Assembly's version of the text was final. In effect, the new procedure gave the Council increased opportunity to air its views, in return for the abolition of its former power to consign bills to limbo (p. 148). The new system worked well during the rest of the Fourth Republic, all but six bills becoming law before the elapse of the time-limit. The most noteworthy tussle was to be over Guy Mollet's old-age pension fund in 1956, which conservative and Radical elements in the Council tried to emasculate by amendment. The Assembly determinedly upheld its version, and it eventually passed on the fourth reading, three days before the hundred days elapsed.

Like de Gaulle in 1958, Mendès France thought that the pre-war electoral system of *scrutin d'arrondissement* was a better guarantor of stability than the existing *scrutin de liste*; and after he fell from office he argued vigorously for its reinstatement, until the carpet was pulled out from under his campaign by his opponent and former colleague, Edgar Faure (p. 249).

If historians think of Mendès France as Janus, he was Cassandra to contemporaries. His oft-quoted dictum of 1953, 'to govern is to choose', was associated with a readiness to take unpopular decisions; and, as in 1944–5, he believed that the economy required a number of hard, painful resolutions. In order to side-step obstruction in parliament, he asked for special economic powers, which he was accorded by the remarkable majority of 361 to 90 (10 August 1954)—the main opposition coming from conservatives, with the Communists abstaining. The deep suspicion with which he was regarded on the Right was partly the product of his supporters' over-enthusiastic publicity, notably *L'Express*, which presented him as a Popular Front or at least a New Deal figure, which was disturbing not only to the conservatives and the right wing of the MRP, but also to the more cautious members of the SFIO.

Perhaps the most contentious of the measures that he took under his special powers were the decrees of 20 November 1954, aimed at reducing alcoholism and limiting the privileges of home-distillers. The alcohol lobby in France was enormous, involving well over five million people. Prominent among these were the three million home-distillers ('bouilleurs de cru'), fruit-growers who were entitled to distil a given quantity of alcohol supposedly for their own domestic consumption and consequently exempt from tax. Comprising half the adult male population in twenty *départements*, and a quarter in forty others, they represented a sizeable slice of the electorate, and were not to be treated with impunity. In reality, they sold three-quarters of their produce on the unofficial market; and Mendès France proposed to limit tax-free home-distilling to full-time farmers, cutting out the amateur gardeners, holiday-property owners, and other non-professional beneficiaries. Another decree of 20 November subjected bars to stricter controls, earning the hostility of half a million bar-owners and -tenders, while a further measure reduced the amount of surplus sugar-beet that the government was prepared to buy at guaranteed prices for alcohol production. 150,000 beet-producers and the distillers they supplied were up in arms—Poujadism undoubtedly gaining many converts among them (p. 251). Despite the approval of various other sectors of the general public, none of these decrees survived more than a year, neither Edgar Faure nor parliament having the courage to brave the anger of the alcohol lobby. As on other issues, *L'Express*'s well-intentioned publicity, portraying Mendès France as a milk-drinking crusader against alcohol, created panic and resentment among a large proportion of the rural population, including the one and a half million wine-growers, who otherwise had fewer privileges to lose than the other sections of the alcohol lobby.

This apprehension spread rapidly to urban segments of the propertied classes when Mendès France relinquished the Foreign Ministry to take on Economic Affairs himself and simultaneously conferred the Ministry of Finance on the left-wing MRP deputy, Robert Buron (20 January 1955). The Ministry of Finance was usually a reliable brake on social generosity, but even under Buron's predecessor it had already allowed the government to double state loans for the building of 350,000 new working-class homes a year; and 12,000 extra teachers had been appointed in the space of six months to improve the staff:pupil ratio. The Right looked hard for an opportunity to destroy the government—and victory came in less than three weeks.

NORTH AFRICA

It was both ironic and a sign of the changing preoccupations of French politics that it should be North Africa that gave the Right this opportunity. Ironic, in that Mendès's North African policies did not represent a significant departure from the theoretical objectives of previous ministries. Both Mendès and his Minister of the Interior, François Mitterrand, affirmed their commitment to 'Algérie française' in parliament (12 November 1954), and the next two months saw French troops there increased from 57,000 to 83,000. Similarly, Mendès's pledge in Carthage that France would respect Tunisia's internal sovereignty (31 July 1954) was coupled with an insistence that the two countries follow a common international policy. And his Moroccan policy was nothing if not cautious. He eschewed the prospect of restoring the old Sultan, preferring to try to establish a working rapport with his successor, justifying his circumspection on the traditional grounds that the feudal nature of Morocco did not lend itself to the same policies of rapid evolution as Tunisia. Moreover, reports of ill-treatment in Moroccan prisons brought protests from François Mauriac and other liberals who were normally enthusiastic supporters of Mendès.

But what disturbed the Right about Mendès's North African policies was that they threatened to make the government's theoretical aims a reality. The Right had hitherto relied on the men on the spot to make a nonsense of these aims; but Mendès replaced several of them with individuals whom he considered more responsive to government intentions yet unlikely to offend *colon* opinion. Although Mendès inherited a seeming impasse in Tunisia (p. 229), the change of atmosphere was shortly to be reflected in the fact that the still-imprisoned Bourguiba described Mendès's Carthage speech as a significant step in the direction of independence. It was precisely this growing optimism that caused René Mayer and other North African Radicals to denounce the direction affairs were taking, their great fear being that the rapid evolution of Tunisia towards independence would encourage large-scale insurrection in Algeria (pp. 230–1). Indeed, the newly formed Front de Libération Nationale (10 October 1954) launched a campaign of co-ordinated violence in Algeria at the beginning of November. Messali Hadj's MTLD had recently been suffering from deep differences of opinion between its older and younger generations, and its proscription by the government had strengthened the hand of the militants. Nor did Mendès's appointment of Jacques Soustelle as Governor-General of Algeria (25 January 1955) necessarily reassure settlers, who feared a rapid escalation of unrest and an eventual sell-out in North Africa. Thought of as a liberal, with roots in the Comité de Vigilance des Intellectuels Antifascistes in the 1930s, Soustelle seemed in 1955 to augur reform and evolution, despite his Gaullist toughness. It was the settlers' misgivings that gave Mendès's enemies the chance they sought. On 5 February 1955, the Assembly refused to place its confidence in the government's North African policy by 319 votes to 273—the majority consisting of conservatives and some twenty Radicals who were as much opposed to Mendès's economic programme as to his colonial policies. Completing the opposition was the predictable personal animus vote of the MRP and the Communists.

Continuing his role of Cassandra to the last, the defeated Mendès took the unprecedented step of castigating the Assembly for its irresponsibility. While many regretted his passing—some seeing it as the death of the Fourth Republic's last chance to fulfil its early ideals—few realized that it was the end of Mendès's ministerial career. For the next twenty years, the return of Mendès to the Matignon was to be a hope held by a large segment of opinion both in France and abroad.

EDGAR FAURE TAKES OVER

One clear lesson of Mendès's fall was that colonial issues were intruding increasingly into French domestic politics—if only as a weapon to settle scores over other issues. The strong settler presence in North Africa meant that it could not easily be shaken off like Indo-China in 1954, when the cost of defending it grew beyond what the taxpayers and parents of conscripts would stand. Moreover, in a multi-party system, the colonial lobby could be a useful secondary plank in a party's electoral and financial support, giving the *colons* an influence beyond their numerical strength. This factor was to become particularly important in the context of the hung parliament of 1956; but it was already evident in 1955.

The next Premier, Edgar Faure (23 February 1955–24 January 1956) knew how to conciliate. His rivalry with Mendès within the Radical party aroused the sympathy of the MRP, whose support he rewarded with four important ministries, while a further four portfolios went to the Gaullist Républicains Sociaux. The conservatives received five ministries, including the Quai d'Orsay which went to Pinay; and the Radical nucleus kept four for itself. Faure's rivalry with Mendès France gave him a somewhat Salieri-like reputation in the annals of the Fourth Republic, eclipsing the similarity of many of their aims and overshadowing Faure's very real achievements in the matter of economic growth and North Africa's evolution towards independence. Indeed, it was his attempts to assert a realistic Moroccan policy which lost him the support of the Right.

The Tunisian issue, by contrast, was resolved with relatively little acrimony. Faure entered into personal discussion with Bourguiba, recently released from prison, and a formal agreement was signed with the Acting Premier, Tahar Ben Ammar (3 June 1955)—after which Bourguiba was able to return home in triumph to secure his political inheritance. France retained a say in Tunisian foreign policy and a military base in Bizerta. And it was a measure of the real issues behind the defeat of Mendès France that Faure's consummation of Mendès's work in Tunisia should be given overwhelming approval in parliament by 538 votes to 44 (9 July 1955).

Morocco was a different matter. There a liberal solution was solidly opposed by a powerful combination of French business interests, administrators, and police, prepared to resort to organized terror to protect their entrenched positions. The fact that the murder—by police agents—of the liberal newspaper-proprietor, Jacques Lemaigre-Dubreuil, still went unpunished was only one example of this impunity. Incensed by this situation, Faure appointed the liberal Gaullist, Gilbert Grandval, as Resident-General (20 June 1955); but he was systematically obstructed by the French forces of order, blatantly supported by two of Faure's own Gaullist ministers, Koenig and Triboulet, until the weary Faure replaced Grandval with the traditionalist Boyer de Latour, former Resident-General in Tunisia.

Nationalist activism escalated, however, and in the United Nations there was growing criticism of French policy. Faced with a worsening situation, Faure took the step that Mendès had feared to take. Reversing the situation that Laniel and Bidault had complaisantly allowed to materialize, Faure persuaded the puppet Sultan, Arafa, to abdicate in favour of the former Sultan, Ben Youssef, whom the die-hards had deposed in 1953. This long-overdue reparation of a major mistake was the key to the deadlock. Ben Youssef emerged from exile in Madagascar to sign a somewhat shadowy Tunisian-type agreement with Faure on 6 November 1955, and ten days later returned in splendour to Morocco as Mohammed V.

The response of the Right in France was much more bitter than it had been over Tunis. Koenig and Triboulet resigned in disgust (6 October 1955), and a large number of Gaullist and conservative deputies who had voted for the Tunisian settlement came out in open opposition to its Moroccan counterpart, albeit without affecting the outcome (9 October).

The evolution of Algerian affairs under Faure offered nothing in the way of hope—but not for want of good intentions on the government's part. Under Soustelle, plans were made to expand education, public-works building, and agrarian reform, while Arabic was given obligatory status in Muslim schools. Particularly imaginative was Soustelle's creation of the Sections Administratives Spécialisées—the Képis Bleus—some four hundred army detachments, designed to give agrarian and medical advice as well as military protection to distant Muslim villages. Soustelle had also made various attempts to establish contact with the FLN, through intermediaries such as Germaine Tillion and Vincent Monteil; but the spring of 1955 saw a hardening of FLN attitudes.

A decisive and tragic change in Soustelle's policies came about on 20 August 1955, following widespread Muslim attacks on Europeans, including the killing and mutilation of women and children. The immediate effect was to convert Soustelle to the view that there could be no negotiation before order had been fully established. With intransigence on both sides, the next month found the majority of Muslim moderates in the Algerian Assembly denouncing the French policy of integration (26 September 1955), while the hitherto circumspect Ferhat Abbas threw in his lot with the FLN. All was set for disastrous conflict—and a recent major discovery of oil at Edjelé dramatically raised the stakes.

As with the two preceding governments, these issues of empire were to be major factors in the demise of the Faure ministry—even if the connection was less direct than before. The government's domestic record was not without merit. Substantial increases in industrial performance and in governmental financial stability were offset by bitter shipyard strikes in the autumn of 1955, requiring large wage-rises to end them. On the other hand, the government had reason to be pleased with the Renault agreement of 15 September 1955, in which the work-force gave a two-year undertaking not to strike in return for a yearly 4 per cent pay-rise, plus confirmation of their recently won social benefits, including three weeks' holiday with pay.

The Algerian situation, however, was ultimately the government's undoing. Its increasing gravity persuaded Faure to bring the French legislative elections forward from June 1956 to January (20 October 1955)—the aim being to strengthen the French position by confronting the Algerians with a new parliament and government, freshly armed with the people's mandate. Yet it was interpreted by many Mendès supporters as an attempt to pre-empt their plan to introduce *scrutin d'arrondissement* before the next election, while deputies of all parties were furious with Faure for confronting them with the expense and uncertainty of an early election. His days were now numbered.

The forces of vengeance, however, miscalculated their response and unintentionally defeated him by an absolute majority instead of a simple one (29 November)—thereby handing him an unexpected constitutional weapon (p. 145). The outcome was that Faure was able to strike back by declaring a dissolution of parliament (2 December), since this was the second such defeat within eighteen months. Stunned by his audacity in resorting to a sanction unused since 1877, the Radical party promptly expelled him. But no French party was strong enough to be able to discard able men for long—only the PCF, roaming in the wilderness, could afford gestures of this sort—and Faure was shortly to be back in the party ranks. Even so, France was now faced with an election.

13
Impasse
1956–1958

A major worry of the established parties was how the new force of Poujadism would fare in the 1956 elections, and what effect it would have on their own fortunes. Poujadism was born in the summer of 1953, mainly among the lower-middle classes, notably shopkeepers and small farmers. By contrast, the social strata above them continued to look to the established conservative parties and to the bodies that represented bourgeois economic interests, the CNPF, FNSEA, and CGPME. Politically, they preferred Pinay to de Gaulle; and it was the relative weakness of organized labour, split between the Communists and the Socialists, that left them without a stronger inducement to create a sturdier instrument of political influence. Only the lower-middle classes of shopkeepers and small farmers, faced with the economic competition of larger units, attempted to create such an instrument—and the resultant Poujade movement had very limited success.

In July 1953, Pierre Poujade marshalled his fellow shopkeepers in the Lot against the growing investigations of tax inspectors—an operation that brought into being the Union de Défense des Commerçants et Artisans four months later. Poujade's links with the extreme Right were strong and varied. Son of an Action Française adherent, he had been a member of Doriot's youth movement in the 1930s and then of Vichy's. This was balanced, however, by his subsequent escape to North Africa, whence he joined the RAF—thereby enabling him to stand as an impeccable Gaullist in post-war local politics.

The nation of Napoleon was now a land of small shopkeepers—one for every fifty-four inhabitants in 1956, compared with an average of one for seventy-one elsewhere in Europe. Their number had grown substantially in the past decade, partly and paradoxically as a result of inflation, which had enabled shopkeepers to sell at high prices, in advance of expected price-rises, but pay off their debts to wholesalers and other creditors at the level that was current at the time of the initial purchase. Restricted travel facilities during the 1940s made customers increasingly reliant on local shops—a situation that had been reinforced by the rationing system which obliged customers to be registered with their local dealer for certain basic goods. The levelling out of inflation in the 1950s, the derestriction of food supplies, and the general improvement in communications removed most of these temporary advantages, while the expansion of the economy encouraged the growth of larger retail businesses which could undercut the small dealer. It was in the context of this clouding perspective that more meticulous tax inspection was particularly resented.

The strongholds of Poujadism were in the south-western half of France—what was later to be called the French desert—where rural depopulation restricted the small retailer's market. Shopkeepers, *bouilleurs de cru*, and impoverished peasants, especially in wine-growing areas, provided the electoral forces of Poujadism. Most were renegades from right-wing parties, notably the Gaullists; others came from the Radical party, and a significant number from the PCF, a simple case of exchanging one form of protest for another. It was backward-looking, the nostalgia of the little man for the times that favoured him, spiced with a sentimental appeal to the 'fraternité française' of the Revolution, the First World War, and the small-town society of the Third Republic. It was a bundle of negatives, asserting 'enough is enough', attacking parliamentarians as a class, but not necessarily parliament itself. 'Sortez les sortants' was a clear enough battle-cry, but the demand for the calling of an 'estates general' was little more than a slogan, with no consecutive programme attached to it. Yet, for all the rhetoric, Poujade did not favour *coups d'état*—something which neither his opponents nor many of his supporters sufficiently appreciated. Its policy of disrupting political meetings mainly took the form of vocal demonstrations rather than violence. Particularly resented was the Dorey amendment of 14 August 1954 to a finance bill, making resistance to tax inspection liable to imprisonment.

In the meantime, Poujade found sympathizers among wine-growers and stock-breeders, who were demonstrating against foreign imports and against Mendès France's measures to restrict alcoholism. The fact that Mendès France was a Jew compounded his eligibility for Poujadist attack, the little man against the (metaphorically, if not physically) big Jew becoming yet again a feature of current political rhetoric. The early months of 1955 found Poujade in his shirt-sleeves haranguing crowds against the Dorey amendment, and putting in a provocative appearance in the public gallery of the National Assembly with a similar sartorial insouciance. The PCF quickly recognized the troublesome potential of Poujadism and gave it various forms of encouragement, until the autumn of 1955, when the proximity of the election predictably made the party and Poujade sworn enemies.

THE 1956 ELECTION AND THE MOLLET GOVERNMENT

Many publicists claimed that the 1956 election represented a choice between Mendès France and Pinay. This was not how the deputies intended it to be; and, as always, they got their way. But, also as always, they got it through the traditional negative means of preventing something happening, without being able to secure a clear-cut alternative that they positively wanted. In practice, the election became a confrontation between Faure and Mendès—an absurd situation, given the proximity of their views on a wide variety of issues. Instead of a broad Centre majority, flanked by hostile extremes, as had emerged from the 1951 election, the political fault-line of the 1956 campaign ran through the Radical party. The expelled Faurists took over the Rassemblement des Gauches Républicaines (RGR), and headed an alliance of MRP, conservatives, and Gaullists—while opposite them stood the self-styled Republican Front of Mendès's Radicals and Mollet's Socialists, backed by a number of left-wing Gaullists. The Poujadists and Communists lay beyond the pale on either side.

The splitting of the Centre made vote-catching alliances more difficult, with the result that only eleven *départements* saw an alliance scoop the pool by getting an absolute majority—compared with thirty-nine in 1951. While the Right got ten of these, the competition of the Poujadists prevented it getting more. Consequently, the mechanics of *départemental* proportional representation were brought into play on a much wider scale than in 1951, causing the percentage of seats gained by each party to bear a much closer relationship to the percentage of votes it gained in the country at large. Thus the Communists, who obtained around 25.9 per cent of the vote in both the 1951 and 1956 elections, won 26.8 per cent of the seats in 1956, as against a mere 17.5 per cent in 1951 (Table 13.1).

The newcomers to politics, the Poujadists, split into three occupational sections in order to compete more directly for the votes of aggrieved sectors of the economy; but they then reunited as electoral alliances in many *départements* so as to exploit the advantages of the electoral system. This stratagem in fact contravened the 1951 electoral law, which forbade parties to present more than one list in the same constituency; and it resulted in the confiscation of eleven of their fifty-one seats. They could nevertheless take comfort from their surprisingly large poll (11.6 per cent), in which they drew votes even from the Communists.

The Mendès Radicals doubled their share of the poll to nearly 10 per cent, their chief gains being in the industrial north-east and Paris basin, where Mendès's appeal to an important section of the working class was clearly demonstrated. Gaullism was among the victims of this success. But the Gaullist drop from 21.7 per cent to 4.4 per cent of the poll reflected not only the inroads of Mendèsism and Poujadism, but more particularly the withdrawal of de Gaulle into memoir-writing at Colombey-les-Deux-Églises. De Gaulle had declared his departure from politics in July 1955, leaving his following to vegetate without formally dissolving it.

As in 1951, the election resulted in a hexagonal parliament. But the Centre alliance which had tied four of its six sides loosely together in the early 1950s was now split down the middle. In this stalemate situation, the weight of seats

Table 13.1.
National Assembly Election of 2 January 1956 (Results for Metropolitan France only)

	% of vote	Seats		
		Total	%	Total plus overseas deputies[a]
Communists	25.8	146	26.8	150
Socialists	15.8	89	16.4	99
Radicals, UDSR, etc.	14.3	70	12.9	92
MRP	11.3	71	13.0	82
Conservatives, various independents, etc.	16.8	100	18.4	100
Gaullists	4.4	17	3.1	22
Poujadists	11.6	51	9.4	51
		544		596

Note: The percentage of electorate casting valid votes was 79.6.

[a] Elections were not held in Algeria, because of the crisis.

was nevertheless with the Left; and since the Socialists' eighty-nine deputies were the indispensable element in any left-wing coalition, the task of forming the next government fell to Guy Mollet rather than to Mendès France—despite the fact that a recent opinion poll showed that 27 per cent of the public wanted Mendès France as Premier, and only 2 per cent preferred Mollet.

Whether a Mendès ministry would have saved France from the humiliation of the next eighteen months is a matter for endless speculation. Of those who believe that it would have done so, many express regret that a Centrist Mendès–MRP alliance did not emerge in the period preceding the 1956 election; and they point out that once again the clerical issue had deflected the Republic from the path of salvation. While it is true that Mendès was an opponent of the Barangé and Marie laws, the religious issue was only one of several factors in the failure of the emergence of such an alliance. The MRP leadership was slow to forgive Mendès over the EDC and his Bidault-baiting. But more importantly, Mendès increasingly saw a coalition with the Socialists as the most promising way of realizing his policies, and he was scarcely likely to be drawn into an unequal competition with Edgar Faure for the support of a party that had mixed views on a number of these policies.

The animosity that the MRP felt towards Mendès re-emerged when Mollet was forming his cabinet. Predominantly Socialist, with several Radicals, it was expected to include Mendès as Foreign Minister; but the MRP made it clear to Mollet that they would oppose the government if this happened. Mendès turned down Mollet's offer of the Finance Ministry, having no desire to be left with the bill for the Socialists' inflationary social reforms; and he was left instead with a token seat in the cabinet and no portfolio.

It was widely expected that the Mollet government would set about the repeal of the Marie and Barangé laws; but in practice, Socialist local authorities were reluctant to see the abolition of laws that provided indirect subsidies for public as well as private education. Thought was given to confining the subsidies to the public sector, but the hesitations of the UDSR, and then the Suez crisis of October 1956, postponed practical pursuit of the matter.

Even if anyone had expected the new government to emulate the Popular Front's social programme of twenty years earlier, the differences and resemblances between 1936 and 1956 militated strongly against it. The hostility of the Socialists towards the Communists ruled out a repetition of the euphoric alliance of 1935–6, while the oppressive shadow and mounting cost of the Algerian war were to erode the material foundations of any far-reaching social programme, just as the Depression and the demands of rearmament had done in the 1930s. The 300 milliards extra expense entailed by the war saw the government's annual budgetary deficit grow from an average of 650 milliards in the early 1950s to over 1,100 milliards by 1957. Even so, following the lead of the Renault works, the government extended the Popular Front's fortnight's holiday with pay to three weeks (28 February 1956). It likewise improved the level and mechanism of state pensions to both the old and chronically ill— despite a rearguard action by the Conseil de la République (p. 244), which resented the consequent 10 per cent increase in income tax and creation of extra indirect taxes. Working-class housing was also given close attention— HLMs receiving top priority in the government's target of 320,000 new houses in 1956. But, predictably, financial problems derailed other reforms that were set in motion, notably the compulsory arbitration of works disputes, as well as extended rights for *comités d'entreprise* and the refunding of a higher percentage of prescription charges. However, the presence of no less than fourteen schoolteachers in Mollet's cabinet guaranteed priority to René Billère's education bill of 25 July 1956, which, among other provisions, would have raised the school-leaving age to sixteen, had the government survived longer.

The economic record of the Mollet government was likewise a chequered one. Although industrial production grew by 10 per cent in 1956 and 9 per cent in 1957, agricultural output slackened in the same period, giving overall GNP a growth rate of 5.8 per cent in 1956 and 5 per cent in 1957, which still compared well with the average of 4.3 per cent in the previous five years. Other countries, however, were doing better; and the French trade balance noticeably worsened, with exports covering only two-thirds of imports in the winter of 1956–7.

Moreover, it was in the context of these disturbing international differences that the Treaties of Rome were put to the National Assembly. Although they were passed by 342 votes to 239 (10 July 1957), Mendès France was among those who opposed them on the grounds that the French economy was not strong enough to sustain the competition that the market would entail.

Yet economic and financial difficulties did not inhibit the government's defence programme, which exemplified the perennial fact that French Socialist governments have always been among the most attentive to the needs of the armed forces. Despite the Algerian war's demands on conventional forces, it was under Mollet's government that preparations were made for developing nuclear weapons, and by July 1957 the Saharan test site at Reggane had already been chosen—even if the completion and explosion of France's first nuclear bomb was to come only under the Fifth Republic.

The choice of Reggane demonstrated the government's commitment to 'Algérie française'—the issue that was to break the Mollet ministry just as it had broken its two predecessors, albeit indirectly. Mollet was likewise committed to the principle of non-negotiation with the FLN while they were still under arms. But he was convinced that only substantial reform could make such a policy successful—a common electoral roll and major agrarian reform being among his priorities.

To insulate French officials from the hostile pressures of settler influence, Mollet transferred the functions of the Governor-General to a specific Minister for Algerian Affairs, who would divide his time between Algiers and Paris. General Georges Catroux was an attractive choice, combining liberal views with intimate experience of North Africa. But when Mollet visited Algiers on 6 February 1956, the government party was pelted with eggs, tomatoes, and sods of earth by the die-hard settler element. What most alarmed Mollet about this 'jour des tomates' was the *petit peuple* nature of the demonstrators, making him aware that the *colon* mentality was not just an attitude of big-business men and large landowners, but probably true of the bulk of the European population in Algeria. Fearing widespread civil disturbance, he capitulated, replacing Catroux with an unambiguous 'Algérie française' enthusiast, Robert Lacoste, whose Socialist vision of Algeria was largely confined to extending influence from the rich settler interest to the less well-heeled. As for Mollet himself, he too was now fully committed to the disastrous principle of pacification first. The Algerian Assembly was dissolved (12 April 1956) and elections indefinitely postponed, while the decrees of 26 March and 25 April 1956, which permitted the redistribution of land above fifty hectares, were forgotten.

All this was in sad contrast to the productive work that was being done in other parts of the empire. Gaston Defferre's *loi-cadre* of 23 June 1956 generalized universal suffrage throughout the *territoires d'outre-mer* and based their assemblies on a common voting-roll—the very reform that was withheld from Algeria. Unhappy at the way affairs were shaping, Mendès France resigned

(23 May 1956); but Defferre, Savary, and Mitterrand were the only other ministers to express misgivings.

In the meantime, Mollet extended military service from eighteen to twenty-seven months, effectively depriving the civilian economy of 200,000 men; and by July 1957 there were 450,000 troops in Algeria, with more to come. Acts of violence in Algeria were running at the rate of 3,000 a month by the end of 1956—the more spectacular of these sparking off acts of counter-terrorism, when gangs of European ultras destroyed houses and murdered residents in Muslim sections of the towns. Furthermore, there was continued fighting between factions in the independence movement itself—particularly between the FLN and Messali Hadj's Mouvement National Algérien, which favoured attempting a negotiated settlement with France. Ironically, the growing intransigence of the FLN owed much to the French capture of Ben Bella and several other FLN notables when the Moroccan plane which was carrying them from Rabat to Tunis was tricked into landing at Algiers on 22 October 1956. Not only did it make room for the promotion of more violent men, but it offended neutral as well as pro-FLN sentiment in other countries. In addition to Morocco and Tunis, the FLN enjoyed the sympathy of Nasser's Egypt, Yugoslavia, and India, to say nothing of the more circumspect approval of the Eastern block. Indeed, large quantities of Egyptian weapons were making their way into Algeria, and were to be a major issue in the abortive Suez affair of November 1956.

SUEZ

Colonel Nasser's nationalization of the Suez Canal on 26 July 1956 presented France with a tempting occasion for striking back. Since the canal had hitherto been controlled by an international company, Nasser was not only appealing to Egyptian nationalism by taking over installations based on Egyptian soil, but he was also securing sources of future revenue. Britain's traditional reflexes of 'defending the routes to India' could still be jerked into life—even if the old specifics of empire were now replaced by concern for the movement of oil and eastern trade in general. It was, however, Nasser's prestige as a leader in the Arab world that principally troubled Britain; he might create instability in an area in which Britain still regarded herself as having wide interests.

Israel was a particularly useful stalking-horse for Britain and France. Israel feared that Egypt might use her newly established control of the canal to disrupt Israeli communications with the east, including her oil imports from the Persian Gulf. A secret Franco-British-Israeli meeting at Sèvres (22–4 October 1956) planned the forthcoming attack on Egypt. The general justification—for world consumption—would be the need to guarantee international freedom of movement through the canal. The British government, however, was less confident than the French of having the bulk of domestic public opinion on its side; and it was therefore anxious that the attack should seem to begin purely as an Israeli–Egyptian conflict—obliging Britain and France to intervene in the interests of free navigation. Hence the subsequent denial by British ministers that any collusion with Israel had taken place. The denial was maintained for decades in the teeth of all the evidence—and to the puzzled amusement of the French, who had no need to maintain such a patent fiction once the operation had taken place. For *Le Canard Enchaîné* it was just a question of 'ou il faut coloniser le canal ou canaliser le colonel'.

The attack took place as envisaged, with an Israeli advance (29 October), followed by an ostensible peace-keeping Anglo-French intervention on the agreed pretexts (5 November). Egyptian air-bases and other installations were bombarded from the air and from the sea, while Anglo-French paratroops made successful drops on Port Fuad and Port Said. The pretexts fooled no one. Bulganin threatened London and Paris 'with the terrible means of modern destruction' (5 November), while Eisenhower put pressure on the pound and made it clear that America would not supply Europe with oil to make up any deficit arising out of the blockage of the canal.

Faced with the wrath of the superpowers and the United Nations, the allies halted the attack and truculently listened to the strictures levelled at them from all sides. The results of the expedition for France and Britain were abysmal. Nasser emerged from the affair strengthened rather than weakened; and the canal was now blocked. Only Israel showed tangible gains—her government refusing to evacuate the territories she had occupied. Whether this was in her long-term interests is still debated; but the resulting increase in Middle East tensions was to cause future embarrassments for France, especially under the Fifth Republic. Yet, whereas Eden felt obliged by the weight of hostile opinion to plead ill health and resign, the French Assembly expressed its confidence in the Mollet government by 325 votes to 210 (20 December). Mendès France went along with the bulk of public opinion, his only reservations being on the timing and execution of the attack. Even the Communists, in retrospect, felt a secret gratitude for the affair, in that it distracted attention from the Soviet suppression of the Hungarian uprising on the previous day (4 November).

PARTY ATTITUDES TO THE ALGERIAN WAR

While it would be hard to think of any year since 1946 as a good year for the Communists, few would dispute that 1956 had been a particularly rough one for them. Khrushchev's celebrated denunciation of Stalin had taken Thorez by surprise—not only had Thorez been a close friend of Stalin, but Thorez had been Khrushchev's guest during his wartime Russian residence. He initially denied that such a denunciation had taken place, and subsequently tried to play down its significance—despite Waldeck-Rochet's plea for rapid de-Stalinization within the PCF. Then came the Soviet suppression of the Hungarian uprising—perhaps the greatest test of French left-wing sympathies since the Nazi–Soviet pact. Apologists sought to justify it by emphasizing Hungary's geographical position in the bulwark of satellite states on which Russia relied for defence against a possible Western attack. They also pointed out that the recently appointed Khrushchev felt an acute need to prove to the sceptical Stalinist old guard that he could be as firm as their dead master when it came to defending Russian interests—despite his leanings towards *détente* and his long-term aim of spending less on armaments and more on raising living standards in Russia. Khrushchev's tough stance was firmly supported by Thorez, regardless of the misgivings and open opposition of fellow-travellers and sympathetic intellectuals. Sartre broke with the Mouvement de la Paix, while Picasso was among the party members who formally protested. Even so, only one Communist deputy resigned from the party, compared with a quarter of the Communist deputies and senators during the furore which followed the Nazi–Soviet pact of 1939—when admittedly France's future safety was also at stake.

By contrast, the CGT refused to endorse Russian behaviour in Hungary, and some of its constituent unions outrightly condemned it. This independence of attitude grew in 1957, with the CGT cultivating better relations with the leadership of other unions and joining in mutual demonstrations against the Algerian war—much to the discomfort of the PCF, which felt obliged to discourage it.

In practice, the PCF was extremely chary about committing itself to unambiguous condemnation of the war. It was conscious of the growing racial antagonisms among the urban workers in France, and of the popular feeling against making concessions to Algerian nationalism. Even more embarrassed was the SFIO. Young Socialist intellectuals, led by Michel Rocard, denounced 'national-molletisme', while highly respected champions of human rights, such as Daniel Mayer, were suspended from the party for voicing their misgivings. The MRP was similarly divided, ranging from hard-liners like Bidault to proponents of a negotiated settlement such as Robert Buron. There were no such misgivings on the part of the Poujadists, who had publicly espoused the cause of 'Algérie française' in November 1955, at a time when the retail lobby was beginning to look inadequate as an electoral surf-board. Paradoxically, what was taking the puff out of Poujadism was the resumption of inflation in 1956–7, re-creating the fiscal conditions of the 1940s that had worked to the small retailer's advantage (p. 250).

The Church, like the MRP, offered a wide spectrum of views on the Algerian question. At one extreme, Georges Sauge and other *intégristes* identified the war with the age-old struggle of the Cross against the Crescent, and their publications enjoyed the patronage of General Weygand and other stalwarts of *la France d'autrefois*. There were even chaplains attached to parachute regiments who were prepared to justify torture in extirpating the FLN. This *exalté* minority, however, were sharply at odds with the much more circumspect outlook of the clergy in general. The Archbishop of Algiers was well aware that his future might rest in an independent Algeria; and his denunciations of torture and settler violence were to earn him the nickname of Mgr Mohammed Duval. Indeed, his cathedral was the scene of counternationalist terrorism, including the planting of bombs in the confessionals. On the other hand, the most courageous critic of collaboration in the 1940s, Archbishop Saliège of Toulouse, was a firm supporter of 'Algérie française', as were many of the older generation of clergy who feared what independence might mean for Catholic proselytism.

Torture

The systematic use of torture in Algeria was fast becoming a major issue—as were the government's blatant attempts to muzzle the Press in Paris and elsewhere. The police regularly confiscated whole issues of reputable newspapers in order to stifle reports on the discreditable activities of the forces of order. If torture and summary censorship were largely initiatives by the army and the police, the government did little or nothing to stop them—despite official reports of extensive police torture as early as March 1955.

It began as the reaction of desperate officers trying to identify and locate the planters of bombs in Algerian city bars, cafés, and shops, where Muslims, including many women and children, outnumbered Europeans among the victims. The initial purpose was not so much to extract 'confessions' as information on other bomb-planters still at large, and on forthcoming planned attacks. But since the whole campaign of FLN violence was inextricably interwoven with its general strategy and resources, any information on its activities was increasingly regarded as justification for torture. French senior officers issued instructions that 'the moral and physical integrity' of prisoners under torture was 'to be respected'. But these theological niceties were not always appreciated by NCOs and junior officers, chosen for their toughness in battle rather than for their moral subtlety. General Jacques Massu, who, with his Tenth Parachute Division, was entrusted with the pacification of the city of Algiers (7 January 1957), sought to soothe public unease at these methods by claiming to have submitted himself to the standard electric-shock treatment. The thought of the hatchet-faced Massu undergoing the inquisitorial ministrations of his respectful subordinates would not have been without humour if the issue had not been so grim.

As the reports of surviving victims and junior witnesses testify, the official image of a quasi-clinical series of carefully graduated electric shocks bore little relation to the activities of exasperated interrogators, faced with obstinate or sometimes plain innocent victims with no information to give. Beatings, burning with matches, sexual assault (real and simulated) were all part of what went on—uneasily obscured in the public mind by the countervailing image of the mutilated corpses of Europeans kidnapped by the FLN.

According to the Secretary-General of Police in Algeria—who resigned in protest in September 1957—at least 3,000 of those taken for questioning were never seen again, some dying under torture, some finished off to prevent their lodging unofficial complaints, and others summarily executed or shot 'attempting to escape'. Officers who protested against this state of affairs were arrested—including a general. As many writers pointed out, few people at the Liberation would have imagined that in ten years time a Socialist government would be turning a blind eye to practices reminiscent of the Gestapo and the Milice. Yet the uncomfortable fact remained that Massu's methods in Algiers succeeded in bringing down the monthly total of acts of terrorism from 4,000 in January 1957 to 1,500 in May 1958.

Even so, there seemed no foreseeable end to the war, except at a greatly increased cost of men and money that the government was reluctant to take to a tight-walleted parliament. The resultant stalemate angered both Right and Left, and accounted for the eventual defeat of the Mollet government—ostensibly on a matter of finance (21 May 1957). The Left wanted negotiations for peace, and the Right wanted a more vigorous prosecution of the war—and neither side was prepared to provide the government with the authorization or the means to pursue the other's prescriptions. Given the composition of parliament, the same problem would face any incoming ministry; and under the existing rules of the game no solution was possible.

Jacques Soustelle recognized this impasse as both a challenge and an opportunity for the Gaullists. The Gaullists would change the rules with a new constitution that would strengthen the executive; and armed with these new powers, the government could then impose an effective solution to the Algerian problem. Quite at what point Soustelle determined on a wrecking policy towards Mollet's successors is not altogether clear. It was much more specific than that employed by the RPF in the early 1950s; and, following his recent experience as Governor-General of Algeria, it was as much directed towards realizing his vision of Algerian integration as towards bringing de Gaulle to power.

BOURGÈS-MAUNOURY AND GAILLARD

The new government was largely a reshuffle of Mollet's Socialist–Radical coalition, skewed towards the Centre, and headed by Maurice Bourgès-Maunoury, a Radical who had been Minister of Defence. As such, its Algerian policies were very much in the mould of Mollet's. Indeed, its proposals for new legislative institutions had begun gestation under Mollet. While affirming the traditional fiction of Algeria's identity with mainland France, this *loi-cadre* envisaged the creation of eight to ten provinces, each with an assembly and executive with certain limited powers of self-administration. The difficult issues, such as whether there would be a common voting-roll, were left to the administrative decrees that would implement the law.

Following vigorous liaison-work by Soustelle among its potential opponents, this anodine document was rejected by 279 votes to 253 (30 September 1957). With the Poujadists and Communists automatically voting against it, it was easy for Soustelle to build a wrecker's majority out of those who feared that even this modest level of fragmented self-government might give Algerian nationalists a dangerous toe-hold. Soustelle could also take pleasure from the fact that it took parliament five weeks to find a successor to the defeated government—after roundly rejecting the contrasting claims of Pinay and Mollet.

Its choice eventually fell on the youngest man to occupy the Matignon so far, the thirty-seven-year-old Radical, Félix Gaillard, who had distinguished himself by his imaginative work at the Ministry of Finance under the previous ministry. One of his more ingenious achievements had been a semi-disguised devaluation of the franc of 20 per cent (12 August 1957), accentuated in the case of exports and diminished in the case of imports, thereby enabling French goods to compete advantageously with foreign produce abroad, while preventing French consumers from having to pay a high price for what foreigners sold in France. Faced now with the double task of securing financial stability and placating the opponents of the Algerian *loi-cadre*, Gaillard broadened the cabinet rightwards to take in MRP, conservative, and Gaullist ministers—notably Pierre Pflimlin (MRP) for financial and economic matters, and Jacques Chaban-Delmas (Gaullist) at Defence, where he was to combine his ministerial functions with the systematic subversion of the regime.

A heavily diluted version of the *loi-cadre* struggled its way through the Assembly (28 January 1958), to be further weakened in the Conseil de la République; but whatever tart satisfaction the government could take from this was almost immediately overshadowed by the Sakhiet affair. A French reprisal raid on the Tunisian base of an FLN unit resulted in the destruction of the nearby village of Sakhiet, killing seventy peasants, many of them women and children (8 February 1958). There was an international outcry, with America threatening to withdraw a forthcoming loan to France unless Gaillard accepted a mediated settlement of the affair. Since the loan was Gaillard's main safeguard against having to put up taxes and risk alienating parliament, he readily agreed to 'the good offices' of America and Britain; but it soon became clear that the mediators wanted to extend their remit to embrace the Algerian question itself. Although Gaillard openly resisted this extension, the issue was enough to give Soustelle the chance he was seeking. January had seen the first Saharan oil making its way to France; and dark rumours had long been circulating that American, British, and Italian business men were plotting to exploit the oil resources of an independent Algeria. Accusing Gaillard of letting foreigners order French affairs, Soustelle drummed up a majority of 321 votes against the government on 15 April and plunged France into her most serious ministerial crisis since the war.

Parliament rejected Bidault's and then Pleven's bid for investiture, until it was eventually the turn of a compromise candidate, Pierre Pflimlin (MRP), who was proposed in desperation by Mollet and Pinay from opposing sides of the house. Pflimlin, however, was suspected by the Right of wishing to open negotiations with the FLN; and the news of his intended bid for office was the signal for the opponents of the regime to resort at long last to direct action.

14
'La République des Dupes'

WRITING four years before de Gaulle came to power in 1958, Herbert Lüthy declared in a political obituary: 'General de Gaulle was and remained a politician of catastrophe; he thought in terms of catastrophe, he judged men and institutions by their behaviour in catastrophe, and apocalyptics like Malraux who formed his entourage did nothing to counter this mental bias based on his profession and his experience.'[1]

In 1958 it seemed that France was approaching just such a catastrophe. While de Gaulle was to be the ultimate victor of this situation, his victory was by no means an obvious likelihood in the spring of 1958. Although an opinion poll in January had put de Gaulle as first choice for Prime Minister in a wide field, this was with a mere 13 per cent of the opinions expressed. And when, during the ministerial crisis of that spring, President Coty privately sounded out de Gaulle on whether he would be willing to try to form a government, de Gaulle wearily replied that the people did not want him (5 May 1958). Nor did his candid comments to journalists about the likelihood of Algerian independence suggest someone who counted on coming to power on the backs of settler and military discontent.

His pessimism was understandable. None of the three main areas of current discontent was clearly committed to him. The political preferences of the settler interests were for whatever government would guarantee their continued presence and influence in Algeria. Some looked to Poujade or to one or other of the generals—active and retired—who were favoured by right-wing groups in Paris. Others felt that it was wiser to back more widely respected figures with established track records in government or administration. Hence the widespread enthusiasm for Soustelle—especially since his initial commitment to improving Muslim conditions had been overlaid by his emphasis on defeating the FLN. The main drawback to Soustelle was his allegiance to de Gaulle, who was disliked by many settlers with wartime Vichy sympathies who suspected de Gaulle of over-liberal attitudes towards the Muslim population.

As for the army, it did not have a tradition of close involvement in politics, despite the exceptional activities of individuals such as Bonaparte, Boulanger, and de Gaulle himself. Its ethos was very different from the armies of Spain or Latin America—or pre-war central Europe. It had, nevertheless, developed a strong tradition of loyalty to itself, partly in compensation for the lack of a firm focus of loyalty in the State. Although the political animosities of the Dreyfus era were past history, the ministerial instability of the Third and Fourth Republics and the drab nature of their presidencies did not provide a stirring expression of national identity. Moreover, the humiliations of 1940 had been followed by defeat in Indo-China, where once again the army accused the politicians of inadequate support and ambivalent policies. Indeed, the recent Affaire des Fuites had raised suspicions that men in government circles had been providing the Viet Minh with military information. A further outcome of the Indo-Chinese war was the assortment of ideas half-learnt from the Communist forces, notably in the realm of psychological warfare, where emphasis was put on the creation of complete commitment in the civilian population as well as among the armed forces. French officers contrasted the single-minded determination of the Viet Minh rebels and their civilian sympathizers with the vacillating politics of Paris—drawing the conclusion that victory in Algeria depended on an identity of views between army and government. And if the government refused to see sense, the government would have to be changed—or so a number of Vietnam veterans were claiming.

De Gaulle was not necessarily their first choice. A number of colonels and young generals thought of him as belonging to a pre-war world, knowing little of warfare in the 1950s, where nationalist and Marxist ideologies were as much the enemy as the guerrilla tactics pioneered in post-war Asia. On the other hand, for the older or more discipline-conscious senior officer, de Gaulle had always seemed something of a maverick, over-ready to reject the collective wisdom of his superiors and over-confident in his personal initiatives. Yet, whatever their political preferences, their prime concern was to avoid yet another humiliating defeat for the army. Settler interests were a secondary matter, many soldiers feeling indifference or contempt for the determination of the *colons* to cling on to their privileged position.

To appreciate the exasperation of both the settlers and the army in the spring of 1958, it has to be recognized that the military methods used in Algeria seemed to be bearing fruit. Apart from the substantial reduction in terrorist attacks (p. 260), the construction of the Morice Line—three hundred kilometres of air-patrolled electrified fence running along the Tunisian border—was seriously impeding the movement of Tunisian-based FLN troops, limiting their operations to a quarter of their normal level.

The troubled waters of Algeria made attractive fishing for the third main element of discontent, the committed right-wing opponents of the regime in metropolitan France—much as Northern Irish loyalism was to be a convenient cudgel for several right-wing politicians in England. Poujadism, although in decline in mainland France, had active adherents in Algeria, notably in the so-called 'Group of Seven', a dangerous cluster of hard men who were to play a vital role in the events of 13 May. They kept an open mind on who was to lead a revolutionary government, some favouring the ebullient, perspiring Poujade, others preferring General René Cogny, an able veteran of the Indo-Chinese war who was more likely to reassure the army and settler interests. Parallel to the Group of Seven in Algiers was the 'Grand O' in Paris, consisting of a variety of right-wing dissidents, some dating back to the 'Cagoulle' of the 1930s which had been responsible for a series of brutal political murders. The importance of the Grand O lay in the membership of various retired generals, General Paul Cherrière—whose comfortable corpulence and ponderous manner had earned him the nickname 'Babar'—and the President of the Indo-Chinese war veterans, General René Chassin.

The prevailing uncertainty and disagreement about who should lead a renovated France was both a source of hope and despair to Gaullist observers. Among the optimists, Soustelle, Chaban-Delmas, and Léon Delbecque played key roles in turning the current confusion to de Gaulle's advantage. The quick-witted enterprising Delbecque was an employee of Gaillard's Minister of Defence, Chaban-Delmas. He had been shuttling between Paris and Algiers, acting as a liaison between the various Gaullist groups, scenting the air, and preparing plans for the eventual propitious moment.

Just such a moment seemed to be offered by the parliamentary programme for 13 May, when Pflimlin would seek investiture. Soustelle had two alternative plans in mind. In the unlikely event of Pflimlin offering him a place in his forth-coming government, it would be open to the Gaullists to work from inside the government. Otherwise, if no portfolio was forthcoming, Soustelle would tele-phone Delbecque as soon as parliament had voted on Pflimlin's investiture; and, if the new government was accepted, Delbecque would launch a take-over of the government offices in Algiers, with Soustelle flying out to join it. De Gaulle was certainly aware that his supporters were laying active plans on his behalf, but he was anxious not to be open to subsequent accusations of complicity—so he distanced himself from the detail of what was afoot, leaving it to his aides, the discreet but hard-minded Olivier Guichard and Jacques Foccart, to observe as much as they thought proper.

The Group of Seven, however, determined to pre-empt the situation. A few hours before Pflimlin's bid for investiture, they took advantage of a public wreath-laying ceremony in Algiers to lead the crowd in an attack on the main government offices (13 May). Both the Commander-in-Chief in Algeria, General Raoul Salan, and the military commander of Algiers, General Jacques Massu, appear to have been genuinely taken by surprise. Even more surprised was Léon Delbecque, who, upstaged in his plan to take over govern-ment headquarters, thereupon worked hard to steer the occupation in a Gaullist direction. Despite his late arrival on the scene, he made himself an immanent centre of interest by announcing the immediate arrival of Sous-telle, the most evocative name one could utter in a gathering of desperate Europeans. In the meantime, Salan and Massu had chosen to parley with the occupiers rather than attempt to evict them; and when they agreed to their leaders' demand for a Committee of Public Safety, Delbecque was invited to be its Vice-President.

The role of Salan and Massu in these events was an ambiguous one. The Paris government, when informed of the take-over of the building, delegated the powers of Minister for Algeria to Salan—Lacoste himself being currently on a visit to Paris. It also swallowed the fact that Massu had become President of the Committee of Public Safety, tamely accepting his claim that it was the only means of maintaining order. Thereafter, however, Massu and Salan behaved with increasing independence; and, indeed, Massu strongly urged the government to invite de Gaulle to form 'a government of public safety' in Paris. Even so, when parliament entrusted Pflimlin with the hazardous task of government—by 274 votes to 129, with the Communists abstaining (14 May)—the new Prime Minister immediately confirmed Salan's provisional powers in Algeria. Admittedly, he took the sound precaution of putting Soustelle under house arrest in Paris, thereby depriving the Gaullists of their trump card—or so he thought. Yet this apparent set-back to Gaullist hopes did not deter Salan from shouting out 'Vive de Gaulle!' to an assembled crowd on 15 May—a portentous step that was apparently instigated by the ever-resourceful Delbecque.

15 May was even more notable for de Gaulle's announcement that he was prepared to form a government, if invited. Faced with this bold initiative, Pflimlin's Deputy-Premier, Guy Mollet, responded by publicly asking de Gaulle if he was prepared to condemn the Algiers junta and go through the normal procedures of seeking investiture—including withdrawal if he was rejected. De Gaulle side-stepped the first question by reminding the government that it was they who had delegated authority to Salan, but he answered the second question affirmatively (19 May). While the humour of de Gaulle's evasion was not lost on the general public, it was to be the main sticking-point in the weeks to come—especially when the Algiers committee moved further along the road to open rebellion.

The catalyst in this development was Soustelle's arrival in Algiers on 17 May, after a deft escape from Paris in the boot of a car. Soustelle's enormous popularity with the settler community was an acute embarrassment to Salan and Massu, who, as his reluctant hosts, found themselves drawn into his insurrectionary plans rather more quickly and decisively than they would probably have wished. Not only were they persuaded by Soustelle and Delbecque to envisage putting pressure on Paris if the government did not make way for de Gaulle, but by the end of the week the generals were making plans for an attack on Corsica. This violent escalation was greatly welcomed by an important segment of Salan's military staff, who, as early as the second day of the Algiers occupation, had been discussing the possibility of an airborne occupation of Paris. This would be done with paratroopers from Algiers and Toulouse—where General Miquel was thought to be sympathetic—and would be backed by Colonel Gribius's armoured units, currently based at Rambouillet and Saint-Germain-en-Laye. Similar secret discussions were taking place in military circles in Paris, where the Deputy Chief of the General Staff, General Maurice Challe, had been arrested for providing the Algiers garrison with extra transport planes. The arrest had provoked the immediate resignation of the Chief of the General Staff, General Paul Ély, on 16 May; and it was henceforth clear that the government could no longer be sure of the loyalty of the army, either in Algeria or in the metropole. At least nine generals in key positions in Paris, Toulouse, Lyon, Bordeaux, Dijon, and Germany were prepared to resort to armed pressure to change the government—depending on a clear lead being given by Algiers or by de Gaulle himself. And, as in Algiers, it was often the colonels in the generals' entourage who were keenest to take action. What some of their seniors feared was the creation of a civil-war situation, reminiscent of Spain in the 1930s, with an insurgent army confronted by the trade-unions, supplied with arms by the Communists.

Despite these reservations in metropolitan France, Algerian-based troops occupied Corsica on 24 May—evoking fond memories of September 1943, when pro-Gaullist forces from Algeria likewise took Corsica as a stepping-stone to the eventual conquest of France. The whole episode was especially humiliating for the Paris government, since the CRS detachments sent to Corsica capitulated on arrival, without offering even token resistance. Moreover, the government did not dare to send military units, for fear that they would side with the invaders from Algiers. With Corsica painlessly in the insurgents' hands, both de Gaulle and the government had to take with extreme seriousness reports that the Algiers forces were poised to implement 'Operation Resurrection'—their plan to occupy Paris on the night of 27 May, should parliament refuse to hand over government to de Gaulle. De Gaulle immediately offered to have a secret nocturnal meeting with Pflimlin in the park-keeper's lodge at Saint-Cloud on 26 May; but the encounter rapidly foundered on de Gaulle's refusal to issue a personal condemnation of the Algiers junta. De Gaulle thereupon took the initiative. Disregarding the inconclusive nature of his meeting with Pflimlin, he announced the following morning: 'Yesterday I set in motion the proper procedure for the establishment of a republican government capable of ensuring the unity and independence of the country', and he called upon the army to remain obedient to its superiors.

The problem remained, however, of whether parliament would accept a Gaullist government and the constitutional reforms that de Gaulle considered essential to national recovery. Despite the firm line that Pflimlin had taken with de Gaulle, Pflimlin was increasingly disposed to make way for him—a step which President Coty strongly favoured. Embarrassingly for Pflimlin, however, the Communists had joined with other sections of the National Assembly to demonstrate their support for him by 405 votes to 165. Even so, he resigned in what he saw as the interests of public safety (28 May)—despite taunts of 'Daladier trente-quatre!'

It was at this point that President Coty played a decisive role. There were various reports that Operation Resurrection, postponed from 27 May, was now scheduled for 30 May. Coty decided to threaten parliament with his own resignation if the Assembly refused to invest de Gaulle. The sharp edge to the threat lay in the uncomfortable fact that in the event of a presidential vacancy, the President of the National Assembly would become caretaker President; and the man in question was André Le Troquer, an implacable opponent of a capitulation to de Gaulle. The fear was that Le Troquer would then invite a Popular Front-type figure to form a government of resistance, which might lead to a confrontation between the army and the unions.

In the various party discussions that nervously considered the President's threat (29 May), it rapidly became clear that a majority could probably be found for investing de Gaulle. The decisive weight lay with the Socialists, where a narrow preponderance (77/74) were prepared to follow Mollet and Auriol into accepting the General. Conversely, opposition to de Gaulle included erstwhile admirers such as Mendès France, who refused to give way to army pressure—as did Mitterrand and a number of other stalwart liberals.

In the meantime, the Algiers junta postponed Operation Resurrection yet again until the outcome of the investiture debate was known on 1 June. The degree of complicity between de Gaulle and the junta still remains a hotly debated subject. Argument still centres on a visit made to de Gaulle on 28 May by an emissary of Salan, General Dulac. Dulac subsequently claimed that de Gaulle was critical of the meagre number of paratroopers envisaged for Operation Resurrection—while regretting the need to consider such a contingency at all: 'I do not want to make an immediate appearance, so as not to seem to be coming back simply as a result of this forcible act. After a few days, I want to be called upon as an arbiter . . .'. According to Dulac, de Gaulle also wanted paratroopers to be sent to Colombey to prevent any attempt to kidnap him.

Whatever the truth of these assertions, the precautions proved unnecessary: the Assembly invested de Gaulle by 329 votes to 224 (1 June). Moreover, it gave his government full powers for six months (2 June), and then, by an even larger majority (350 to 161), it entrusted him with the elaboration of a new constitution. While the Assembly stipulated that the constitution should be democratic, it agreed that it should be put to a referendum rather than put through the parliamentary hoops stipulated by Article 90 of the 1946 constitution. Already the Gaullist ethos of 1945–6 was reasserting itself, with its direct appeal to the people over the heads of their accredited representatives.

The relative ease of these dealings with parliament owed something to the composition of the cabinet that de Gaulle presented. It was very broadly based, containing three representatives from each of the main parties (except the

Communists); and it included such diverse figures as Mollet, Pflimlin, and the reassuring Antoine Pinay at Finance, thereby appeasing the social and democratic conscience as well as the guardians of financial orthodoxy. Moreover, the only full-blooded Gaullists within it were the respectable literary André Malraux and the normally dangerous Michel Debré, who had reportedly been rendered innocuous by lumbago during the subversive activities of May. Soustelle was very pointedly left out—until the initial parliamentary hurdles were safely cleared. Otherwise, the most remarkable feature of the cabinet— and a foretaste of the future—was the inclusion of non-political experts: Maurice Couve de Murville at the Quai d'Orsay, Émile Pelletier at the Interior, and Pierre Guillaumat at Defence. The age of the technocrats was about to begin.

IMMEDIATE TASKS

The most pressing problems facing de Gaulle were the Algerian question and the new constitution. The constitution was rapidly put together in the summer of 1958, published on 4 September, and accepted by a large majority in the referendum of 28 September. Its principal architect was the future Prime Minister, Michel Debré, who had been primarily responsible for the Bayeux proposals of 1946 (p. 141). Mollet, Pflimlin, Pinay, and Reynaud participated in the preliminary discussions, but it thereafter went to the referendum without consideration by parliament, which was still in suspension pending the forthcoming November elections. There was relative indifference among the general public concerning its precise provisions, the referendum being popularly regarded as largely a matter of confidence in de Gaulle. The PCF was the only major party to campaign *en bloc* for rejection, the position of the other parties corresponding to the divisions that characterized the investiture vote on 1 June (p. 268). The non-Communist left-wing parties were split internally, while the MRP and the Right were solidly in favour; even the Poujadists and other right-wing extremists could not prevent many of their supporters from joining the 66.4 per cent of the electorate in France who voted 'yes'. In fact, the total 'noes' (17.3 per cent) amounted to a million less voters than had habitually voted Communist under the Fourth Republic.

The method of electing the new parliament was established by a simple ordinance of 13 October, there being no Assembly in session to approve it. The drafting committee had debated whether to follow the preference of Debré, Pinay, and Soustelle for a *scrutin de liste* system, where the list with most votes got all the seats for the *département*—either by an absolute majority on the first ballot, or a simple majority on the second. Or, alternatively, whether to adopt Mollet's and Pflimlin's plea for a return to something resembling the Third Republic's *scrutin d'arrondissement*—a solution favoured by the non-parliamentary members of the cabinet. De Gaulle associated systems of *scrutin de liste* with the subservience of candidates to the parties that nominated them, and he resolved the debate in favour of an updated version of the pre-war two-ballot *scrutin d'arrondissement*—or *scrutin uninominal* as it was now called, since the new single-member constituencies were no longer subdivisions of the old administrative *arrondissements*. But, in order to eliminate the smaller parties and to make for greater stability, only candidates who had stood in the first ballot could henceforth stand in the second. Moreover, they had to have received at least 5 per cent of the initial vote—a stipulation that was subsequently raised to 10 per cent in 1966, and then to 12.5 per cent in 1976. It was also decided to reduce the number of deputies representing metropolitan France from 544 to 465, leaving 67 seats for Algeria, 4 for the Sahara, 10 for the Départements d'Outremer, and 6 for the Territoires d'Outremer—the empire thereby holding 16 per cent of the total.

The first ballot of the elections themselves (23 November) saw well over a third of the electorate change its political allegiance—a striking testimony to the force of the recent crisis on the public imagination (Table 14.1). The Gaullist gains were spectacular: with 20.3 per cent of the votes on the first ballot, they finished with 42.2 per cent of the seats, compared with a mere 3.1 per cent in the previous parliament. As before, their strength lay in northern

Table 14.1.
National Assembly Elections of 23 and 30 November 1958
(Results for Metropolitan France only)

	% of vote on first ballot	Seats gained after both ballots		
		Total	%	Total plus overseas deputies
Communists	18.9	10	2.1	10
Socialists	15.7	40	8.6	47
Radicals, etc.	8.2	37	8.0	40
MRP	10.8	55	11.8	64
Gaullists	20.3	196	42.2	206
Conservatives, various independents, etc.	24.2	127	27.3	129
Others	1.9	0	0.0	81
		465		578

Note: The percentage of electorate casting valid votes was 75.2.

France, but 1958 was notable for their gains in those industrial areas that were more usually associated with the Left. By contrast, the conservatives held their own against the Gaullists rather more successfully than anticipated, increasing their share of the vote from 16.8 to 24.2 per cent on the first ballot, and thereby putting themselves temporarily ahead of the Gaullists. They had undoubtedly benefited from *scrutin uninominal*, with its premium on grass-roots connections, and they had also gained from de Gaulle's refusal to allow the Union pour la Nouvelle République to use his name in its title in case it detracted from his intended role of an above-party President. As a consequence, a number of conservatives who were broadly sympathetic to de Gaulle were able to pick up extra support from those voters who were unclear about the party distinctions.

Although the MRP predictably lost seats to the UNR—and also to the conservatives—they also gained compensatory votes from the wreck of Poujadism, as well as from the general national swing to the Right. On the other hand, the MRP was increasingly prone to centrifugal tendencies within itself—exemplified notably by the rightward drift of George Bidault's personal following, the Démocratie Chrétienne (thirteen seats).

The fortunes of the Left were a very different story. Both the Radicals and the Socialists were severely handicapped by their disarray over support for de Gaulle. While the Radicals lost nearly half their former voters on the first ballot, the Socialists initially retained their accustomed electoral support, but lost half their seats in the run-off—the Communists refusing to adopt a common strategy on the second ballot.

The collapse of the Communist presence in parliament—from 26.8 to 2.1 per cent—had no parallel in the party's history. While they were the obvious victims of the national panic of 1958 as well as of their own intransigence, the Communists were also beginning to be affected by the rising standard of living of the working classes, who found the slogans of the PCF increasingly remote from their own circumstances and aspirations. Even the rural bulwark of Communism, the traditional protest vote (p. 45), resorted to abstention in many cases, with some choosing to vote Gaullist. How *scrutin uninominal* worked in the atmosphere of 1958 was reflected in the average number of votes required to put each successful party candidate in parliament. For the UNR, it was a mere 19,000; for the conservatives, 23,000; the MRP, 46,000; the Radicals, 76,000; the Socialists, 79,000; and the PCF, a staggering 380,000.

Notable casualties of the election included Mendès France, Mitterrand, Le Troquer, Laniel, and Teitgen; and, indeed, a striking feature of the new Assembly was the number of new and relatively young faces—only a quarter of the deputies had sat in the previous parliament. Another remarkable feature was its overwhelmingly middle-class nature—only seven deputies were workers, the lowest number since 1885. At the same time, the traditional phalanx of lawyers, journalists, and teachers was eroded by an influx of business men and management personnel—a current that was also to find its counterpart at government level.

With de Gaulle as a candidate, the presidential election on 21 December 1958 was little more than a formality. The electoral college, which was made up more or less by the same 80,000 individuals who elected the senators (p. 402), gave him 78.5 per cent of their votes, against the token opposition of two left-wing candidates. Secure in the Elysée, he handed over the premiership to the devoted, if abrasive, Michel Debré, the St Sebastian of the regime as de Gaulle was to call him, in recognition of his readiness to suffer the arrows intended for his master. None were to be sharper than those from the Algerian settlers—and all the more painful, given Debré's personal sympathy for 'Algérie française'.

De Gaulle's Algerian and colonial politics were intimately connected with his domestic and foreign policies. Not only was he convinced that colonialism had had its day, but he saw it as a grave embarrassment to French foreign policy, where France was continually encountering the strictures not only of the Third World, but also of the United States, Russia, and other substantial countries. It was his intention to make France a leading spokesman of the middle-rank powers in world politics (p. 310). Not only was such a policy far more important to France's future than the short-term psychological and material losses that would accompany the abandonment of empire, but it would itself provide part of the cure for these losses. The army would find itself linked to a bold, forward-looking role on a far larger stage than France's Afro-Asian possessions, and the prestige of this role would be further enhanced by de Gaulle's development of French nuclear power. Already the shape of things to come was indicated by de Gaulle's note to America and Britain (24 September 1958) proposing a re-structuring of the Western alliance so that France, as a nuclear power, would have parity with her partners. Predictably, the demand received an unenthusiastic reply from Eisenhower; but it was a preparatory affirmation of French determination, correctly interpreted by its recipients as a warning shot.

THE ALGERIAN NETTLE

Even before this, however, de Gaulle had recognized that everything was contingent on settling the Algerian impasse. Immediately after parliament had given him what he wanted in June 1958, de Gaulle went to Algiers, where he performed what critics have called the greatest confidence trick in post-war France. Addressing an ecstatic crowd from the much-used balcony of the government building (4 June), he began with deliberate ambiguity: 'Je vous ai compris'—taken by his cheering listeners to mean 'I share your aims', but seen in retrospect as signifying 'I have got the measure of you'. Eschewing any mention of 'Algérie française', he announced that 'from this day forth France considers that there is only one category of inhabitant in the whole of Algeria ... French citizens in the full sense with the same rights and the same duties'. Once again, the crowd responded enthusiastically, understanding this as a commitment to Soustelle's formula of full civil equality for Muslims, but within the framework of a French Algeria. Yet however permanent 'from this day forth' might sound, it only specified the initiation of such a state of affairs, without guaranteeing its perpetuity. Similarly, when at Mostaganem he finished his speech with 'Vive l'Algérie française!', this was not necessarily inconsistent with giving the population the eventual opportunity of deciding whether it wished to remain French or not—or so de Gaulle's apologists subsequently argued. Yet, few of his hearers at the time appreciated the open-ended ambiguities of this patriotic affirmation. Even more explicit, or so it seemed, was his declaration earlier in the day at Oran: 'France is here with her vocation. France is here forever'. De Gaulle's apologists were later to justify these apparent hostages to fortune by saying that the French *mission civilisatrice* did not necessarily require political sovereignty to sustain it.

Nevertheless, self-determination and independence were to come much sooner than de Gaulle probably anticipated in June 1958. And his forthcoming programme of massive economic aid and negotiation with the FLN suggested that he hoped that even a self-governing Algeria would remain within the French Union. It is simplistic, therefore, to portray the deliberate ambiguity of these speeches as no more than a cynical disguise for an intended policy of desertion. At the same time, his speeches of June 1958, both in Algeria and at home, spoke of the ballot-box ultimately determining the 'conditions' of the relationship between France and Algeria—which, if insufficiently appreciated at the time, left open a wide range of options.

De Gaulle thought it prudent in the short term to put Algerian civil administration in the hands of the military, and to confirm and regularize Salan's position by making him Delegate-General. The co-option of Soustelle to the cabinet as Minister for Information (July 1958) was also something of a mixed gesture. His hands were kept off the levers of power in Algeria and elsewhere, but his voice as spokesman of government policy would have a reassuring ring for those who had brought de Gaulle to the Élysée.

De Gaulle then embarked on a rapid tour of Black Africa and Madagascar (August 1958), during which he made it plain that in the forthcoming referendum on the French constitution the population would in effect be choosing between secession or association with France in the federal community. Since the bulk of these territories were heavily dependent on France for their economic progress, the assumption was that most of them would opt for membership of the community. This was how the referendum was understood in Black Africa—where only Guinea, under the influence of Sekou Touré, voted 'no' and, *ipso facto*, for secession and deprivation of French economic aid.

In Algeria, the referendum was held on a common voting-roll, with women voting for the first time. Thanks to army propaganda and close organization of the referendum, 96 per cent of the 79 per cent who voted did so affirmatively. Reassured by this demonstration of support, de Gaulle's Constantine Plan of

3 October 1958 envisaged a five-year development programme that would create 400,000 jobs and ensure that at least two-thirds of the country's children received adequate schooling. It also stipulated that a tenth of the places in the public services should be reserved for Muslims—a modest proportion that reflected both the low level of educational progress made in Algeria and the entrenched position of the Europeans.

The administration of Algeria reverted to normal civilian control on 12 December 1958, Salan relinquishing his civil functions to a new Delegate-General, Paul Delouvrier, and his military command to General Maurice Challe. Salan himself was kicked upstairs to a couple of honorary sinecures, which, as intended, were the ante-room to his retirement. However, the prospect of a heavily hostile vote at the next session of the United Nations decided de Gaulle to promise the Algerians self-determination (16 September 1959)—but only after four years of 'peace', during which the annual death-toll had to remain below two hundred. Of the alternatives he offered—secession, assimilation, or self-government in close relationship with France—he indicated his preference for the last. Even this modest offer, however, was against the advice of half the cabinet, and the announcement set in motion the first of two major military plots against him.

Although the offer was approved by the Assembly by 441 votes to 23 (31 October 1959), it had aroused the opposition and resignation from the UNR of a dozen deputies, including Léon Delbecque, whose cool calculation in May 1958 had been a major factor in bringing de Gaulle to power. Other parliamentary opponents included Jean-Marie Le Pen, Jean-Louis Tixier-Vignancourt, and Delbecque's rival on 13 May, Pierre Lagaillarde of the Group of Seven, who was now provoked into re-establishing contact with his fellow conspirators, notably Joseph Ortiz. Ortiz's Front National Français had close links with the quasi-fascist Jeune Nation—their Celtic-cross graffiti still surviving on many walls decades later. Together with disaffected elements in the army, they looked to a trio of generals—Edmond Jouhaud of the air force, Marie-André Zeller, a retired general who had studied methods of psychological warfare in Indo-China, and Raoul Salan, whose recent Algerian career and current presidency of the veterans of the Indo-Chinese war made him a key figure.

The plan was to demand a government commitment to 'Algérie française'. In the expected event of a refusal, a coup would be launched on 15 October and power temporarily invested in the trio of generals. The scheme ran into a series of difficulties, however. Ortiz was doubtful whether he could get his militants ready for action as soon as 15 October, while a number of key military personnel, including Generals Massu and Gracieux, indicated that they would have nothing to do with such a venture. It was also clear that no support could be expected from the Commander-in-Chief in Algeria, General Challe. Among politicians in metropolitan France, Soustelle was opposed to the whole enterprise, while Léon Delbecque was equally hostile, despite his anger at the evolution of de Gaulle's Algerian policy. Speculation surrounded Bidault's potential role in what was rumoured to be afoot, but there is little evidence to indicate active involvement. Indeed, the whole episode has remained shrouded in considerable mystery, since it was the government's policy to play down its significance in case the publicity augmented its supporters. Suspected military figures were largely left alone—though under discreet observation—much as Spain was to do with many army conspirators in the spring of 1981. The fact that the plotters called it off also meant that there was relatively little concrete evidence for prosecution, had the French government been disposed to resort to exemplary punishments.

The dangers of not taking action, however, were to be demonstrated in April 1961, when the unmolested central figures of October 1959 were to come out into open rebellion, together with others who had refused to participate on the earlier occasion. Even before this, a foretaste of the hazards to come was demonstrated in the so-called Affaire des Barricades of January 1960, when Joseph Ortiz and various colleagues led bands of armed ultras in an occupation of the Plateau des Glières in Algiers (24 January 1960), demanding a new government and commitment to 'Algérie française'. But General Challe and his subordinates remained loyal to the government, and their cautious handling of the situation paid dividends in the voluntary dissolution of the occupation, albeit after 19 deaths and 141 serious injuries.

Both the army and the settlers were dismayed by what they saw as the growing cynicism of de Gaulle in 1960–1. Receiving Algerian deputies (16–19 January 1960), de Gaulle told one of them: 'L'intégration? Une connerie. L'armée a fait l'affaire Dreyfus, elle a soutenu Pétain, maintenant elle est pour l'intégration . . . Les musulmans ne seront jamais des Français.' And when one of the Muslim deputies evoked the likely fate of the pro-French Muslims in an independent Algeria—'mon général, we will suffer!'—de Gaulle's short reply was: 'well then, you will suffer'. In the same vein, when Roger Duchet on another occasion asked de Gaulle about the future of Saharan oil, de Gaulle contemptuously replied: 'Ah! le pétrole, c'est ce qui vous intéresse. Eh bien, monsieur, on essaiera de vous le garder!', as though he were addressing some selfish old woman who wanted him to rescue her umbrella from a burning building.

The issue of whether Algeria should be offered self-determination was put to France and Algeria in a referendum on 8 January 1961, de Gaulle making it clear that he would resign if it resulted in a negative vote. Many Algerians followed Ferhat Abbas's advice and abstained, while Thorez exhorted Communists to vote against de Gaulle as a gesture of defiance; the result in metropolitan France was 56 per cent in favour and 18 per cent against. Strengthened by this mandate, de Gaulle startled both the army and the settlers by publicly saying that Algeria was a liability to France and that rapid disentanglement was the best policy (11 April 1961).

The blow to the army was all the greater, in that Challe's strategy against the FLN had produced results. If his policy of evacuating peasants from vulnerable areas created discontent, it did at least leave the French forces free to shoot at everything that stirred. Not surprisingly, Challe's replacement by the unswervingly obedient General Crépin (April 1960) had already fuelled ill will; and this latest move by de Gaulle seemed to shrug off all the army's labours, as well as leave its Muslim auxiliaries, the *harkis*, a potential prey to nationalist vengeance.

The mainspring of the impending *putsch* was a group of colonels, Argoud, Broizat, Gardes, Godard, Lacheroy, and Vaudrey, all of whom had been transferred from Algeria to less sensitive posts elsewhere. The colonels needed more senior figures, however, to secure the following of the rank and file, and to ensure civilian acceptance of whatever regime might emerge from a *putsch*. The triumvirate of Salan, Jouhaud, and Zeller was still available—all of them now in retirement, with Salan enjoying the relative privacy of residence in Spain, where Lagaillarde, Susini, and Ortiz had also sought refuge. But it was the addition of Challe—also in retirement, but bitterly incensed by de Gaulle's announcement—that gave the conspiracy a popular and respected leader.

The spearhead of the *putsch* was the First Parachute Regiment of the Foreign Legion, which occupied the strategic buildings of Algiers in the small hours of 22 April 1961. Nine other regiments joined Challe in the course of the day, but others rejected Challe's telephone appeals. A major factor here was the refusal of the conscripts to obey officers who were anxious to respond to the summons. Moreover, de Gaulle's personal appeal of the following day, heard on countless transistor radios in Algeria, confirmed the conscripts in their resolve. Recognizing that no further progress could be made, Challe surrendered two days later. In Paris itself, Debré summoned residents to block the road with their vehicles to prevent a recrudescence of Operation Resurrection; and the enthusiasm of their response drove an infuriated de Gaulle to berate Debré for causing 'ce tumulte grotesque que vous organisâtes sous mes fenêtres'.

De Gaulle's victory over the dissidents strengthened his hand in going for a swift settlement of the Algerian question. Secret negotiations with the rebel Muslim Gouvernement Provisoire de la République Algérienne had been taking place intermittently since June 1960. But a major sticking-point continued to be sovereignty over the Sahara, with its rich oil and natural gas deposits. Eventually, France gave way on the issue of territorial sovereignty (5 September 1961), in the expectation that French oil companies would continue to enjoy special privileges. Even so, the negotiations continued to be a source of continual recrimination among the Muslim rebels themselves, with the hard-liners accusing Belkacem Krim of not being firm enough in his demands.

In France, opinion was increasingly in favour of a rapid settlement. With the war taking 10 to 15 per cent of the entire French budget, the public found it increasingly convenient to remember that the bulk of the European settlers were of non-French stock. Moreover, because the 1960s were 'hollow years', as a result of the impact of the Second World War on the birth-rate, the presence of half a million troops in Algeria was proportionately a much greater drain on the French work-force—the cost in lost labour being in the order of some £400 million a year. The implications of a rapidly growing Muslim population for the balance of French politics was also invoked more often; and it is significant that de Gaulle delayed implementing the election of the President by universal suffrage until Algeria and Black Africa had been shuffled off the electoral roll. As for the army, its dissidents were unlikely to attempt a repetition of the April 1961 débâcle. With over half of the defence expenditure absorbed by the hole-in-the-corner war in Algeria, career soldiers realized that de Gaulle was right in seeing the army's destiny as lying in the infinitely more respectable role of upholding France's position on the international stage—even if army leaders suspected that the navy and the air force would be the main beneficiaries in a world of nuclear deterrence.

A final settlement was not reached until 18 March 1962, when the Évian agreements listed a series of proposals to be put to referendums in France and Algeria. Algeria would receive full independence, while the Europeans would have three years in which to opt for Algerian nationality or remain as foreign residents without the privilege of dual nationality. In this three-year period they could not be dispossessed without compensation, but thereafter they would be subject, with the Muslim population, to whatever laws the new Algerian state chose to enact. Various provisions were made for the guarantee of European representation on the political institutions that would govern Algeria in the transitional period. French troops would be withdrawn in stages during these three years, while the naval base of Mers-el-Kebir would initially be leased to the French for fifteen years, with renewal open to negotiation thereafter. On the difficult issue of Saharan oil, French oil companies were to continue to enjoy their concessions and to receive preferential treatment in the allocation of newly discovered resources during the next six years.

In fact, de Gaulle had progressively given way on virtually every sticking-point: his demand for a cease-fire before agreement; his refusal to recognize the GPRA as sole negotiator; his insistence on four years of 'peace' before self-determination; the question of sovereignty over the Sahara; and his insistence on dual nationality for the European settlers—all of these were in turn abandoned. Metropolitan France accepted the agreements by 64.86 to 6.65 per cent (8 April 1962), while Algeria accepted them by 76.8 to 0.25 per cent (1 July 1962).

The three-month delay in holding the Algerian referendum was caused by the escalation in violence by the desperate European die-hards. The activities of the Organisation Armée Secrète had assumed serious proportions in September 1961, when uncaptured *putsch* members of the previous April joined with militant settlers in a series of terrorist outrages, designed to provoke Muslim reprisals and to demonstrate what might come should Algeria be given independence. Some forty schools were destroyed in the weeks following the Évian agreements, together with the library of Algiers University and many economic and social amenities. The principal accredited leaders of the OAS were Generals Salan and Jouhaud, who nevertheless found it increasingly difficult to control the excesses of their followers. Indeed, when they were arrested in the spring of 1962, they appealed for an end to the violence—which had been rendered pointless by the Évian agreements. The nominal leadership of the OAS passed to Georges Bidault—de Gaulle sarcastically commenting: 'At last, some good news!'—but Bidault later denied exercising any executive role in the OAS, and was to spend the next four and a half years in Brazilian exile. The other respectable hope in the firmament of the ultras, Jacque Soustelle, had likewise packed his bags for foreign climes (August 1961) after resigning from the cabinet early in the previous year.

Despite the short-term guarantees of the Évian agreement, 85 per cent of the European population had left the country by the end of the year. Not only did these immigrants pose an employment problem for France—few of them being prepared to accept the menial jobs that their Muslim predecessors had traditionally taken—but they were to strengthen the militant right-wing organizations that extended OAS violence into metropolitan France. Even so, the liquidation of the Algerian question removed the principal *raison d'être* of right-wing extremism; and the economic boom of the 1960s was quickly to resolve the employment problems of the new arrivals.

On the French side, the eight years of fighting had entailed the death of 17,456 soldiers—5,966 'accidentally', as a result of mistaken identity, mishandling of weapons and vehicles, and the numerous other mishaps that befall nervous and inexperienced conscripts. As for the Muslims, 141,000 had been killed by the French security forces, and a further 16,000 by the FLN. The FLN additionally abducted and probably killed another 50,000 Muslims, and 12,000 were killed in internal settlings of scores within the FLN. At least 30,000 *harkis* were killed in independence reprisals, many with appalling cruelty. In Alistair Horne's words: 'army veterans were made to dig their own tombs, then swallow their decorations before being killed; they were burned alive, or castrated, or dragged behind trucks'.[2]

The liberation and return to Algeria of Ben Bella resulted in his becoming Algeria's first President in September 1962. Claiming that the Évian agreements had to be revised, he proceeded to collectivize the farms of Europeans who had stayed behind—and demanded that 50 per cent of the oil and gas profits of foreign companies should be given to Algeria. Despite these breaches of the Évian settlement, French aid to Algeria continued, accounting for nearly a third of the total overseas-aid budget in 1965. After Ben Bella was himself ousted in 1965, his successor, Houari Boumedienne, was to take over majority holdings in French oil and gas companies in 1971—leaving little of the legacy of empire that the French believed they had retained at Évian.

15
'La République des Citoyens'

In a much-quoted phrase, de Gaulle spoke of the need for France to 'marry her century'; but 'the problem is to carry this through without France ceasing to be France'. Yet, when he spoke about 'the French' rather than 'France', there was much that brought him almost to despair. Shortly before his death, he told Malraux that the great weakness of the French was their powerlessness to believe in anything.[1] They were both blessed and cursed with a corrosive intellect which cast doubt on everything and destroyed enthusiasm for any cause for fear that it might turn out to be empty. They had too much intelligence, and too little instinct.[2] Following Bergson, he claimed that it was always the weakness of intelligence to be embarrassed by reality, because the untidy inconsistencies of reality do not correspond to the expectations of intelligence. And it is here that instinct and intuition come to the rescue, recognizing the nature of a situation when the intelligence is merely bewildered by it.

De Gaulle himself was a powerful combination of principle and pragmatism. His principles were simple and few—and tenaciously held. But being simple and few, they allowed him unusual latitude and flexibility in their application. He recognized the power of material circumstance and the fundamental importance of economic forces. But he attached an equal importance to psychological factors, which he saw as consistently underrated by the bulk of politicians. He believed that self-respect was ultimately the only durable form of happiness, either for an individual or a collectivity—and that self-discipline was the key to its achievement. This partly reflected his own personality, upbringing, and harsh experience. He had been wounded three times in the First World War, and had made five attempts at escape from German captivity—his conspicuous tallness frustrating each attempt. At the same time, as an officer and leader of the Resistance, he had seen for himself the degree of self-sacrifice that people were capable of when highly motivated—especially when with members of a larger collectivity, be it a regiment, a Resistance cell, or the nation. Indeed, for him, self-respect found its highest fulfilment in service to the State.

De Gaulle saw the State as the crystallization of social bonds, which, like the family, was an inescapable expression of human nature, corresponding to the deepest levels of its instinct. It was the tragedy of French politicians that they seemed incapable of appealing to these instincts, partly because they were too blind and too cynical to recognize their existence: 'For today, as always, it is ideas which lead the world'[3]—a remark which echoed that of another prag-matic patriot, Georges Clemenceau.

But if principles determine the ends, the means of attaining these ends depend on circumstances. In the matter of means, there was no absolute truth, 'only the circumstances': it was these that determined the choice of method, provided that it did not conflict with the principle at issue. De Gaulle's prag-matism was often misunderstood. He was continually charged with opportun-ism—especially by doctrinaire politicians of the Left, who saw principles as determining the methods as well as the ends. But de Gaulle professed always to have an eye on the long-term implications of tactical moves, especially their repercussions on personal and collective self-esteem, and how they would appear to future generations: 'In the last resort, decision-making is a moral question'.[4]

How far his Catholicism affected his thought is hard to assess. A practising Catholic and assiduous church-goer, there was little in his private or public life that suggested that he did not take it seriously. But a perceptive colleague once described him as basically a monist or pantheist, seeing Catholicism as the human expression of this inner reality. And it is tempting to see parallels with Bismarck's Protestantism, which A. J. P. Taylor described as owing more to 'the God of the Old Testament ... the God of Battles ... He believed that he was doing God's work in making Prussia strong and unifying Germany'. And like Bismarck—and most Catholic kings—he was at pains to see that the Church kept to its place in the civil order of things. It is true that Catholic private educa-tion emerged from his presidency with far-reaching gains (p. 296). Yet he saw these concessions as meeting a long-standing grievance of a substantial section of the community; and their prime purpose was to help create a national con-sensus—even if they triggered off a short-term rumpus in militantly secular circles.

Although de Gaulle held that self-esteem found its highest expression in service to the nation, his patriotism had little in common with that of Action Française. He did not share the Maurrassian view that France was the prime embodiment of civilization, the only heir to the Graeco-Roman tradition. Although as a young man, de Gaulle had attended meetings of a group of Action Française disciples, he found Maurras's ideas too cerebral and remote from the realities of the modern world. His father had been a monarchist and a former teacher in a Jesuit school, but de Gaulle's own respect for the royal line arose from its links with the nation's past. Despite private exchanges of courtesies, de Gaulle did not easily forget the Comte de Paris's ambivalent record during the Second World War (p. 106); and in any case, he believed that monarchy in France had had its day.

De Gaulle was deeply conscious of the French contribution to civilization. The intellectual distinction of France was both the glory and the curse of her people, advancing thought and culture on the one hand and creating deep internal divisions on the other: 'ça, c'est notre génie'. But his patriotism sprang from his belief in the nation-state. He recognized the same rights and aspirations in other nations—which is why he never regarded Germany or Russia as an implacable enemy. Indeed, his belief that Russian aspirations were national rather than ideological made it easier for him to regard Russia as a country like any other—a country with which France could usefully have a working relationship when it suited her, or not when circumstances directed otherwise. It was also his respect for the force of national concerns and loyalties that made him sceptical of the future of overseas empires.

PRESCRIPTION AND THE NEW CONSTITUTION

The prime task of government in de Gaulle's view was to bridge the deep internal divisions of the French people and to give them back their self-esteem after the humiliations of the past thirty years. The implication of this for constitution-making pointed towards a semi-presidential regime along the lines contained in the Bayeux proposals of 1946 (p. 141). This would ensure a strong, purposeful executive, able to give direction and coherence to the disparate demands of a divided electorate, while democratic safeguards would be provided by a representative parliament with more modest powers than its predecessors.

The election of the President, which the constitution of 4 October 1958 delegated to approximately the same 80,000 or so individuals who chose the senators, not only reflected de Gaulle's determination to free government from the control of parliament, but also his caution when faced with the opportunity of edging France towards his ultimate goal of a popularly elected President. This curious 'electoral college', which was little more than a 'liste de notables' in the old Napoleonic sense of the word, also reflected Debré's view of the President as an arbiter, a constitutional monarch above politics, who was not the leader of a party like the American President—and who therefore ought not to be elected by the general public, otherwise he would inevitably become identified with the political majority of the day.

The President had the customary attributes of Heads of State, but without any nominal right of veto on legislation. However, he could resort to a referendum, a dissolution of parliament, or the assumption of emergency powers under Article 16 without necessarily obtaining the counter-signature of his ministers—although the constitution obliged him to consult the Premier and the Presidents of the two houses of parliament beforehand. The weapon of dissolution, however, could not be used more than once in a twelve-month period. The President chose the Prime Minister and, 'sur proposition du Premier Ministre', the other ministers. This was sufficient to breathe official life into the government—parliamentary investiture no longer being required.

It was still incumbent on the Premier to get his legislation through parliament; but de Gaulle made it clear that if parliament withdrew its confidence from the government, he would not hesitate to call an election.

During the first four years of the new Republic, Debré's concept of an arbiter President was upheld, in that it was the Prime Minister who took charge of all spheres of policy, except defence, foreign affairs, Algeria, and the Communauté, which were collectively regarded as the President's 'domaine réservé'. This was to change markedly in the following four years, however, when de Gaulle, effectively head of the parliamentary majority, took over all spheres of government from the more compliant Georges Pompidou (14 April 1962–11 July 1968). Whereas Debré had had a respectable parliamentary base, Pompidou had none; and as de Gaulle's *directeur de cabinet* during the General's brief tenure of the Matignon in 1958, Pompidou was very obviously his master's man, whatever his personal strengths. From 1966, however, de Gaulle increasingly allowed domestic matters to slip back into the hands of the Prime Minister, as he became preoccupied with foreign policy and increasingly aware of his personal limitations in the socio-economic sphere—where his measures were encountering considerable opposition. The violent upheaval of 1968, however, was paradoxically to reverse this trend. Despite the fact that it was Pompidou's cool appraisal of events that enabled the Gaullists to emerge strengthened electorally from the crisis, he was then replaced by Maurice Couve de Murville (11 July 1968–20 June 1969), who had little experience of domestic affairs, and was therefore quite content to see de Gaulle resume control of general policy.

This toing and froing of responsibility was also reflected in the conduct of government meetings. As had been the case under the Fourth Republic, it was the custom for the *conseil des ministres* to be chaired by the President, while the Premier chaired the larger *cabinet*. After 1962, however, when de Gaulle took over responsibility for most spheres of government, *cabinet* meetings became rare, and the *conseil des ministres* became the usual forum for government business. But, since many discussions required the views of junior *cabinet* figures, it became de Gaulle's practice to invite them to attend relevant meetings of the *conseil des ministres*, over which he presided, rather than call meetings of the *cabinet*, where he did not. The conduct of the weekly meetings of the *conseil des ministres* left no doubt as to where decision-making lay in French government. Routine meetings were merely a matter of report, in which the President and the relevant minister would outline the decision that in effect they had already taken. On major issues, the President would invite each minister in turn to give his views; and then at the end, often without further discussion, the President would announce his decision. Otherwise, ministers were expected to mind their own business and only comment on other spheres of government when expressly invited to do so. When Pinay raised the matter of de Gaulle's cavalier attitude to NATO in November 1959, de Gaulle icily asked: 'Monsieur le Ministre des Finances is interested in problems of foreign policy?' And when Pinay persisted, de Gaulle merely said: 'Thank you, Monsieur Pinay. Gentlemen, the meeting is adjourned.'[5]

On the other hand, de Gaulle was relatively sparing in his recourse to the emergency powers accorded him by the constitution—their use mainly being confined to the periods of threatened subversion in 1960–2. His use of referendums, by contrast, was a major cause of controversy throughout his presidency. The constitution expressly confined referendums to constitutional matters, or issues with constitutional implications. But de Gaulle repeatedly used them as political weapons, even if he perfunctorily dressed them up to look like institutional proposals. The constitution also insisted that referendums should be held on the initiative of parliament or the Premier, not the President.

Yet, in reality, it was de Gaulle who personally decided on each of the referendums that took place under his presidency—with a compliant if sometimes hesitant Premier tamely acquiescing. Pompidou was particularly unhappy about the referendum of October 1962 on electing the President by universal suffrage, since it violated the requirement that the terms of the proposal should be discussed in advance by parliament. Indeed, when Pompidou himself became President in 1969, he was extremely wary about using referendums at all, feeling that he lacked the personal prestige to risk ruffling parliament's self-esteem.

Although the Prime Minister was responsible to parliament, in practice it was the President who appointed and changed him. The changes of Premier in the 1960s were all presidential decisions, sometimes running counter to the prevailing feeling in parliament, as when Pompidou was replaced with Couve de Murville in 1968 (p. 328). Conversely, when Pompidou resigned in October 1962 as a result of the only successful censure motion under the Fifth Republic, de Gaulle demonstrated his authority by promptly reappointing him. With parliamentary investiture no longer a necessity, de Gaulle and his Premiers had a much freer hand in their choice of ministerial colleagues. In Debré's government (8 January 1959–4 April 1962), only fifteen of its twenty-eight members were parliamentary figures; most of the others were specialists from public administration. With the new Republic *in situ*, Debré could afford to pass over Mollet and Pflimlin—who had provided de Gaulle with a ministerial gangplank from the Fourth Republic—and keep only Pinay from among the leading bastions of the old regime. It became a commonplace of journalists to compare the Gaullist era with the Second Empire, and describe the burgeoning rule of experts and technocrats as the advent of the new Saint-Simonianism.

This change in the type of minister appointed was reinforced by a novel feature of the constitution: the incompatibility of functions. Members of parliament who were appointed to ministerial office had to relinquish their seats to a substitute, whose name featured on the ballot-paper at the time of the member's initial election. This rule reflected not only de Gaulle's insistence on the sharp distinction between legislative and executive function, but also his bitter experience as Prime Minister in the Liberation era, when his cabinet colleagues appeared to show more loyalty to their parties than to the government. Ministers, of course, still had the right and duty to speak and answer questions in parliament on their spheres of responsibility. Yet the remoteness of ministers from the electorate, and from the rough-and-tumble of parliamentary politics, was eventually realized to be something of a disadvantage; and by the 1970s it was increasingly expected that non-parliamentary ministers should acquire political respectability by contesting a seat at the next election—if only to relinquish it to their substitutes in the event of election and reappointment to the next cabinet. Moreover, the utility of certain old parliamentary traditions was to become sufficiently recognized for ministers to be encouraged to develop provincial fiefs, where, as mayor of the principal town, they could emulate their Third and Fourth Republican predecessors and sink deep roots at local level.

The new type of Fifth Republic minister provoked a division of response in the civil service. In general, the senior grades sympathized with the regime's technocratic, *étatist* principles, and shared a fellow-feeling with their former colleagues who had become ministers. Indeed, the Club Jean Moulin was an influential forum, where senior civil servants and rising *énarques* discussed current problems with ministers and academics. Some *fonctionnaires*, however, felt nostalgia for the Third and Fourth Republics, when the brevity of governments *ipso facto* gave the civil service importance as a firm element of continuity beneath the shifts of the political surface. The governments of the Fifth Republic, by contrast, were much longer lived, and the civil-service background of many of the ministers deprived the bureaucrats of their traditional mystique as sole repositories of a hallowed corpus of administrative knowledge and wisdom.

Running parallel with these misgivings, was a growing fear among the ministerial staffs that the expansion of the President's body of Élysée advisers was rapidly creating an occult counter-government that was overriding the proposals of the ministries—but without a corresponding responsibility to parliament or public. This disquiet, however, was largely unjustified, in that the total number of advisers never exceeded forty, most of them talented civil servants seconded from the *grands corps*, but without the back-up facilities that would make them serious rivals to the ministries.

Parliament

The tighter controls which the Gaullist constitution placed on parliament stemmed from the assumption that no French government would enjoy an overall majority. It was the unexpected triumphs of the Gaullists and their allies in successive elections that made many of these controls seem unnecessarily restrictive. It had been a broad feature of earlier constitutions that silence on an issue was to be understood in a permissive sense as far as parliament was concerned—but in a prohibitive sense when it concerned the executive.

The new constitution, however, explicitly outlined those areas in which parliament could legislate, the implication being that it had no right to legislate on other matters. Another major advantage to the executive was that parliament could no longer control its own timetable, the government now having the right to stipulate when its bills should be discussed.

Critics made much of the fact that parliamentary sessions were now reduced to five months in the year, averaging just under 150 days or about 500 hours. While this was substantially less than in Britain, where the time allowed to the Opposition raised the overall average to 167 days or 1,528 hours, it was more than that of many European countries—for example, Italy (137 days), Sweden (123), Denmark (111), Ireland (84), and Austria (39). But although the constitution recognized parliament's right to convene an extraordinary session, de Gaulle claimed that it was his right to refuse it if he thought it unnecessary— and he did so in the spring of 1960. Indeed, it was symptomatic of the new dispensation that the great issues of the day, such as Algeria, were discussed very little in parliament. These, in effect, were largely a matter for dialogue between the President and the people—the President announcing his intention, and the people saying 'Yes' or 'No' in referendums. Characteristically, the Évian agreements were never put to a parliamentary vote; after a debate they went straight to a referendum.

Even the nature of parliamentary legislation was changed. It had been the traditional practice in France for much of the peripheral detail of new proposals not to be included in the parliamentary bills themselves, but to be left to administrative regulations that were issued afterwards. The Fifth Republic carried this practice much further. Laws on a whole range of issues were now limited by the constitution (Article 34) to the enunciation of broad principles, thereby transferring responsibility for the specific provisions of such laws to civil servants and legal experts. At the same time, the increased range of matters now dealt with by government ordinance greatly reduced the number of minor bills handled by parliament, cutting its overall legislative output by more than half. With much of the structure of the new regime still to be erected, it was perhaps inevitable that there should be little room for private members' bills in the early years of the Fifth Republic. In 1959 such bills accounted for less than 2 per cent of enacted legislation, gradually rising to 22 per cent by 1967.

The great rock that had interrupted the flow of parliamentary business during much of the Fourth Republic had been the budget. Discussion in the Assembly was now limited to a maximum of forty days, after which it must pass to the Senate. If the whole parliamentary procedure was not completed in seventy days, the government would then implement the budget by ordinance. This provoked a predictable outcry that the new system destroyed parliament's sacrosanct open season for creating mayhem, obliging deputies to digest twenty kilos of documentation in a few weeks. But Debré indignantly retorted that the British Parliament disposed of the budget in a single afternoon. (Like Cavour a century earlier, Debré entertained many convenient misapprehensions about the Mother of Parliaments.)

The constitution also provided the government with various gagging devices to rule out unwelcome bills and debates. Resort to the inadmissibility clause (Article 41) depended in practice on how secure the government felt. The reassuring majority of 1962 saw its use in the Assembly drop from six times a year in 1961 to less than one a year in the rest of the decade; but Giscard d'Estaing felt the need for it more often in the 1970s, when it was used on average four times a year during his presidency. More frequently brandished was the prohibition of bills and amendments that would affect the level of government revenue and expenditure (Article 40). The government axed over a hundred such bills in 1959 and sixty in 1967–8—symptomatically, periods when the government was unsure of its majority. Stifled amendments ran into thousands. Another useful weapon, if deputies were in a nit-picking mood, was the *vote bloqué*, which enabled the government to confront them with a take-it-or-leave-it vote, bypassing article-by-article discussion. An average of seventeen bills a year were pushed through in this way when Pompidou was Prime Minister.

It was also open to government to request parliamentary permission to legislate by ordinance during a fixed period (Article 38), such legislation requiring subsequent ratification by parliament if it was to remain on the statute book. This facility was resorted to seven times during the first two parliaments (1959–67), and twice during the period 1967–71—the main issues that were treated being Algeria, the EEC, agriculture, social security, alcoholism, and prostitution. Indeed, if all these exceptional methods are taken into account, 15 per cent of de Gaulle's successful legislation was passed in one or other of these ways.

If the powers of government over parliament were strengthened, those of parliament over government were correspondingly pruned. *Interpellations* disappeared altogether; and although the circumstances obliging a government to resign had a superficial similarity to those of the Fourth Republic, there were significant differences. Bills that were made an issue of confidence were automatically deemed adopted unless a motion of censure was laid down within twenty-four hours and passed by an absolute majority of the house. Moreover, in motions of censure, abstentions and absences were counted as positive votes for the government—a procedure that was unique to France. Deputies who wished to bring down the government, therefore, had to come out into the open and vote for the censure motion—whereas, under the Fourth Republic, cautious deputies could hedge their bets by limiting their wrecking tactics to abstention. Making a bill a matter of confidence now became a convenient way of converting abstentions into support, and bringing to heel disobedient back-benchers. Given these general rules and the government's majority in the Assembly, it is scarcely surprising that there was only one successful censure motion among the twenty-five proposed in the first two decades of the Fifth Republic. And even that—the famous vote of 1962—merely saw the defeated Premier reappointed by his unrepentant master.

Armed with these constitutional weapons and a solid phalanx of support in parliament, the average life of a government grew from six months under the Fourth Republic to three years in the 1960s, with Pompidou heading most of them. And above them stood the real ruler, the President, impregnable for seven years, unless he indulged in patent misdemeanours. He was the true author of policy, yet only his Premier carried responsibility for it before parliament. The disconcerting novelty of stable government to the old Republican personalities of the past was exemplified in Vincent Auriol's speech of November 1959, in which he declared that the absence of ministerial crises called into question whether France was still a free country. Yet if cabinets were more secure, they were subject to much internal reshuffling. In the first decade of the Fifth Republic, there were twelve Ministers of Education (as against fourteen in 1944–58) and eleven Ministers of Information. But most of these changes were as a result of decisions of the President or Premier, rather than of pressure from parliament.

Not everything was negative in the government's handling of parliament. Debré introduced question time on Friday afternoons; but the choice of day was guaranteed to render it a largely harmless affair, with most members wanting to set off to their constituencies where many exercised municipal functions. It was the government's intention to discourage the proxy voting that had been such a feature of the previous regime. Part of a deputy's salary was made dependent on attendance; and the new system of electronic voting, in which each deputy had an individual key, was designed to make proxy voting more difficult. In practice, however, deputies merely supplied colleagues with a duplicate key, and it was a not uncommon sight to see a handful of deputies in a semi-deserted Assembly, turning a succession of keys, like night-porters illuminating an office-block in preparation for the nocturnal incursion of cleaners. Even the debates on such major issues as regional and Senate reform found less than 15 per cent of members in attendance. Not that debates were much fun under the Fifth Republic. Many members read their speeches, despite house regulations to the contrary, and there was a dull sense of ultimate pointlessness when the government's position was so secure.

Government control over parliament was also strengthened by the drastic curtailment of the system of standing committees. The 19 old committees of 44 members each, roughly corresponding to the various ministries, were replaced by 6, 2 of which had no less than 121 members each; and inevitably the wider remit and size of these committees reduced their expertise and capacity for informed discussion. Moreover, when bills passed from the committee to parliament, it was the government's original draft that was debated; and it was then incumbent on the committee members to propose once again the amendment that they had adopted in committee. Furthermore, the bill was steered through parliament by the relevant government minister, not the *rapporteur* of the committee as under the Third and Fourth Republics.

Turning to the upper house, the Council of the Republic reverted to its prewar shape—not only resuming its old title of Senate, but also its nine-year mandate, with a third of the senators being replaced every three years. Debré and his fellow constitution-makers assumed that the Senate would be a potential ally against an impetuous Assembly, rather than a brake on government; and it was with this expectation that the caretaker duties of the President of the Assembly were transferred to the President of the Senate, in the event of the President of the Republic resigning or dying in office. The *navette* system of the last years of the Fourth Republic was retained—but with important modifications which substantially increased the power of the government over legislation.[6]

If these various adjustments seemed to assure a quiet life for the government, the outcome of the senatorial elections in April 1959 was something of a disappointment. To guarantee a completely fresh start, and a healthy influx of Gaullists, all the seats were up for contest, not just a section as in previous and subsequent elections. But the base of the electoral pyramid rested as before largely on the municipalities; and the municipal elections of the previous months had seen the re-election of many of the old-established local *notables*, with no particular allegiance to Gaullism—with the result that the personnel of the Senate remained relatively unchanged. A number of crusty champions of Republican democracy who had been defeated in the Assembly election made a come-back in the Senate—including Edgar Faure, François Mitterrand, Gaston Defferre, and Jacques Duclos—and de Gaulle was to find the Senate more of a truculent grouser than a co-operative ally. In Chapsal's words, it became 'le temple de la critique et de la mauvaise humeur'—in many ways the ghost of the Fourth Republic.

President, People, and Television

These constitutional changes also reflected much of de Gaulle's deep-seated suspicion of parliamentary representatives, whom he considered representative of pressure groups and self-interest rather than the national interest. Far from being intermediaries between people and government, he saw them as obstacles to an effective rapport between the public and those who ruled in its name. Not only did they disperse and deflect the opinions of the electorate, but they were incapable of acting in the national interest in moments of crisis—as the recent past had demonstrated. They had abandoned parliament's powers to

Pétain in 1940; they had cast aside de Gaulle's proposals for a more effective regime in 1946; and they had shown themselves incapable of dealing with the Algerian question in 1956–8. While assemblies were essential for democratic debate, they were no good for decision-making and action. It was therefore part of his overall concept of government that not only should parliament be kept in its place, but the President should cultivate direct contacts with the public. Apart from the elaborate use of referendums, there were his visits to the provinces and the hand-shaking with the crowd—'crowd baths' as his nervous security aides called them—which set a pattern of behaviour that other Heads of State were to imitate in their 'walkabouts'. More important were his television broadcasts, which were very different from the radio talks of Pinay or Mendès France. De Gaulle's 'charisma' was not of the kind that this overworked term popularly suggests. Neither his voice nor his manner was the type to rivet crowds. Nor did he charm or fascinate people in conversation. He was impressive in a cold, forbidding sort of way. As he himself wrote: 'Nothing heightens authority more than silence'. And André Malraux said that in conversation 'his silence was a question ... Despite his politeness, one always seemed to be reporting to him'. What people admired in de Gaulle, and what inspired able and intelligent men of widely divergent political views to follow him, was his combination of courageous integrity on the one hand, and his clear-headed ability to appraise the requirements of a situation on the other. The public watched his television appearances because of who and what he was, rather than because of the performances themselves. His initial appearances were at times of great crisis—January 1960 and April 1961—when viewers were transfixed by his direct appeal to them to be prepared to fight against insurrectionary elements. But the impact, once made, left a lasting impression. De Gaulle was careful to use television sparingly; his appearances were something of an event—and viewers brought to them the remembered excitement of the earlier ones. At the same time, his style and delivery were a careful blend of simplicity and ambiguity. The language was comprehensible to all, but what it signified in terms of specific intent was often deliberately left unclear, lest his options be prematurely limited.

With the number of television sets in France rising from one million to three million in the first five years of the Gaullist regime—compared with twelve million in Britain—government control over the Office de la Radio et Télévision Française became tighter. Premiers, such as Guy Mollet, had blatantly censored programmes during the last years of the Fourth Republic, when the conduct of the Algerian war was an embarrassment; and with the ORTF now under the control of the resurrected Ministry of Information, opposition leaders were given scant coverage. Although the Gaullists' opponents were allowed a greater airing in 1966, the public was kept in the dark about what was happening in May 1968—with the result that the ORTF staff staged an initially successful protest, and then went on strike, the whole episode ending in dismissals and yet tighter government control. With only one channel until 1964—and no commercially owned competition until the 1980s—the public had no option, short of switching off the set, but to watch what the government allowed it to see.

The development of a strong rapport with the views of the population was an important element in de Gaulle's edging towards his prime constitutional goal: the election of the President by universal suffrage. His decision to take the final step in 1962 stemmed from two factors. His popularity was high after the settlement of the Algerian war—twice as high as in the previous year—and public fears that an assassin's bullet could leave France leaderless and vulnerable to another *putsch*, assured him of an attentive and sympathetic audience should he chance his arm with a bold constitutional proposal. What gave him a particularly strong leverage was the Petit-Clamart assassination attempt of 22 August, when an OAS hit group pursued his car, firing a hundred or so rounds, which only the skill of his driver prevented from being fatal. A stunned public took little persuading by de Gaulle that it might suddenly find itself having to look for a new President with sufficient prestige to continue the work of unification and stability that he had initiated. If there was no one whose personality and achievements could match his own, then the new President must be able to command respect through the forces that he represented. The current mechanism that elected the President, and in a different fashion the senators, would scarcely confer prestige on an individual who did not already possess it, whereas a President elected by universal suffrage could claim to represent the people in a way that parliament could not hope to emulate, since it was merely an aggregate of disparate opinions, with no member—no matter how large his majority—representing more than 0.2 per cent of the national electorate.

An alternative solution to the sudden-death problem would have been for de Gaulle to groom a Vice-President on the American model. But de Gaulle had consistently opposed the idea of a dauphin who would automatically succeed in the event of the President dying in office. There was always the danger that senior party members would start to look to the rising sun, and de Gaulle intended to share the firmament with no one. But, more disinterestedly, it was questionable whether a President who owed his position to misfortune could exercise the same authority as one who had been elected by the will of the people specifically for the presidency. The method de Gaulle proposed was broadly similar to that of the parliamentary elections: there would be two ballots, with an absolute majority required for election on the first, or a simple majority on the second. Unlike the Assembly election, the second ballot would be confined to the two leading candidates.

Such a change to the constitution, however, could not legally be put to a referendum until it had first been approved by both houses of parliament. Moreover, this approval was not subject to the normal legislative procedure, by which the government could ultimately give the final verdict to the Assembly, in the event of the Senate being recalcitrant. And de Gaulle had every reason to suspect that the Senate would be difficult—given its political composition and the fact that it was the senators' own electoral base that was being indirectly disparaged by the proposed change. The Assembly, too, was disposed to be belligerent; and the issues were never more evident than in an exchange between Paul Reynaud, highly respected relic of the Third Republic, and the Gaullist benches (4–5 October 1962). A decade earlier, Herbert Lüthy had described Reynaud as 'the most brilliant mind and most ambitious politician of the French right and a perpetual prophet of disaster, both in the pre-war and post-war periods, whose mistake it was always to be right'.[7] This was an assessment to which many of Reynaud's parliamentary colleagues would have subscribed; and there was solid backing for him when he protested: ' "For us republicans, France is here in parliament and nowhere else." "No! France is in the people!" "The representatives of the people gathered here *are* the nation, and there is no higher expression of the will of the people than our vote after a public debate." '

A vote of censure was carried by an absolute majority, isolating the UNR and its close allies. The deputies nevertheless realized that they were on uncomfortable ground, since they were open to the obvious accusation that they were denying the public a voice in the matter. And the government, knowing this, promptly dissolved parliament—confident that the referendum would make the deputies look foolish. Moreover, lest there should be any wavering, de Gaulle broadcast to the nation that if the 'yeses' were 'faible, médiocre, aléatoire', he would resign.

If the gamble came off, the success was not spectacular. The 'yeses' on 28 October amounted to 62 per cent of those voting, but this was still under half of the electorate. Even so, it was enough to persuade the Constitutional Council to drop any idea of challenging the legality of the referendum; and it put heart into the Gaullists and their allies as they entered the legislative elections of the following month.

THE BIPOLARIZATION OF POLITICS

The parliamentary elections of November 1962 not only endorsed de Gaulle's bold constitutional step, but they marked the beginnings of the bipolarization of French politics that was to transform the country's political life. As everyone anticipated, the referendum threw its shadow right across the electoral campaign, causing the Socialists, Radicals, MRP, and conservatives to move together in what was loosely termed 'le cartel des Non', with the Communists tagging on as awkward outsiders. The adherence of the left-wing parties was predictable enough, but the addition of the MRP and conservatives followed agonizing soul-searching by many of their supporters. The MRP had been broadly sympathetic to de Gaulle's Algerian policy, and this had been the prime reason for their continual presence in the government during the early years of the regime. They had been increasingly impatient, however, with the conservatism of the government's social policy, and had serious misgivings about what they saw as its unparliamentary tendencies. But it had been essentially de Gaulle's attitude to European co-operation that had caused the resignation *en bloc* of the MRP ministers from Pompidou's first cabinet on 15 May 1962. The conservatives, for their part, had broken with de Gaulle principally over what they saw as his sell-out in Algeria, but also because of the government's refusal to continue the traditional feather-bedding of farmers.

Leading the government forces was André Malraux's Association pour la Cinquième République, an *ad hoc* umbrella, which enabled a number of free-range candidates and crypto-Gaullists from other parties to stand beside the UNR without being swallowed by it. Should anyone doubt the nature of the confrontation, de Gaulle told the nation on 7 November that the conflict would be between the 'Oui' majority and 'the parties of yester-year'. Although he was careful not to endorse the UNR by name, the arbiter President of Debré's imagination was already slipping into the role of party leader. Even on the first ballot, the Socialists chose not to challenge a number of conservative opponents of de Gaulle. But even restraint on this scale was not enough to prevent the UNR emerging with the highest poll of any party since the war (35.5 per cent). The greatest losers were the conservatives, whose 'Non' label lost them much of the support they had picked up in 1958, when they had been part of the mounting pro-de Gaulle tide. A number of them also paid the penalty for too openly professing their nostalgia for the dead and increasingly unpopular issue of 'Algérie française'—a mistake that was likewise made by a large proportion of the extreme Right who were eliminated on the first ballot. The MRP was another party that paid heavily for deserting de Gaulle when his popularity was in the ascendant. On the other hand, the Communists positively benefited from the very consistency of their opposition to de Gaulle; and a number of Socialist voters who had been offended by Mollet's initial flirtation with the General switched their allegiance to PCF candidates. With 21.9 per cent of the vote, the Communists could not be ignored; and Mollet felt obliged to counsel voting for them on the second ballot wherever it might exclude a Gaullist. Indeed, in over a quarter of the constituencies the second ballot was a straight fight between a Communist and a Gaullist—a situation which tended to improve the Gaullists' successes still further, since many 'moderate' voters of the 'cartel des Non' could not bring themselves to vote for a Communist (Table 15.1).

The UNR, with 233 seats (including overseas supporters), were only eight below an absolute majority; but between the first and the second ballot, Valéry Giscard d'Estaing had led a phalanx of conservatives into the Gaullist camp, their thirty-five elected members forming the new Républicains Indépendants (RI) and thereby assuring de Gaulle of the absolute majority that his own party

Table 15.1.
National Assembly Elections of 18 and 25 November 1962
(Results for Metropolitan France only[a])

	% of vote on first ballot	Seats gained after both ballots	
		Total	%
Communists	21.9	40	8.6
Socialists			
PSU	2.0	2	0.4
SFIO	12.7	65	14.0
Radicals, etc.	5.8	39	8.4
MRP	8.2	31	6.7
Gaullists, etc.	35.5	256 (see text)	55.0
Conservatives, various independents, etc.	13.6	32	6.9
		465	

Note: The percentage of electorate casting valid votes was 66.6.

[a] The overseas deputies now numbered only 17.

had come so near to achieving. Yet the Left, too, could take some satisfaction from increasing their representation in parliament, even if it was a very slim shadow of what they had had under the Fourth Republic. The Assembly was now largely made up of a disciplined governmental majority and a left-wing opposition, which in the years to come was increasingly obliged to seek ways of sinking its differences, if it was to come anywhere near challenging the government. Bipolarization was beginning to emerge.

Institutional Factors

Although few contemporaries were fully conscious of the great change that was coming about in French parliamentary politics in the 1960s, most historians would now see the bipolarization of the 1960s as a major turning-point in post-war France. It was partly a product of the change in institutions, which obliged the parties to think and act increasingly in terms of two large opposing alliances. But it was also accelerated by the nature of Gaullism, which embraced a wide diversity of opinion on socio-economic and religious issues, yet put a premium on discipline and co-operative action, obliging its opponents to attempt similar strategies, albeit with chequered success. But, underlying these surface developments, were fundamental changes in public opinion itself, where the three 'Cs' of the colonial, clerical, and constitutional issues were fast being smothered by the relentless spread of 'class'—the socio-economic issues that had long dominated the politics of most of France's neighbours.

On the parliamentary level, the two-ballot *système majoritaire* not only increased the chances of governments having a working majority, but in practice it often confronted the electorate with just two candidates on the second ballot, a Gaullist and a left-wing alternative—thereby forcing it in the last resort to choose on what were fundamentally issues of 'social order' and the redistribution of wealth, albeit embroidered with other matters of political principle. Not only did the system oblige the various constituency parties to sink their differences and rally behind the two opposing finalists, but habits of co-operation increasingly flowed over into parliament itself, despite the disparate nature of constituency alliances where regional differences imposed a variety of tactics.

These consequences were to be even more marked in the new system of presidential election when it was put to the test in 1965. Here, the contest was explicitly confined to two candidates on the second ballot, and the resulting alliances were taking place on a national scale. And in all but one presidential election during the Fifth Republic, the ultimate contest was fought on issues of social order and social justice, whatever banners the protagonists chose to wave ostensibly. The parties became increasingly identified in the public eye with the opposing candidates, so that the bipolarization of the presidential election tended to affect the legislative elections and parliamentary life in general. The process was much more self-evident in the case of the UNR, since, unlike all other parties, its *raison d'être* had always been the support of an individual rather than the expression of a package of principles and group interests. Yet other parties soon found themselves being pulled by the new rules of the game into being support troops for a presidential candidate. Prophets and wishful thinkers pointed to the bipolarizing effect of the American presidential system, where the politics of a vast continent were dominated by two parties, each of which contained a wide spectrum of views—many in mutual contradiction—and whose only unifying principle was support for the party's presidential candidate.

The long-term process of bipolarization was to be reflected in the evolution of the governmental majority. During de Gaulle's presidency, it was essentially composed of the Gaullists and Giscardians. Under Pompidou, it was widened to embrace a section of the Centrists (including the rump of the MRP), while under Giscard d'Estaing it was extended to include the Radical Réformateurs as well. The Left, for its part, was also seeking to cannibalize the Centre. The Communists assumed that Gaullism would disappear with de Gaulle, and regarded their first priority as the elimination of the Centre. Indeed, the attacks of the Right and the Left on the Centre during the 1960s were all too reminiscent of the Russo-German partitions of Poland in the eighteenth century prior to their bloody confrontation in the world wars of the twentieth.

The Waning of Old Issues—Religion

The bipolarization of public opinion itself along a socio-economic axis was a gradual process, but owed much to the happenings of the 1958–62 period. The evaporation of the colonial issue was an obvious case in point. But so, too, if less obviously, was the waning of the religious issue. The assimilation of Catholicism into the ranks of Republican respectability had started well before the war and had gathered pace in the post-war decade. But the schools question had rekindled animosities in the 1950s, especially among militant defenders of the secular ideal. Paradoxically, it was the major concessions of the Debré law of December 1959 that ultimately dampened them down. The Debré law offered financial aid to private schools under two types of contract. Under the *contrat simple*, the State paid the salaries and half (eventually all) of the national insurance of those teachers in private schools whose qualifications and teaching were on a par with those in state schools. The *contrat d'association*, on the other hand, went much further and provided state contributions to the running costs of the school as well; but in return the school had to accept a greater degree of state supervision. Within four years, two-thirds of French private schools had signed contracts, 95 per cent of them *contrats simples*. But in the years to come an increasing number of private secondary schools saw the huge financial advantages of *contrats d'association*, which enabled them to cover the mounting costs of laboratories and libraries, as well as reassuring parents that academic standards were guaranteed by the State. For them, the *contrat d'association* eventually became the norm, whereas in primary schools, with their modest running costs, the more flexible *contrat simple* was preferred.

Concessions of this magnitude would have been inconceivable under the Fourth Republic, and their generosity largely sprang from the government's need to hold together a large right-wing majority that might split on the Algerian issue. Debré was in no sense a clerical, but he wanted to have done with the schools question while hostility was distracted by the Algerian war and the general upheaval of assimilating new institutions. The need to conciliate the Right, however, resulted in the concessions going much further than he had intended and provoked the resignation of the Minister of Education, André Boulloche. The bishops, remembering bitter lessons from the past, kept a low profile throughout the affair, and the bill passed the Assembly with the remarkable majority of 427 to 71 (24 December 1959).

Paradoxically, the very size of the concessions gave them an air of semi-finality which discouraged a continuing fight against them—until such time as the Left itself should enjoy a majority in parliament. And by that time the point at issue was to become a largely secular debate on the rights and wrongs of buying a separate education for one's children—with religion playing a secondary role (p. 373). The Barangé and Marie laws, by contrast, had offered such small concessions that they had been seen by their opponents as the thin edge of a growing clerical wedge—and had therefore created a continuing demand for their repeal.

The 1960s, in any case, were to witness a rapid fall in church attendance and in other signs of religious commitment. This was a phenomenon that affected all the Christian Churches in the developed economies, and were part of the much broader ground swell of doubt and criticism that questioned most traditional forms of authority and constraint in a decade when age, experience, and past esteem were no longer regarded as providing teachers, parents, and the time-honoured pillars of society with a pre-emptive right to guide younger generations—or, indeed, anyone. This disaffection was particularly evident in the Catholic Church, where the personal price of observing its teaching was greater—regular Mass-going, no artificial contraception, celibacy for the clergy, etc. At the same time, the Second Vatican Council (1962–5) was exercising an ambivalent influence on those sections of the Church that were conversant with what was going on. On the one hand, it was creating interest and enthusiasm among both conservatives and liberals, and sustaining the commitment of waverers who had despaired of seeing significant change in the Church. But it also contained the seeds of bewilderment and disappointment. Traditionalists were upset by the liturgical changes and what they saw as the abandonment of attitudes and practices which they had defended against the criticism of non-Catholics. While few of these were driven into leaving the Church, many of them turned their backs on what was creative and innovative and retreated into a routine world of personal devotion. Conversely, many liberals who expected major changes from the Council were disappointed by the modest nature of its proposals. They were then assured that the major changes would come when the broad declarations of the Council were processed into detailed policy by the various post-conciliar commissions; but they were even more disappointed by the feeble results and the apparent reluctance of the new Pope, Paul VI (1963–78), to proceed in the adventurous spirit of his predecessor. The period of waiting for the reports of the commissions was initially accompanied by wishful speculation, in which the recommendations of the more progressive wing in each body were given wide publicity and acted on in some circles as though they were already agreed policy. Whatever its many virtues and welcome innovations, the effect of the whole exercise, for conservatives and liberals alike, was to weaken unity and authority; and those marginal Catholics who had been delaying departure to see whether the Council would make the Church a more congenial haven, left—some more noisily than others. The main exodus was to come after *Humanae Vitae*, the papal encyclical of 25 July 1968, which reasserted traditional Catholic teaching on birth-control. It was widely known that a majority of the Pope's advisory committee on the matter favoured change, and change was widely expected. Many practising Catholics had long been using methods that the Church officially condemned; and others, anticipating a liberalization of church teaching, had started to adopt them. Once again, authority and unity were in question—and this time on an issue with a direct and serious bearing on family life. Regular Mass-going in France dropped from well over 20 per cent of the adult population in the early 1960s to less than 15 per cent by the end of the decade.

Running parallel with this drop in the number of practising Catholics, was the fragmentation of their political allegiance, accelerating the decline and eventual disappearance of the MRP as their prime political spokesman. With the Gaullists in power and responsible for the greatest financial concessions to Catholic education since the *ancien régime*, the MRP no longer seemed necessary for the defence of Catholic interests—even if MRP and conservative pressure were important factors in persuading Debré to be much more generous to Catholic education than he intended. Increasingly, the MRP became identified first and foremost with agricultural interests and the European idea—although the schools question was one of several stumbling-blocks in its negotiations with the prospective Socialist presidential candidate, Gaston Defferre, in 1964. Its transformation into the Centre Démocrate in the mid-1960s more or less signified the end of its career as the accredited defender of confessional interests. Thereafter, what was left of this role was split up and parcelled out to other parties with a record of Catholic support—like the patients' list of a retired doctor.

Conversely, there were committed Catholics who looked to the Left and were attracted by the SFIO and PSU. Symptomatically, the Confédération Française des Travailleurs Chrétiens split in 1964, the minority struggling on under the old Catholic banner, while the majority took the secular title of the Confédération Française Démocratique du Travail and developed links with the SFIO and PSU.

Politics in the mid-1960s

Paradoxically, the constitutional issue also began to wane in the 1960s, as politicians increasingly came to see the virtues of the new system for the party in power—which one day might be themselves. Not all shared Mendès France's conviction that the Left could never win a presidential election based on universal suffrage; and many felt that his stern refusal to be a presidential candidate in 1965 and 1969 deprived France of her best chance of having a left-wing government before 1981. He justified this refusal in terms of not wishing to condone an unparliamentary system; but colleagues argued that he was merely condemning France to perpetual right-wing government. Both Defferre and Mitterrand were ready to stand as presidential candidates in 1965; and opposition to the constitution increasingly took the form of proposed modifications to the President's term of office and to his emergency powers, rather than to the principle of a popularly elected President. 1969 was to find even the Communists putting up a presidential candidate. All of which left socio-economic issues as the main axis of political division.

The logic of this situation was that de Gaulle should be opposed by a Socialist in the 1965 presidential election. But a remorseful Guy Mollet claimed that this would endorse the legitimacy of an unparliamentary institution, and suggested that Albert Schweitzer should be invited to transform the post into that of a venerable figure-head—holding, no doubt, an out-patients' clinic at the Élysée, and giving the occasional organ recital at the nearby Madeleine. Gaston Defferre, however, took the situation more seriously, and proposed to legitimize his own candidature by campaigning on the platform of a five-year presidency, coinciding with the life of parliament, and a cut in the President's emergency powers. Throwing his hat into the ring on 12 January 1964, his schemes gradually expanded to envisage a broad Left–Centre alliance, stretching from the MRP to the left-wing Socialist dissidents, with the hope that the Communists might also give it their votes, albeit with no quid pro quo from Defferre. The MRP were not keen, however, since they distrusted Mollet's anticlericalism and contemplated running their own candidate—while the left-wing Socialist clubs and the Parti Socialiste Unifié wanted closer links with the Communists. The Socialist clubs had drawn hope from the death of Maurice Thorez in the summer of 1964, after he had relinquished the secretary-generalship of the PCF to Waldeck-Rochet; the change of leadership might possibly lead to a thaw in Communist attitudes. Sensing impasse—and the continuing hostility of Mollet—Defferre withdrew in June 1965, leaving the field open for the clubs to see if they could come up with something more plausible.

Paradoxically, the clubs had started life among those Socialists who felt unable to follow Mollet's support for de Gaulle in 1958. Like all splinter groups, they claimed to embody unity and truth, demonstrating this with the formation of the Parti Socialiste Unifié in April 1960. Despite their proud possession of Mendès France, they were more reminiscent of a debating society than a party, and were to remain unrepresented in the Assembly until 1962. It was Mendès France who suggested to Mitterrand that he should take up Defferre's mantle and try for the presidency. Having had his eye on it since 1962, he needed no persuading. Mitterrand was still regarded as belonging to the largely defunct UDSR. But, symptomatically, he had attempted to show an interest in the embryonic PSU in 1959 (when it was still the Parti Socialiste Autonome). And in June 1964, he had brought the remnants of the UDSR into a loose *entente* with a number of the left-wing clubs to form the so-called Convention des Institutions Républicaines. Putting himself forward in September 1965 with the backing of the newly formed Fédération de la Gauche Démocratique et Socialiste, he rapidly built a left-wing alliance, stretching from his Radical friends to the Communists, but without advance pledges on policy. He made no attempt to elaborate a specific social programme, and put most of his emphasis on ending the arbitrary nature of Gaullist government. But even here it was not clear exactly how he proposed to make the presidency more responsive to parliament, as he promised. Given the parties that supported him, it was not surprising that his main electoral backing came from traditional left-wing strongholds in the southern half of France.

In the meantime, de Gaulle had remained aloof from it all. Announcing his candidature a bare month before polling day, he disdained to use the television time accorded to him and restricted his campaign to reminding the electorate that it was 'moi ou le déluge'. He would almost certainly have won straight away on the first ballot (5 December 1965), had not the MRP decided to put up Jean Lecanuet on a mixed programme of Kennedyesque 'youth at the prow' and more help for farmers. Less serious as a contender was Maître Jean-Louis Tixier-Vignancourt, who masked his appeal to the *pieds-noirs* immigrants by claiming that he would appoint the coloured President of the Senate, Gaston Monnerville, as his Prime Minister. Softening his Poujadist stance of the 1950s, he also made much of his 'European ideals' and 'defence of Western values'. As expected, the second ballot was a straight contest between de Gaulle and Mitterrand, with Lecanuet advising his erstwhile supporters to abstain or vote for Mitterrand. It made no difference: de Gaulle won with 54.5 per cent of the vote in metropolitan France against Mitterrand's 45.5 per cent.

The presidential election of 1965 strongly influenced the legislative elections of 5–12 March 1967, creating in effect an electoral campaign that lasted sixteen months. While the vagueness of Mitterrand's presidential platform had concealed the socio-economic issues that underlay what seemed superficially to be a straightforward democracy versus paternalism struggle, the parliamentary elections brought them into the forefront of public debate. Mitterrand promised equal pay for women, more jobs, earlier retirement, more houses and schools, and better social security benefits for the farming community. Indeed, it was widely recognized on both sides of the political divide that social progress had failed to keep pace with economic growth. Prices and unemployment had risen in 1966, occasioning a bitter tussle between the left-wing Gaullists, who wanted greater expenditure on social welfare, and the Giscardians, who thought it electorally safer to woo the Centre and play for economic and financial stability.

The Giscardians were currently irritated by Pompidou's dropping of Giscard in favour of Debré at the Ministry of Finance in the government changes of 8 January 1966, and it was with some reluctance that they accepted the Gaullist insistence on single Gaullist–Giscardian candidates for the first ballot of the 1967 elections. The Left, for their part, were uncertain what to do with Mitterrand's FGDS of 1965. Gaston Defferre wanted to transform it into a party that would replace the old formations, whereas Mollet wanted it to be no more than an electoral umbrella, under which each party would be free to maintain its own candidates on the first ballot. Mitterrand successfully proposed that the constituent parties should retain their separate identity, but should agree on single candidates on the first ballot; and—a very significant innovation—he persuaded the Fédération to form a shadow cabinet. This was yet another symptom of bipolarization. Under the previous regime, the identity of the future Prime Minister had been a lottery; and the concept of a shadow cabinet, formed in advance of the event, would have been thought a presumptuous waste of time. Now, at last, it seemed that the voter could make real choices between alternative ministerial teams, instead of awaiting the outcome of back-stairs inter-party bargaining. The 'République des députés' was giving way to the 'République des citoyens'. Equally encouraging was the Fédération's electoral pact with the Communists (20 December 1966), envisaging single candidates on the second ballot but without commitment to a common programme.

In the meantime, the Centre had sought to protect its dwindling numbers by giving concrete expression to Lecanuet's presidential following of 1965. The resultant Centre Démocrate (2 February 1966) embraced the MRP, the non-Giscardian Independents, and an assortment of unattached members from the Centre Left. The Centre Démocrate, with 15.4 per cent of the vote on the first ballot, did marginally less well than Lecanuet had done in the presidential election of 1965 (15.8 per cent); and although comparisons with the legislative elections of 1962 are difficult—given the changing formations and allegiances—the Centre lost well over a quarter of its seats. The combined Left—the Fédération, the PSU, and the PCF polled a slightly smaller percentage than in the 1962 election, yet mutual strategy and good discipline in the second ballot enabled it to emerge with a gain of nearly a third in terms of seats, leaving the Gaullist–Giscardian alliance with the barest of absolute majorities—a margin of one

Table 15.2.
National Assembly Elections of 5 and 12 March 1967
(Results for Metropolitan France only)

	% of vote on first ballot	Seats gained after both ballots	
		Total	%
Communists	22.5	72	15.3
PSU	2.2	4	0.8
FGDS	19.3	117	24.9
Centre Démocrate	15.4	44	9.4
Gaullists and allies	38.3	233	49.6
		470	

Notes: The percentage of electorate casting valid votes was 79.2.
There were additionally 17 overseas deputies.

(Table 15.2). Given the composition of the Centre, however, the government was confident that it could carry its legislation on most issues.

A remarkable feature of the election was the return of the Communists in strength. Although their share of the vote on the first ballot (22.5 per cent) was little more than in 1962, they nearly doubled their seats (from 8.6 to 15.3 per cent), thanks to left-wing discipline on the second ballot. As always, however, there were left-wing voters who could not bring themselves to vote for a Communist, and this favoured government candidates in those constituencies where the second ballot was a straight fight between a government and a Communist candidate. Indeed, the other remarkable feature of 1967 was that over 70 per cent of second-ballot contests were straight fights between government and left-wing candidates. Bipolarization was unmistakably a fact of life.

The Pursuit of National Unity

The counterpart to de Gaulle's search for political stability was his attempt to create national unity; each was a facet of the other, and each buttressed the other in a mutual system of support. De Gaulle's pained awareness of the divisions within French society inclined him to concentrate his energy on those spheres of government that united rather than divided Frenchmen. Foreign policy was an obvious example, since it focused attention on the common interests of the French as against those of other peoples. And in practice de Gaulle was to achieve a high level of consensus here—with even the Communists finding much to admire, however tacitly. It was true that the MRP and a section of the Independents were to find him insufficiently pro-European and were alarmed by his distrust of NATO. There was also disquiet in other parties about the cost of the nuclear *force de frappe*, but much inner pride about France's elevation to nuclear status.

Another sphere which de Gaulle found conducive to unity was that of French cultural prestige. Once again, it represented an emphasis on French achievement as against that of the foreigner. Prizes were awarded to characteristic examples of indigenous attainment, not always happily. The Ministry of Culture was given to André Malraux (January 1959–June 1969)—'ce repris de justice' as Bidault liked to call him, evoking his alleged connection with the disappearance of seven statuettes from a temple near Angkor Vat in 1923. His saturnine countenance, dangling cigarette, and bleak assessment of the human condition were a chastening reminder of earlier decades, when French literary eminence in the world was represented by the familiar photographs of truculent but respected faces, their desks littered with paper-knives and overflowing ash-trays. The Grand Prix du Président de la République was given to Charles Munch's somewhat ephemeral performance of Berlioz's *Symphonie Fantastique* with the Orchestre de Paris, while far finer performances with less prestigious French orchestras were passed over in silence. The cleaning and renovation of public buildings and street façades created an air of prosperity and pride in the nation's past, while the *roi soleil* ethos of the regime made a similar impact. Like Mary Magdalene's critics, elements on the Left argued that the money would have been better spent on the poor; but the financial cost of cultural chauvinism was relatively modest, and the results were plainly and pleasingly visible to all. Indeed, when the Left eventually came to power in 1981, Mitterrand was to go rather further than his patrician predecessor in subsidizing the arts—even if the emphasis was on diffusing its benefits through society rather than on cutting a dash internationally. Not that the provinces were neglected under Malraux. He established a dozen or so municipal Maisons de la Culture, for which the State shared the cost with the local authority. But his purpose was to bring national and international culture to the natives, rather than to encourage local initiative and talent based on regional themes. 'In ten years' time, this hideous word "provincial" will have ceased to exist in France', he announced in 1966, when opening the Amiens *Maison*. The *Maisons* were grandiose, expensive to run, and unavailable to amateur companies. On the other hand, Malraux's appointment of Michel Landowski as head of a new music department in 1966 was to result in a musical renaissance in the provinces in the 1970s, when Malraux's successors were more generous with funding. Not only were new orchestras created and life injected into the provincial opera companies, but local *conservatoires* started belatedly to compensate for the inadequate provision of musical education and activity in French secondary schools.

De Gaulle's emphasis on economic growth, apart from its obvious material benefit, was intended, like his cultural policy, to play a unifying role. Everyone wanted economic growth—except environmentalists, who had not as yet become a significant force in French politics—and once again economic growth was an achievement that was measured against foreign rivals. If the franc was in effect devalued by 17 per cent *vis-à-vis* foreign currencies, this fact was obscured by striking off a couple of noughts and giving the new franc a value roughly equivalent to the highly respectable German mark or Swiss franc (28 December 1958). The liberation of French prices from their Italian-style trail of noughts did much for domestic morale and prestige abroad. Indeed, in the course of the 1960s, the French franc was to become a respected haven for savings and investment.

But economic growth also had its divisive aspects when it came to apportioning the cost of achieving it. Although both employers and workers welcomed the higher profits and wages that increased productivity promised, there was inevitable disagreement as to how these rewards were to be divided between the two sides of industry—and even more disagreement about who should bear the brunt of the initial hard work and self-restraint. Similarly, there was often disagreement between shareholders and management on whether profits should be distributed or ploughed back into further expansion. And state help to industrial growth in the form of subsidies or tax concessions was likewise often a bone of contention with other taxpayers. But it was the restraint of wages and workers' benefits that was the most divisive aspect of the economic growth of the 1960s. Paradoxically, the collective goal of increased national productivity created a greater initial sense of common identity than the distribution of its benefits among the various sections of society did—which may have been a subsidiary, perhaps unconscious, factor in de Gaulle's reluctance to make expansion more appetizing by passing on some of its rewards *en route*. Apart from the potential disagreements already listed, improved social benefits were likely to arouse opposition from employers, who, under the French system of social insurance, bore a heavy percentage of the contribution. And while all workers wanted better wages, the question arose as to whether those industries that had contributed most to economic progress should enjoy the largest increases—or whether increases should be standardized across the board, on the premiss that society was an interdependent team.

De Gaulle personally, and the UNR as a party, saw themselves as a focus of national identity that transcended these divisions of interest. Shortly after the foundation of the UNR, Chaban-Delmas declared: 'L'UNR doit être à la Cinquième République ce que le parti radical a été aux bonnes années de la Troisième République, le parti qui, pouvant gouverner à droite ou à gauche, assure l'équilibre, le parti au pouvoir.' The paradox of the Fifth Republic was that although de Gaulle was seeking to pursue a policy of national consensus, it was essentially government from the Right rather than the Centre—despite the fact that the Third and Fourth Republics had been largely governed from the Centre. And this paradox was nowhere more evident than in the economic and financial policies that his ministers pursued, which belonged unmistakably to a conservative tradition that served to sharpen the political division of France along Left–Right socio-economic lines. This was particularly true of financial policy, where Antoine Pinay, Jacques Rueff, and the professed goal of balanced budgets were all redolent of an old-style orthodoxy that was viewed with some amusement by neighbouring governments, until their own overheated economies in the mid-1960s caused them to look to France with more respect. The trade-unions resented the fact that wages were no longer directly linked to the cost-of-living index—and, indeed, the SMIG was now the only surviving relic of the ratchet-mechanism which had kept wages in step with price-rises—and which Gaullists were currently condemning as the source of the country's inflation.

In fact, inflation worried the general public rather less than it worried the government. The French had lived with periodic bouts of inflation since the First World War, and short-term personal advantage tended to take precedence over long-term collective advantage in the priorities of many. With inflation running at an average of 5 per cent in 1958–63, the Debré government had issued instructions that wage settlements should be confined to 4 per cent a year. Yet private enterprise was giving increases that averaged 12 per cent in many sectors, since a healthy rate of turnover was enabling them to do so without markedly increasing prices. The subsequent disparity between wage increases in the private and public sectors was a major factor in the miners' strikes of February–April 1963, which de Gaulle made a trial of strength. This was notably a period when he was extending his personal control of government to economic and financial matters; and his prime fear in 1963 was potential overheating of the economy. However, his resort to the old *belle époque* ploy of requisitioning labour (3 March 1963) was largely ignored, since the miners were well aware that the courts and gaols could not cope with disobedience on such a massive scale. Moreover, the affront to the right to strike generated a lot of public sympathy for them, particularly among the bishops, with even the conservative *Figaro* speaking in their favour. De Gaulle's presidential popularity-rating sank to its lowest ebb—42 per cent in May 1963—and he had to settle on much more generous terms than he intended.

The government's austerity policies in a time of economic expansion aroused resentment in other quarters, and Pompidou in January 1965 eventually conceded that the policy of cooling the economy was inhibiting growth. Moreover, with the prospect of an impending presidential election, the government affected a more benign countenance and made a belated attempt to stimulate consumer spending.

INTEREST GROUPS—PAST AND PRESENT

While socio-economic issues predominated in politics, the ability of particular interest groups to influence government was arguably lessened under the Fifth Republic by the fact that both the President and the parties that supported him were based on much wider electoral bases that had been the case during the multi-party stalemate of the Fourth Republic. At the same time, many of the powerful lobbies of the Fourth Republic were in decline. The electoral clout of agriculture was being continually eroded by the growth of the industrial and service sectors of the economy, while the colonial lobby was a mere ghost.

All that now remained of the old colonial empire were the four Départements d'Outre-Mer (Martinique, Guadeloupe, Réunion, and Guiana), the seven Territoires d'Outre-Mer (French Polynesia, New Caledonia, French Somaliland, the Comoro Archipelago, St Pierre and Miquelon, the Southern and Antarctic Territories, and Wallis and Fortuna), and the condominium of the New Hebrides. The disappearance of the rest had been the result of France's own improved generosity in the wake of the Algerian war. The stark choice of 1958 between penniless independence and subsidized dependency had been softened by a third alternative: self-government within a loose framework of formal links with France (December 1959). This had been a national consequence of the similar alternative offered to Algeria three months previously (p. 274). The offer had been swiftly taken up by Madagascar, Senegal, and the former AEF (Congo, Gabon, the Central African Republic, and Tchad); but, disappointingly for France, the other colonies had opted for complete independence, with nothing but 'special relations' with France, entitling them to the prospect of French economic help but giving France little diplomatic advantage in return. Bowing to reality, de Gaulle dissolved the Community Senate early in 1961 as no longer serving a useful purpose. While the new terms of independence no longer left the errant children penniless on the street, as initially threatened in 1958, French aid to the former empire rapidly fell from 2.33 per cent of GNP in 1959 to 1.16 per cent by 1967. At the same time, the percentage of French trade conducted with her empire and former possessions fell from 32 per cent in 1958 to 14 per cent in 1966—the fall reflecting French membership of the EEC as well as the changed relationship. So, not only was the colonial lobby deprived of its quantitative bargaining power but, with the disappearance of the French Maghreb, the dispersed and disparate nature of what was left deprived it of any coherent thrust.

As for French farmers, their effectiveness as a political lobby was weakened by their internal animosities. The pressures and opportunities of the EEC widened the division between the progressive elements in French agriculture—principally based in the north-eastern half of France—and the more backward farming sectors of the west and south. Symptomatically, Edgard Pisani's bill encouraging the retirement of older farmers encountered conservative opposition in the Assembly (July 1962), but met with the approval of a substantial number of young farmers. Indeed, the backward sectors increasingly felt that both parliament and the agricultural unions were being dominated by the voices of the progressive element, and that consequently the most effective way of expressing themselves was to resort to direct action, blocking roads and dumping surplus produce on the steps of public buildings. That times had turned against them had already been made clear in July 1960, when the notorious privileges of the *bouilleurs de cru* were swept away.

It was true that the declining influence of pressure groups on parliamentary parties was partly compensated for by a greater readiness on the part of government departments to have informal discussions with them—but little was given away, except in election years. At such times, a numerous following was a very strong card. The VAT concessions to small businesses in 1965 exemplified the fact that ten impecunious voters were electorally worth far more than a single rich one, however welcome his financial contributions to party funds. At the same time, however, loquacious and strategically placed cogs in the machinery of day-to-day life could sometimes extract concessions out of all proportion to their actual numbers—notably taxi drivers and country lawyers, whose daily chat provided much of the copy of local and foreign journalists, and whose good opinion was therefore worth courting by governments.

Table 15.3.

Professional Backgrounds of Members of Government, 1958–1976 (% Representation of Various Groups)

Survey date	De Gaulle 1958	Debré 1959	Pompidou 1962	Pompidou 1966	Couve de Murville 1968	Chaban-Delmas 1969	Messmer 1973	Chirac 1976	Barre 1976	Average
State employees (inc. teachers)	69.5	52.0	43.3	50.0	58.0	43.5	66.0	62.0	64.0	55.0
Business men	4.3	16.0	16.6	25.0	13.0	12.8	10.5	5.4	8.3	11.4
Lawyers	8.7	8.0	16.6	14.8	13.0	15.4	2.6	5.4	2.8	9.9
Journalists	8.7	8.0	13.3	3.5	6.4	7.7	—	8.1	8.3	6.8
Political party officials		4.0			3.2	5.1	8.0	2.7		4.2
Farmers		4.0	6.6				2.6	2.7		2.0
Doctors						5.1		2.7		1.3
Shopkeepers	4.3	4.0		3.5	3.2	5.1	2.6	2.7	2.8	2.7
Men of letters	4.3	4.0	3.3	3.5	3.2		2.6			2.0
Miscellaneous						5.7	5.2	8.1	5.6	3.5

Source: Vincent Wright (ed.), Continuity and Change in France (London, 1984), 119.

As in all advanced Western countries, however, it was labour and industrial capital that were the main contenders for government attention—and around whose rival claims politics tended to gravitate. While Debré could be relied on to take a firm line with employers and unions—and, indeed, with all pressure groups which offended his Jacobin, *étatist* principles—his admirers were less confident about his successor, Georges Pompidou. If they warmed, like everyone else, to the Pompidolean image of the benign manipulator, cigarette on lower lip, soothing ruffled sensibilities, they also remembered that he was a former director of Rothschilds Bank. His business connections were also noted by the unions, who complained that living standards failed to keep pace with the expanding economy. They also pointed out that the proportion of business men in the cabinet had grown from 4.3 per cent in 1958 to 25 per cent by 1966 (Table 15.3). If this was still modest compared with the number of former civil servants and other state employees who held most ministerial portfolios (50 per cent in 1966), the complaints were symptomatic of a mounting feeling at shop-floor level that the prosperity of the 1960s was something being cooked up by government and business for their own consumption. Admittedly, these feelings were partly the result of lack of communication between management and employees in the huge, impersonal work-units of the economic boom. But this lack of contact was itself a pointer to trouble ahead—all of which was to come much sooner than anticipated, in the explosion of 1968.

16
De Gaulle's Foreign Policy

IT is often assumed that de Gaulle's advent to power represented a major change in the aims of French foreign policy; and in some ways it did. But in substance de Gaulle's policy remained consistent with the traditions of French post-war policy, and it was mainly in its range and style that it differed. His period of office coincided with a growing nuclear balance in the world, when American predominance was not as absolute as it had been during much of the Fourth Republic. It also coincided with the relatively flexible attitudes of the Khrushchev and Kennedy era, which seemed to offer middle-sized nations some opportunity of wriggling out of the vice-like grip of the bipolarized world of the superpowers and of establishing some freedom of manœuvre. It is arguable that any French government would have attempted to take some advantage of this. Yet de Gaulle was especially disposed to do so.

De Gaulle's foreign policy was closely linked with his domestic aims. It is often suggested that his foreign policy was pursued for its own sake, and that his domestic policies were little more than a method of achieving the material means to fulfil it—like some sporting landlord who runs his estate with the main purpose of finding the wherewithal to enable him to hunt five days a week. 'Depoliticisation in domestic affairs, plus a foreign policy';[1] 'de Gaulle's nationalism was a kind of permanent quest with varying content but never any other cause than itself',[2] exemplify an attitude reminiscent of historians' comments on the Austro-Hungarian Empire, which was often described as merely an organization for the pursuit of foreign policy. There is in fact a valid analogy between de Gaulle's France and the Austro-Hungarian Empire. The Habsburg monarchy pursued a foreign policy which was primarily intended to offset the internal divisions of the empire, and in the same way de Gaulle was seeking to transcend the internal divisions of French opinion by channelling energies into national ambitions that would remind Frenchmen of what they had in common rather than what divided them. De Gaulle's celebrated utterances on the matter have already been cited (p. 119). These were echoed again, shortly before his coming to power: 'You see, the truth is like this; the French no longer have any spur to action.'[3]

Yet, given the expense of de Gaulle's policies and the arguable fact that they failed to achieve their ultimate aims, it is tempting to agree with his critics, who claimed that French foreign policy was 'made in three admirably synchronized steps: it is prepared at the Quai d'Orsay, decided on at the Élysée and implemented nowhere';[4] or, de Gaulle 'called on France "to be herself" ... in order to console her for the fate of becoming like others'.[5] Yet de Gaulle's foreign policy achieved a remarkable degree of popularity at home. The army was given new aspirations to console it for its loss of prestige in Algeria—even if inter-service rivalry made for mixed feelings among the generals on the nuclear *force de frappe*.

De Gaulle's aim was to encourage a shift towards a multipolar world, where middle-sized powers might enjoy more freedom of manœuvre. It is arguable that in a nuclear age, bipolarization and the balance of terror offered the best hope for the survival of humanity. To encourage the emergence of a multipolar world was to run the risk of nuclear proliferation, where any Third World state with little to lose might decide to invest such resources as it had in acquiring nuclear weapons, so that it could get what it wanted by 'threatening to blow its brains out all over your nice new suit'.[6] De Gaulle might have replied that if that was what other nations believed, they were at liberty to argue the case for the bipolar strait-jacket; but that France in the meantime could only press for what she saw as in the best interests of herself and middle-rank powers in general.

In reality, de Gaulle was in no doubt that France was and ought to be under the American nuclear umbrella. His main concern was that France should not become a mere pawn in the superpowers' game. As many politicians had argued under the Fourth Republic, the danger was that America might not always see it as in her best interests to protect France against a Russian attack with conventional weapons, if such protection might escalate into a nuclear confrontation involving the American mainland. In practice, de Gaulle's short-term policies were based on the assumption that America would attach considerable importance to defending France. But there were many issues on which France might find herself without American support—or, conversely, might be press-ganged into helping America on matters that did not interest her. So a greater measure of independence was desirable.

The first French atomic bomb was exploded in the Sahara on 13 February 1960—the culmination of work set in motion under the Fourth Republic. Later landmarks were the launching of France's first nuclear submarine on 29 March 1967, and the explosion of her first hydrogen bomb on 24 August of the following year. De Gaulle recognized that the development of a respectable nuclear force was a relatively cheap way of achieving a greater measure of independence. For a middle-sized power, it was certainly far cheaper than seeking to achieve it through emulating the superpowers' economic strength and conventional weapons. The problem, as always, was convincing other nations that France was prepared to use them; but in practice, possession of such weapons was usually enough to induce circumspection. However modest the French nuclear deterrent, no superpower would lightly risk the obliteration of even 5 per cent of its cities—while the danger of international escalation enabled

France to threaten the world with consequences far more devastating than her own weapons could achieve.

As indicated earlier (p. 272), de Gaulle inaugurated his foreign policy with a letter to America and Britain demanding parity for France in the counsels of the West. Receiving an unenthusiastic response, he informed NATO that it could no longer count on the French Mediterranèan fleet in time of war (7 March 1959), a withdrawal that was later extended to the Channel and Atlantic fleets (21 June 1963). Analogous limitations were put on the authority of NATO over French air strength (28 September 1960), while on 3 November 1959 de Gaulle made his famous École Militaire speech, in which he spoke of military integration having 'had its day'—a declaration that eventually led Pinay to resign from the cabinet and prepared the way for the MRP secession two years later (p. 293).

De Gaulle's confidence in this policy was greatly strengthened by the Cuban crisis of October 1962. It demonstrated that Kennedy was prepared to be tough, and that Khrushchev was prepared to respect that toughness. This encouraged de Gaulle to presume that a Soviet advance into Western Europe was extremely unlikely—especially since Russia was becoming progressively apprehensive about China. De Gaulle could consequently afford to adopt a more independent attitude towards America, with little risk of Russia seeking to take advantage of it. Yet it was symptomatic of the basic realities of the situation that de Gaulle announced full French backing for the USA during the Cuban crisis, despite the fact that the Americans had mobilized their NATO forces in Europe without prior consultation with their treaty partners.

De Gaulle felt little enthusiasm for international institutions. He preferred the traditional system of what he called 'bonnes et belles alliances' between individual states. Like many Frenchmen, he was extremely sceptical of the ability of the United Nations to achieve effective solutions. He referred to it contemptuously as 'les nations soi-disant unies', 'ce machin', 'ce truc'. He saw it as not only ineffective, but dangerous. For one thing, it was dominated by the concerns of the superpowers and was incapable of coming to fair decisions based on the merits of the individual case. Moreover, French colonial policies had continuously come in for a hard hammering in the General Assembly from the Third World countries and the Eastern bloc. The only worthwhile part of the United Nations for de Gaulle was the Security Council, where, as a member, France could exercise her veto to protect her interests against the hostile proposals of other countries. Indeed, he saw the Security Council as a convenient formal meeting between powers which had already come to prior decisions and agreements through their bilateral representatives. The General Assembly, on the other hand, was, in de Gaulle's estimation, a mere talking-shop, which should be at liberty to debate issues but not to decide them. Admittedly, he made use of the General Assembly in the later 1960s as an instrument for improving French relations with Third World countries—but this was a matter of making the best of a bad job. Symptomatically, France refused to meet her share of United Nations peace-keeping costs until 1972.

De Gaulle was strongly opposed to the various gestures that the United Nations made towards disarmament. His formal attitude was that countries like France were so far behind the superpowers in their level of weaponry that it was up to the superpowers to take the initiative and strip themselves down to the level that they expected the middle-sized powers to observe. For de Gaulle, arms control was merely a device of the superpowers to clear the board of inconvenient minor pieces. He was happy enough to sign the treaties that put the Antarctic (1959) and space (1967) out of nuclear bounds, but these were prohibitions that affected the superpowers much more than France. Similarly, when it came to the vexed question of supplying nuclear information to other middle-sized powers, de Gaulle refused to sign the non-proliferation treaty (1968); but he assured the United Nations that in practice France would observe the treaty as if she had signed it.

Turning to the EEC, it goes without saying that de Gaulle saw no future for the EEC as a federal superstate. Not only would such a development destroy France's sovereign independence, but he did not think that such a federation could have any firm basis in human reality. For him, the nation-state was the only effective mediator between the conflicting desires of the various sections of society. The nation-state had roots in the historical and living reality of a shared experience. It was precisely this that supranational structures lacked. If you spoke of the EEC, all that people could imagine were soulless buildings that looked like a cross between a maternity hospital and an airport lounge—and men in dark suits with lock-covered brief-cases. And it was characteristic of the EEC that it should choose Brussels as its home, lost between two cultures, having, as an English diplomat remarked, all the dullness of the Dutch and the insanitariness of the French. And lest even that should give the organization too human an identity, it was thought necessary to blur the situation by adding Luxemburg and Strasbourg—the latter admittedly bringing some historical tradition to the idea of Europe.

De Gaulle had opposed the Coal and Steel Community in the early 1950s, and had likewise opposed the Treaties of Rome. But, after coming to power, he sought to exploit this embarrassing inheritance from the previous regime as best he could. He recognized its advantages for French agriculture, but in the early 1960s he also saw its potential for creating a special relationship with West Germany, which would be both a means of controlling Germany and a way of achieving greater independence from the United States. He accordingly put forward various confederal proposals for the EEC, notably those of 29 July 1960, which appeared to indicate some commitment to the European idea; but the Benelux countries and Germany were quick to recognize them as an instrument intended to loosen the EEC from American influence. Germany's vulnerable military situation made her far from anxious to slacken ties with America—while the smaller EEC members were still nervous of Germany and looked to America to keep Germany under control. Consequently, the net effect of de Gaulle's proposals was to create deep suspicion in the EEC against any suggestion of greater political or military union among the six. Indeed, Belgium and Holland proposed that Britain should be invited to join as a guarantor of German good behaviour (21 November 1961). This was the last thing de Gaulle wanted, Britain in his view being little better than a stalking-horse for American ambitions; and he was to maintain a steady opposition to British entry throughout his presidency. With the failure of his German stratagem, de Gaulle lost interest in confederal proposals, and his EEC policies were thereafter mainly directed towards strengthening its economic advantages for France.

On 7 March 1966, de Gaulle announced his intention to withdraw from NATO. This was the highest card he wished to play in his game against American pressure, which is why he did not play it earlier. But, having played it, there was little of significance left—short of quitting the Western alliance which he had no intention of doing. There were, it is true, the brief 'tous azimuts' gestures of 1967–8, when France claimed that her nuclear force was multidirectional and was not necessarily pointed at Russia. But this did not constitute withdrawal from the alliance; it was merely a reminder that France was not to be taken for granted. At the same time, de Gaulle was seeking to free France from the financial domination of the dollar by building up French gold reserves.

The failure of his attempt to detach Germany from America in the early 1960s turned his attention to a more general European policy, in which he hoped to negotiate a European *détente* with Russia, which, he claimed, might ultimately lead to the emergence of a European power block stretching from the Atlantic to the Urals, in which Russia would remember her European heritage and occupy a middle ground, with Western Europe against the wild men of the Third World, such as China, and the plutocrats of the New World. This was, admittedly, largely rhetoric, and was not meant to do more than encourage Russia to consider a more flexible attitude to Western Europe. Russia might be induced to tolerate a loosening of the Warsaw Pact in return for a slackening of the Atlantic Alliance. And so there might be a gradual move towards multipolarization. Visiting Russia in June 1966, he spoke of 'détente, entente and co-operation'. Russia, for her part, was pleased by de Gaulle's opposition to American suggestions that West Germany be allowed nuclear weapons, and was also gratified by his consistent support for the view that East Germany would have to accept the Oder–Neisser Line, an opinion he had expressed as early as 1959. De Gaulle was nevertheless careful to avoid suggesting that the division of Germany between East and West was an irreversible feature of Europe.

De Gaulle's designs were made to look somewhat sorry in June and July 1967, however, when the six-day Arab–Israeli war ended with America and Russia more or less imposing a joint solution on the Middle East—thereby demonstrating that the superpowers not only called the tune in the world at large, but were prepared to act together to their mutual advantage. Some commentators claimed that de Gaulle's petulant behaviour in the summer of 1967 reflected his deep disappointment at this outcome—notably his virulent attacks on Israel, and his extraordinary behaviour during his state visit to Canada in July 1967, when he shouted 'Vive le Québec libre!' from the balcony of the Montreal town hall, causing the Canadian authorities to protest. It is perhaps both an asset and a misfortune for the French to have such a convenient phrase as 'Vive'. Its brevity and bravura—and its ultimate ambiguity—make it hard for an orator to resist; yet, while committing him to nothing precise, its very ambiguity gives rise to the worst suspicions in listeners' minds. De Gaulle, however, was not one to let things slip unintentionally. It was perhaps all on a par with what many saw as the degrading spectacle of the French ministers lined up at four in the morning to greet the returning President at Orly airport; de Gaulle believed that the world rates you as you rate yourself.

De Gaulle's desire to loosen Russian control over the Warsaw Pact countries was made to look sadly chimerical in August 1968, when Russia and her allies invaded Czechoslovakia to suppress the liberal 'Czech Spring'. Indeed, the Czech crisis obliged de Gaulle to behave as though France was a member of NATO, with France assisting America in monitoring the movement of Russian ships in the Mediterranean. It likewise demonstrated the dependence of French forces on the information provided by NATO radar systems. Little was left of de Gaulle's bridge-building with Eastern Europe by the time he resigned in April 1969; and, in so far as there was East–West *détente* in the late 1960s, it was largely West Germany that was making the running.

Nor did de Gaulle's record look appreciably better in the Third World. It is hard to gauge the practical consequences of his spectacular tour of Latin America in September and October 1964, or his recognition of Communist China in the previous January. His continuous critique of American policy in Vietnam had been accompanied by suggestions that China be given a role in achieving a settlement (April 1964), and he had been at some pains to present himself as the prime contact man between Europe and China. But it rapidly became clear that China wanted close relations with the EEC as a whole—and, indeed, was prepared to go further and seek American help as well, which was particularly mortifying for de Gaulle. The fact, too, that Paris was accepted as the venue for peace-talks on Vietnam (May 1968) was rapidly offset by the impasse into which they fell. Moreover, the domestic disturbances of May 1968 put a severe financial strain on the government, with the result that de Gaulle had to reconsider his gold policy against the dollar. Swallowing his pride, he was quickly constrained into adopting a much mellower attitude to America.

President Nixon was invited to France, obliging de Gaulle to spend an excruciating two days in February–March 1969 playing host to his beaming visitor and trying to keep Nixon's fingers off the Élysée piano. Unkind Democrats said that de Gaulle's rapid retirement and early death were not unconnected with the episode, while the degree to which he had to pare down his policies was demonstrated in March, when he publicly accepted the American principle of 'graduated response' towards Russia.

Even so, to describe de Gaulle's foreign policy as a failure is to judge its achievements by the very high standards of de Gaulle's own rhetoric, which was not meant to be taken as anything but long-term goals. To argue that de Gaulle failed to replace the bipolarization of the world by a multipolar system is to expect the impossible. At bottom, he was primarily concerned with obtaining greater independence of movement for France—but to make this aim palatable to other nations he had to elevate it to a general level of universal principle and application. While regarding the principle as a desirable goal, he did not consider it realizable except in very gradual stages—especially in view of the timid reluctance of other second-class governments to follow his lead in challenging the dictates of the superpowers. As far as France was concerned, de Gaulle undoubtedly made her a voice to be reckoned with in international affairs—and he succeeded in uniting a broad segment of French public opinion behind his policies. An opinion poll of December 1967 reported over two-thirds of the public as expressing confidence in them, while a parliamentary motion of censure on his withdrawal of France from NATO obtained only 137 votes (20 April 1966).

Europeans, however, saw de Gaulle's influence on the EEC as deplorable. Not only did he seek to frustrate such potential for political evolution as it might have had, but his method of negotiation within it was one of confrontation, threats, and bluff. The battle over the Common Agricultural Policy in the mid-1960s resulted in a settlement which destroyed the EEC's ability to move forward on majority decisions. Like the United Nations Security Council—or the Polish Diet in the eighteenth century—it was always to be hypnotized by the fear of vetoes. The behaviour of Gaullist France in the EEC was like that of a power with grievances, reminiscent of Germany in the inter-war period, whereas West Germany in the 1960s was acting the good European. But bad examples are catching, especially when bad behaviour is seen to be rewarded with appeasement rather than reproof. And some commentators have argued that Germany's increasingly independent attitude in the late 1960s, and her solitary pursuit of an *Ostpolitik*, was to some extent influenced by the lesson of the French example. To such reproaches, de Gaulle would merely reply: 'A state worthy of the name has no friends'. At the same time, de Gaulle created uncertainty in American minds about the desirability of the EEC moving to greater political and military co-operation. Whereas in the 1950s, America had urged Western Europe to create its own military system, she became increasingly hesitant in the 1960s, once de Gaulle had demonstrated that such a development could lead to a weakening of American influence. It has also been asserted that de Gaulle's behaviour soured the attitude of the EEC itself to supranationality, transforming it from one of cautious curiosity to outright hostility—and thereby bringing about a collective retreat from the prospect. However flattered by the accusation, de Gaulle would probably have replied that the retreat was a mere manifestation of mutual good sense, made all the wiser by experience.

17
Challenge and Response
1968–1974

THE events of May 1968 were a legend in their own time. Within months they had engendered a torrent of books, with such titles as *French Revolution*, chronicling and numbering the days of the opening moves as though they were the start of a new revolutionary calendar. In R. W. Johnson's words:

For many, especially the young, it had indeed been bliss to be alive through that utopian spring. Everything, suddenly, had seemed possible. A handful of laughing radicals had created a sort of joyous anarchy and . . . the system had trembled . . . How to be a political leader after May, when all politicians and all authority had been so decisively challenged, so hilariously mocked? How could any mere party programme compare with dream-world slogans such as 'L'imagination au pouvoir'? How could one go back to the mundane tedium of electoral politics?[1]

Yet if the events were a legend in their own time, it *was only* in their time; and in the years that followed, the liturgical sequence of happenings was largely forgotten, except by a dwindling number of participants who periodically celebrated the anniversaries, like the faithful nucleus of an old boys' society, most of whose members had drifted away into family and professional life.

The events of May 1968 were disconcerting rather than destructive; it was their unfamiliarity not their violence that alarmed authority. While the children of the bourgeoisie threw petrol-bombs and set cars alight, the Communist party called for calm and proclaimed itself 'the great and tranquil party of order'. De Gaulle despairingly summed it all up as 'incomprehensible'. In some ways, 1968 was a caricature of 1848. It was an international phenomenon, with an 'intellectual' minority taking the initiative in each country; but, as in 1848, the activities of this minority were only taken seriously by government when they coincided with the parallel but very different demands of the working class. Once it became apparent that there was little common ground between the two currents of protest, the authorities regained courage and sought to drive a wedge between them, and increasingly regarded the intellectual minority as an insignificant factor. As in 1848, fear of disorder caused the bulk of the electorate to side with the government, the whole affair ending with victory for the authorities, and with negligible concessions to reform. A remarkable feature of the May events was the almost complete absence of fatalities. By contrast, the resumption of petrol sales after the strike saw a sudden increase in deaths on the roads over the Whitsun weekend; life was back to normal. Moreover, although ten million people went on strike, the loss of working hours was equivalent to only 3 per cent of GNP—against a background of 4.5 per cent national growth.

As later pages will demonstrate, the mild recession of 1967 exacerbated grievances in industry, reminding workers that social progress had not kept pace with economic progress—and that perhaps the chance for it to do so was now lost. Nor were the rapidly expanding work-units of the 1960s an easy terrain for the effective resolution of these worries. Despite the existence of works committees, dialogue and horse-trading with an increasingly remote management was difficult; and confrontation rather than discussion became the dominant style of negotiation. As outlined in Chapter 1 (pp. 14–15), French labour relations traditionally showed a penchant for this style of bargaining—as, indeed, in other dealings between subordinates and authority. But the size and impersonality of the firms and factories of the 1960s exacerbated it.

In 1968, the State was the target for much of this frustration. This reflected the obvious fact that a sizeable sector of the economy and the greater part of education were under public control—and that most of this control lay with central rather than local government. In this national context, the growing taste for the language of no return inexorably took on the connotation of revolt and repression—with the protesters asserting the need to recast the whole socio-political system. But if 1968 demonstrated this sequence all too dramatically, it also revealed the simple fact that at heart most people were realists—and that, since the whole matter began with relatively modest demands, the granting of some or all of them was sufficient to restore the atmosphere to routine normality.

The initial explosion of 1968 came in the universities. Student unrest was a world-wide phenomenon in 1968, taking in the United States, Mexico, and Japan, as well as Europe—with American policy in Vietnam a powerful ingredient in the transatlantic troubles. Yet an opinion poll among French students in November 1968 revealed that 56 per cent believed that the May upheavals essentially represented anxiety over future employment, 35 per cent felt that inadequate university facilities were primarily to blame, while only 12 per cent saw it as an attempt to transform society.

France had over half a million students in 1968—56,000 more than in 1967, without a commensurate increase in staff and teaching facilities to accommodate them. The government was well aware of the attendant problems, and the 1960s had seen a substantial acceleration in university building and expansion of staff—not nearly enough, however, to meet the demands of the influx of students from the baby-boom of the late 1940s, and the increasing population of school-leavers who wished to enter higher education. Ninety-five per cent of those who passed the *baccalauréat* went on to university or an equivalent institution; and de Gaulle believed that the only effective answer was to follow the British pattern of selective entry, ridding the system of most of the 50 per cent or more who dropped out or failed to obtain degrees, and enabling resources to be concentrated on those who were likely to profit from them. Indeed, in Britain, drop-outs and failures represented only 10 per cent of the total.

Such a policy would also have released funds for increased financial aid to working-class students. In 1964, only 2 per cent of French university students were the children of industrial workers or agricultural labourers—compared with about 20 per cent in Britain, where selectivity enabled the provision of substantial grants for those students who could not otherwise afford to go. But the tradition of many Continental countries, of allowing access as a right of all those students who had passed the *baccalauréat* or its equivalent, was deeply engrained in French democratic sensibilities; and it was only in highly competitive vocational subjects, where training facilities were expensive to provide, that a more rigorous selection was operated. Symptomatically, the Fouchet plan for higher education (November 1967) did not even begin to address itself seriously to this basic problem.

The principal worry of many students in the late 1960s was finding congenial employment. Under the Third and Fourth Republics, a degree had been a guarantee of some sort of position with career prospects. But the influx of numbers undermined this assumption, especially as the newcomers were tending to swell the student rolls of arts and social sciences, where the employment opportunities were fewer. In effect, the universities had reached a point of uneasy cohabitation between two radically different concepts of their function: the traditional role of what Stanley Hoffmann has called 'the temple of discourse'—where the training of the mind and powers of communication were developed—and the conveyor-belt provision of vocational skills, designed to provide employees for the State and the economy.[2] And once again, selectivity at university entry remained the nettle that had to be grasped.

When Fouchet was promoted to the Ministry of the Interior in April 1967, his successor as Minister for Education, Alain Peyrefitte, was inexperienced and anxious not to offend de Gaulle. His advisers were mainly his cronies, and neither he nor they had the confidence to deal realistically and incisively with the crisis that was shortly to overtake them. Neither were they to be given the time to direct their minds to realizing de Gaulle's wish that selectivity be introduced. In the meantime, student militancy was being actively prepared following an international meeting of revolutionary groups in Brussels in March 1967, where it was implausibly claimed that the contradictions of capital were creating favourable conditions for the outbreak of revolution. The new university campus at Nanterre was a fertile ground for such ideas, an *alphaville* of concrete, far from the amenities of Paris, and surrounded by depressing apartment blocks as far as the eye could see. Active among those at Nanterre was Daniel Cohn-Bendit, the son of German Jewish refugee parents who had returned to Germany in 1950. Cultivating close links with revolutionary student groups in Germany, he had failed most of his examinations in his first three years at Nanterre, and, as cynics observed, had little to lose in organizing a boycott of exams 'to prevent youth being sucked into the capitalist system'. Anarchist, Trotskyite, and Maoist groups were the principal participants, whereas the Communist UEC largely remained aloof from the growing unrest, remembering Georges Marchais's withering denunciation in January 1967 of 'small groups of young agitators not representing anything'.

The attempted assassination of the German student leader, Rudi Dutschke, by right-wing elements in Berlin on 11 April 1968, occasioned demonstrations in the Quartier Latin; and, after the authorities closed the Nanterre campus on 2 May, Nanterre militants transferred much of their activity to the Sorbonne, where they were much more likely to attract notice than in the grey wilderness of the *banlieue*. The Rector's appeal to the police to restore order led to successive nights of confrontation in the streets, familiar to all from the television record—as was the violence of the riot police, who hospitalized passers-by and wrecked apartments in which students had taken refuge, beating up the owners, most of whom had been helpless witnesses of what was going on. The police admittedly suffered many injuries from the stones and petrol-bombs that were hurled at them—official hospital statistics showed more police casualties than civilians, but these figures concealed the many students who were given unofficial treatment by medical students and who feared arrest if they went to hospital. The closure of faculties created widespread resentment among non-militant students, and encouraged the parliamentary Left and the unions to make political capital out of the situation. Hitherto, they had been critical of student violence, but 8–9 May found Mitterrand accusing the government of 'treating the students as objects', and both the CGT and CFDT declared their solidarity with them.

Paris students occupied the Théâtre de France at the Odéon on 15 May; and for a month they held interminable debates on a vast range of issues, while Jean-Louis Barrault helplessly tolerated the situation—and was subsequently a victim of the government's punitive displeasure for doing so. These debates, together with those held elsewhere, were a forum for a wide spectrum of burgeoning causes—feminism, the ecology movement, gay rights, etc.—which gave rise to the myth in many quarters that these movements were effectively launched by the events of 1968. It would be more accurate to say that the publicity and euphoria of these debates strengthened the impetus of such causes, enabling them to shorten the distance that still separated them from the more advanced equivalent movements in the United States and neighbouring European countries. Thus, a number of women students, active in the May events, were to be prominent in the formation of the Mouvement de Libération des Femmes in 1969–70 (pp. 341–3). It is likewise hard to assess the role of 1968 as a symptom and factor in the widening gap between youth and the adult world that affected most prosperous countries in the 1960s—some more markedly than France. The affluence of these years had given young people the material wherewithal to create and develop distinctive styles of life, while the sexual liberation of the 1960s and current trends in popular psychology had undermined parents' confidence in their ability to give effective guidance to adolescent children and exercise parental authority of the traditional kind. At the same time, the active concern of the young with alleviating domestic and overseas problems of poverty, disability, and old age was uncomfortably felt by some parents as a sharp if tacit indictment of their own generation for failing to do as much. Yet opinion polls in France, both before and after 1968, indicated less resentment and mutual incomprehension between children and parents than in a number of other countries.

It was inevitable that the May developments in Paris should have their counterpart in many other French universities: and, indeed, secondary education in general also had its demonstrations, sit-ins, and demands that curricula and disciplinary procedures be reformulated. Interviewed by Sartre for *Le Nouvel Observateur* on 20 May 1968, Cohn-Bendit declared that the aim of the militants was not the impossible dream of overthrowing bourgeois society at one blow, but rather the staging of a series of revolutionary detonations, each one of which would send out widening ripples of change, beyond either the power or the wish of the militants to predict or control. Students would give an example, and it was for the workers to take it up in joint action. Although little would come of it in the short run, it would be a brief glimpse of what could be. As he told *Paris-Match*: 'We will begin by destroying—then little by little action will teach us what we must build ...'. Its natural spontaneity precluded the whole idea of leadership or attempting to plan the future; and, symptomatically, Cohn-Bendit played a progressively inconspicuous role in events as they developed in May. Indeed, for much of the time he was out of the country.

As in other countries, the ideological militants of the student movement were overwhelmingly middle class—which was, admittedly, scarcely surprising in the case of France, where working-class students were so thin on the ground. But the militants' total rejection of established social values was only possible for those who started life with a secure base in the society they were rejecting. Indeed, the middle-class nature of the movement was one of the contributory factors that prevented it from receiving working-class sympathy. But, like student militants elsewhere, their criticism of the capitalist economy took some of its flavour from pre-industrial as well as post-industrial concepts. Various of their schemes for workers' co-operatives and communes were reminiscent of early nineteenth-century Utopian socialism, with its artisan and agricultural assumptions. Symptomatically, they mythologized the example of the workers' collectives in Barcelona in 1936–7, which were themselves part-products of the anarcho-syndicalism of Andalusian peasant immigrants with strong Bakuninist roots. Cohn-Bendit himself commented: 'Some people have tried to force Marcuse on us as a mentor: that is a joke. None of us has read Marcuse. Some read Marx, of course, perhaps Bakunin, and of the moderns, Althusser, Mao, Guevara, Lefebvre. Nearly all the militants of the March 22nd movement have read Sartre.'[3]

Worker involvement in the events of May 1968 was an ambivalent affair. On a superficial level, the embarrassment of the State in one sector inevitably encouraged other sectors to take advantage of it to press their own demands— an inherent hazard of any centralized system. As with so much else, de Gaulle was aware of the lack of participatory channels and had sought to remedy them, but with little enthusiasm from his collaborators. Management feared loss of authority in running the enterprise, and the unions feared absorption into the very system they were seeking to confront—with the notable exception of a sizeable sector of the CFDT, which still enjoyed the idealism of recent self-metamorphosis and maintained close relations with the forward-looking PSU. At the same time, middle management felt increasingly like wage-earners within the impersonal structures of expanding enterprise—which was one of the reasons why sections of it joined the protests of May 1968.

The mild recession of 1967 brought wage stagnation and a rise in unemployment. The worst affected were the young, unskilled workers and those on short-term contracts; and it was significant that it was mainly these groups who were most drawn to participate in the events of 1968. They were the least unionized section of the work-force, which was why the unions grossly underestimated the level of discontent and unease in 1967–8. It was true, in any case, that conditions compared badly with those of West Germany and the Benelux countries; and at the start of 1968, a quarter of the wage-earning population received less than £46 a month (worth £140 in 1986). Admittedly, French workers enjoyed four weeks vacation and ten public holidays; yet the working-day had remained unchanged for ten years, and they could justifiably point out that the current forty-eight-hour week left them in much the same position as they had been before the reforms of the Popular Front thirty-two years previously. Moreover, their social security contributions had been put up in 1967, reducing their meagre pay-packets even further.

The self-appointed conscience of the working class, the PCF, was reluctant to back the student movement—other than to make it an occasion for vilifying the government. As Georges Séguy of the CGT remarked on 7 May: 'The French working class has no need of petit-bourgeois leadership. It can find among its own class the experienced leaders and cadres that it needs.' In any case, there had always been a basic tension between students and the PCF, partly arising from the differing purposes to which they put ideology. Both sides conceded that ideology served two necessary functions: to analyse current and future trends on the one hand, and to serve as a rallying and unifying banner on the other. The students were primarily interested in the first function, while the party concentrated on the second. As an analytical tool for social and political planning, ideology had to be subtle and perceptive; but as a banner, it had to be simple and readily understood by the uneducated. This tension was a continuing feature within the party itself. Intellectuals such as Baby, Barjonet, and Hervé had wearily criticized Thorez, Duclos, and Waldeck-Rochet as simplistic and over-dogmatic. As a sympathetic outsider, Sartre sought to bridge the rift by viewing the banner role of ideology as a necessary working-class myth, which did not need to correspond with literal truth as it did when used as an analytical tool. Moreover, Sartre was content to leave the banner role to the political discretion of the party leadership—recognizing, like Althusser, that the future of society depended on the party rather than on intellectuals like himself. Yet it remained a fact that the Union des Étudiants Communistes was characteristically much more sympathetic to Popular Front tactics than the PCF itself—their heroes being Sartre, Gorz, and Togliatti, whose revisionism they found attractive.

But whether students and workers would come together depended on factors that lay far beyond the control of the PCF. And the ability of the government to exploit their differences was a major one among them. Pompidou had been absent on a visit to Afghanistan and Iran during the early riots; but on his return on 11 May, he adopted a conciliatory stance towards the students and determined to play the whole matter very coolly. Events, however, were moving quickly. The Union Nationale des Étudiants Français invited the CGT, the CFDT, and the FO to call a one-day general strike for 13 May; and the CGT, conscious that matters were slipping out of their control, decided to follow the other two unions in complying. On the day of the strike, some 700,000 people demonstrated in Paris; but the CGT went out of its way to try to keep contact between the students and the workers to a minimum.

In parallel with Pompidou, de Gaulle maintained his own chilly stance of business as usual, and departed on a state visit to Romania the following day. But during his absence, strikes and occupations spread rapidly from Sud Aviation, Renault, and other state-owned companies to a number of other factories; and he came home on 18 May to find the railways, postal services, and airlines at a standstill. Indeed, by 20 May a large segment of the private sector was also in the grip of a general strike, and overall numbers escalated within a few days to a staggering ten million. Faced with the paralysis around them, there was now widespread alarm in major cities, as people stocked up with food, petrol, and money in anticipation of a complete freeze of commercial life. And, echoing public dismay, a censure debate launched by the parliamentary Left was lost by a mere eleven votes.

De Gaulle was utterly and uncharacteristically disorientated by the whole situation, which did not correspond to his expectations of human behaviour. 'Shit in the bed' was his verdict; and he exasperatedly suggested that the resolve of the weary riot police should be strengthened with drink, like troops about to go over the top in 1917. Addressing the public on television on 24 May, he spoke of the need for the readaptation of structures to meet current needs, thereby linking the crisis with his own belief in participation. And, pulling out the old familiar stops, he announced a referendum on whether he had a mandate to engage in this reconstruction—warning that a negative response would result in his resignation. Technically speaking, such a referendum was illegal, since it did not concern the constitution—and the Conseil d'État swiftly pointed this out. But public irritation with yet another illegal demand for a national vote of confidence was largely offset by mounting apprehension, following the worst day of violence the capital had seen so far in this month of upheaval. The Bourse was set on fire (24 May), and fighting in the Quartier Latin resulted in more than a thousand injuries. At the same time, the day was also marked by riots and demonstrations in the provinces; the banks closed, and radio and television staff went on strike, compounding apprehension with uncertainty.

But this mounting wave of seeming solidarity with the students was largely illusory. Once the Communists had decided to support the strike in an attempt to control it, they steered it towards specific short-term gains. The size of pay-packets progressively squeezed out the long-term issues of participation and the restructuring of management–labour relations, which had interested the CFTD and many of the non-unionized workers. Moreover, from the CGT's point of view, participation not only ran the risk of unions being drawn into the system—or at least of being accused of a sell-out—but it ran the greater risk of the CGT being outnumbered in the committees by non-union members and other unions. This ostensible Popular Front approach, with the Communists trying to obtain overall command, characterized their policy over the forthcoming weeks. They steered a difficult course, pressing hard for settlement when it seemed that their partners were becoming too independent, and, conversely, encouraging local strikes to keep going when it seemed that the government was about to claim credit for a successful outcome to its mediation.

During these events, the CGT made it clear to the *patronat* that they were interested in reforms not revolution; and 25–6 May saw a joint meeting of unions, employers, and government at the Ministry of Social Security in the Rue de Grenelle. The result was the so-called Grenelle agreements (27 May), which promised an immediate 7 per cent wage increase, followed by a further 3 per cent in October. Workers' contributions to health charges would be proportionately limited, and a firm undertaking was given to investigate how union officials might be afforded better facilities to conduct their business at shop-floor level. The SMIG was to be increased in effect by some 30 per cent—

an offer which was followed in June by raising the SMAG to the same level, which meant a major increase of 56 per cent for agricultural workers. But, despite these substantial concessions, there was no specific agreement as yet on the other crucial issues—the length of the working week, the retirement age, and the matter of strike pay.

When the CGT Secretary-General, Georges Séguy, took the agreement to the Renault workers at Billancourt on 27 May, he was embarrassed, but not surprised, to be met with a hostile reception and loud accusations that he had sold the strike for paltry concessions. Moreover, other state factories followed suit, stubbornly refusing to accept the agreement; and once again the distance between the CGT and the militants, most of them non-unionists, was starkly demonstrated.

The non-Communist Left now attempted to take the initiative. Following a rally of the UNEF and the PSU at the Charléty stadium on 27 May, with Mendès France among the participants, the Fédération took the plunge: Mitterrand announced on 28 May that he would run as a presidential candidate if de Gaulle resigned, and he proposed an interim government with himself or Mendès France at its head. Responding swiftly to this call for readiness, Mendès France declared on the following day that he would be prepared to form such a government, and include the Communists. But the Communists, having ordered their followers to boycott the Charléty rally, were witheringly dismissive of these proposals—the older members doubtless remembering Mendès's cold rejection of their help in 1954 (p. 239). By contrast, Jean Lecanuet declared that the Centre Démocrate was ready to support Mendès France (29 May), provided that he pursued a pro-European policy—another barbed reminder of 1954.

This was the point at which the government seized the initiative. Pompidou believed that the idea of a referendum was a mistake, and that the government would be better advised to call a general election, on the assumption that the violence and inconvenience of the last fortnight would dispose the mass of the electorate to support the government. De Gaulle saw the force of the argument. But on 29 May he greatly disconcerted Pompidou by disappearing, no one knew where. In fact, he had secretly flown by helicopter to Baden to consult with General Massu, who was then commanding the French army in West Germany. General Massu claimed, after de Gaulle's death, that de Gaulle was dispirited and considering resignation, and that his visit to Germany was partly intended to establish a refuge for his family. Massu, according to his own account, managed to persuade de Gaulle that resignation would be regarded by the public as a betrayal, and that he should return to Paris at once and postpone all thought of it, at least until the outcome of the referendum should be known—if he still wanted to hold one. Much has been made of this meeting; but de Gaulle may well have gone to Baden with an open mind as to what he should do. It must be remembered that there had been occasions in the past, notably in the 1940s, when de Gaulle had displayed or affected discouragement in an attempt to bring out the genuine feelings and opinions of colleagues on key issues. A bold front might merely have elicited loyal expressions of encouragement, whereas what he probably wanted to know was whether he could count on Massu and his colleagues in the event of widespread civil disorder. Massu claims that there was no discussion on using the army to suppress subversion; but at that stage it was probably sufficient for de Gaulle to know that he had Massu's good will. Pompidou, in any case, had tanks ready on the south-western outskirts of the capital to deal with any serious attempt at an armed take-over. Whatever the truth of the matter, de Gaulle returned very confident, finally convinced that Pompidou's preference for a general election rather than a referendum was the wiser course of action—a decision which he announced to the nation in a radio broadcast on 30 May.

The announcement, and the confident tone in which it was delivered, was in marked contrast to his performance on 24 May—and signalled to opponents and supporters alike that he intended to fight to win. Over half a million people demonstrated their enthusiasm in the streets, a huge cortège marching down the Champs-Élysées in triumphant celebration of de Gaulle's determined stance; and the Left were now desperately conscious that the tide of events was beginning to turn against them. Mitterrand was already widely accused of opportunism, and his loosely worded proposal of 28 May for an interim government was interpreted by hostile commentators as an appeal for some sort of coup. Similarly, Mendès France was accused of abandoning his long-held pose of high principle now that a serious chance of getting back to power had come his way. But their position was an extremely difficult one. Had they taken a different tack and maintained an attitude of silence or circumspection, they would doubtless have both been open to the greater charge that they had betrayed the Left by failing to give a decisive lead at a time when its chances were good.

With parliament dissolved, pending elections, the immediate concern of the parliamentary Left was to get its electoral campaign in motion. The students and militant extremists found themselves forgotten in the rush to prepare for the contest, while the factories still on strike were too concerned with specific material demands to be bothered with groups whose millennial expectations were remote from their interests. A significant return to work, which began on 1 June, continued steadily over the next three weeks, Citroën car-workers being the last to go back on 25 June. Student militants made various attempts to join in the picketing; but they were continually denounced by the CGT, as well as attacked by the police and CRS, whose snatch squads arrested students making their way to the factories.

The mood of the election was reflected in the new title of the Gaullist party, the Union pour la Défense de la République—a label that was also worn by the Giscardians. The Gaullists not only obtained a large increase in their share of the vote, but they emerged with an absolute majority in the Assembly— without even counting their Giscardian allies (Table 17.1). This resounding success was unquestionably the result of the deep-felt public fear of disorder; and their victory had therefore some of the ambiguities of 1958–62, despite its magnitude. It also represented the capture of electors who were disillusioned with what they saw as the mixture of indecision and opportunism that characterized the Left in May; and, even more markedly, it exemplified the decisive capture of former Centre voters, who were capitulating even faster than in 1967, the Centre no longer seeming to fulfil a clear function in the confrontation politics of the summer of 1968. But if the bipolarization of the 1960s was continuing at an unprecedented pace, the benefit in these elections was almost exclusively to the Right. The Communists' loss of the popular vote was dramatically compounded in their loss of seats, where the breakdown of left-wing co-operation on the second ballot reflected the bitterness and accusations of mutual betrayal that had characterized the opposition parties in the last week of May. The Fédération itself was as heavily defeated as the Communists, while the PSU, despite having three times more candidates than in 1967, lost all its seats, including that of Mendès France. Mitterrand's own parliamentary group of sixteen members, the Convention des Institutions Républicaines, was entirely eliminated except for himself, while his popularity among the general public, according to opinion polls of April and October 1968, fell from 51 per cent to 29 per cent. It therefore came as no surprise when he resigned from the presidency of the Fédération in the following month.

Table 17.1.
National Assembly Elections of 23 and 30 June 1968
(Results for Metropolitan France only)

	% of vote on first ballot	Seats gained after both ballots	
		Total	%
Communists	20.0	33	7.0
PSU	3.9	0	0.0
FGDS	16.6	57	12.1
Centre groups	12.5	31	6.6
Gaullists and allies	46.4	349	74.3
		470	

Notes: The percentage of electorate casting valid votes was 78.6.
There were additionally 17 overseas deputies.

Conversely, it was a great surprise to many that the new ministry should be headed by Couve de Murville (11 July 1968). For all his loyalty, intelligence, and capacity for hard work, this cold fish seemed a poor exchange for the genial, level-headed organizer of victories, whose personal popularity with the population at large had steadily grown in his six years in office—and was still mounting while de Gaulle's was falling in the crisis weeks of 1968. Indeed, it was widely suggested that it was precisely Pompidou's popularity and his cool assessment of events—when de Gaulle was plainly baffled—that prompted de Gaulle to get rid of him and appoint a man with no claims to charisma, so that de Gaulle could reassert his own ascendancy. It has also been suggested, however, that de Gaulle was acting in Pompidou's interest. Six years was a long innings by French standards. He had wanted to replace him with Couve in 1966; and to let him go now, with his prestige at its highest, put him in a very strong position for the presidency when de Gaulle retired. The matter had been complicated by Pompidou's own request for retirement following de Gaulle's disappearance to Baden. When Pompidou had second thoughts, de Gaulle informed him that he had already offered the succession to Couve and that it was too late to change matters. Even so, 'ingratitude' was a widespread verdict on the change of Premier.

Debré took Couve's place at the Quai d'Orsay, while Ortoli, who had briefly taken over Education on Peyrefitte's resignation, became Minister of Finance. Education passed to Edgar Faure, a key assignment, given the promises of major reform that had been made in the previous month. It was a measure of how times had changed that three days later the national holiday of 14 July was celebrated with a general amnesty for 'Algérie française' militants, including OAS activists—a mark of gratitude, perhaps, for the Massu therapy of 29 May. Among the long-lost sons who returned to the metropole were Bidault and Soustelle.

Faure's experience as a parliamentary manipulator stood him in good stead in getting his education reform law of 12 November 1968 through both the Assembly and the Senate—without a single dissenting voice on the final votes. Since both students and parents were dead set against selectivity, the government made no attempt to introduce it. The bill notably laid down guide-lines for the provision of new universities that would help to keep down the national average size to 15,000 students. As a result, France was to have 67 universities by 1972, comprising 720 *unités d'enseignement et de recherche*, which replaced the old faculties. Of the 13 universities allocated to Paris and its suburbs, 2 were designated for experimental methods of teaching and grading. Universities and UERs were henceforth to be governed by elected councils, on which students and teachers would have equal representation; and the councils were to elect their own presidents, chosen from the professoriat for a limited term of office.

To prevent the student representatives from being dominated by militant minorities, a poll of less than 60 per cent would result in a graduated scale of reduction in the size of student representation. But in practice, the left-wing militants boycotted the elections, leaving the moderates and Communists holding the floor. Sadly and predictably, apathy prevailed; and within three years, the proportion of students voting for these bodies fell from just over half to little more than a quarter.

The new law allowed universities greater autonomy in running their affairs; but the Ministry of Education still retained control over those examinations that conferred national diplomas, while the Ministry of Finance continued to supervise all expenditure that was not covered by private appeals. In practice, however, the universities were slow to take advantage of this greater freedom, habit and timidity inclining them to depend heavily on ministry guidance. By contrast, the experimental University of Vincennes was a mixture of exhilarating innovation and anarchy, which initially relied on Communist students to keep order and eject activists who disrupted classes.

The university reforms served to emphasize the absence of any comparable changes in secondary education, despite its being the scene of widespread trouble in 1968. *Lycée* teachers and administrators remained segregated in their hierarchy of castes—as remote from each other as they were from their pupils, despite the cosmetic exercises of school councils. Classes continued to be meticulously structured one-man performances, in which dialogue was kept to a minimum lest the programme fall behind its carefully calculated schedule. As always, passivity and resentment were the in-built hazards of the system.

Equally modest were the consequences of the Grenelle agreements on labour relations. Shop-floor facilities were granted to unions by the law of December 1968, while February 1969 saw the signing of the promised agreement on job security. Profit-sharing, however, was slow to become a reality, with under 250,000 workers participating by mid-1969. But the main grievance, predictably enough, was the government's refusal to allow a succession of post-Grenelle pay-rises, which had been naïvely envisaged by many in the euphoria of May 1968. The so-called 'Conference of Tilsitt' of March 1969 merely offered a 2 per cent rise, plus 2 per cent in the autumn, provoking a one-day national strike on 11 March. The unions, however, were too busy unravelling their internal differences in the aftermath of 1968 to contemplate sustained confrontation with the government; and the fact that unemployment was easing helped to defuse discontent elsewhere.

The government's financial caution was understandable. French currency reserves had fallen by almost half between July 1968 and June 1969, and yet the government recognized the need to give direct help to exporters and to those smaller firms that would be hardest hit by the pledge of higher wages. Various restrictions were put on imports, taxes were raised, and the pace of the nuclear programme was lowered. There was a strong case for devaluation, but de Gaulle dismissed it as a potential blow to French prestige in a delicate period of psychological recovery.

De Gaulle was keen to build on the government's recent electoral success and push through changes that he had long envisaged but had hesitated to attempt. He not only viewed the Senate as an anachronism, but it was increasingly a thorn in his flesh. Its method of election preserved it as an athenaeum of Fourth Republican *notables*, and it was here that de Gaulle felt himself confronted with the old parliamentary élites and intermediaries whom he had fought for so long in the 1940s and 1950s. The changes he envisaged were analogous to those of the Bayeux programme of 1946. One hundred and fifty of the existing seats would be allocated to representatives of socio-economic interests, while the remaining 173 would continue to be elected in the usual way. In effect, this would transfer to the Senate the role of the Economic and Social Council, which in consequence would cease to exist. But this injection of new blood and new functions would entail the loss of the Senate's legislative role, leaving it as a purely advisory body. Moreover, in the event of the President of the Republic dying or resigning, the President of the Senate would no longer become the caretaker President; this role would be filled by the Prime Minister, who, unlike the President of the Senate, was very much the retiring President's man and likely to continue his policies.

While few electors felt much sympathy for the senators, they were nevertheless aware that an attempt was being made to get rid of irksome critics under the guise of participation and modernization; and an opinion poll revealed that only 38 per cent of the public favoured the change. Some voters disliked the vaguely corporatist ring of the proposal, suggestive of faint shades of Vichy and the inter-war dictators. In order to give it a 'participatory' character and link it with his promises of 1968, de Gaulle paired it with a scheme to make the new twenty-one regions, plus Corsica, important units of local government. The remit of the existing Commissions de Développement Économique Régional (CODER) was to be broadened to include non-economic matters. But since these bodies were part-appointed, part-elected, this was more of an attempt to involve local socio-economic forces in modest decision-making, rather than a significant piece of democratic devolution. It was, nevertheless, a comparatively popular proposal, favoured by 59 per cent in a recent opinion poll, which was why de Gaulle was determined to link it with the less-favoured Senate reform in a take-it-or-leave-it package to be put to a referendum—the whole parcel being wrapped up in a threat of resignation if the public rejected it. De Gaulle was doubtless aware of the risk he was taking. But, having steered the regime through its most hazardous crisis to a firm base, he probably found little attraction in staying on unless he could achieve the completion of the constitutional changes he considered desirable. He wanted to win—but if he lost, he would be resigning by his own wish and on his own terms. And he knew that the regime would be in safe hands with Pompidou and his majority.

None of which softened the blow when it came: rejection by 53.2 per cent to 46.7 per cent of voters in a relatively high poll on 27 April 1969. Beaten by his old enemy, the Senate, it was in some ways the last laugh of the ghost of the Fourth Republic. There is little evidence to suggest that this reflected a positive desire on the part of most voters to see de Gaulle go. It was de Gaulle, not the public, who had made it a resigning issue; but the public was not prepared to be blackmailed into accepting something it did not want. There was none of the impending threat of disorder as in 1958–62 or 1968 that had secured large majorities for de Gaulle in the past; and Pompidou was an able and popular heir apparent.

Immediately on receiving the result, de Gaulle issued the briefest of communiqués: 'Je cesse d'exercer mes fonctions de Président de la République. Cette décision prend effet aujourd'hui à midi.' In his remaining eighteen months of life, he gave no interviews and made no public appearances. The last memorable image that he left behind was a film of him walking with his wife along a wind-swept beach in western Ireland—deserted but for the white-mackintoshed photographers, circling like gulls around a pair of plough-horses.

POMPIDOU'S PRESIDENCY

There was much that was unexpected about the presidential election that followed. Although Right and Left continued as always to woo the Centre vote, the post-1968 disarray of the Left resulted in the final contest being between the candidates of the Right and the Centre. Pompidou made it clear that his policies would be more *centriste* than those of de Gaulle, with parliament and business being treated as partners rather than unruly subordinates. This soothing language won over not only the Giscardians, who re-entered the governmental camp, but also various notabilities of the Centre, such as Pleven. Much of the Centre, however, was prepared to stick to its guns. The President of the Senate, Alain Poher, was regarded by many as the hero of the 'Non' campaign in the referendum; and, as caretaker President, he was able to add the publicity and panoply of State to this popularity. He was initially careful in his electoral campaign to steer clear of positive declarations of policy, contenting himself with his hitherto successful role of defender of democratic institutions against the arrogance of the Right. But as the campaign continued, he was increasingly obliged to declare his intentions, many of which were bound to create deadlock between him and a largely Gaullist Assembly.

The Left entered the contest under the worst of circumstances: the Fédération was dissolved, Mitterrand discredited with the general public, the PSU was tainted with *soixante-huitard* activism, and the Communists were isolated after the double blow of May 1968 and the Warsaw Pact's invasion of Czechoslovakia three months later. The SFIO was itself divided: the Mollet camp regarded the defeat of Pompidou as the prime objective, even if it meant supporting Poher, while Gaston Defferre's followers believed that it was possible to capture enough of the Centre vote to put a Socialist in power. Eventually an uneasy compromise was patched up around Defferre's candidature. The PSU, for its part, put up its Secretary-General, Michel Rocard, as a candidate, since Mendès France steadfastly refused to consider a post he regarded as undemocratic. The Communists, conscious of their isolation, also joined in the *chasse au centre* by putting up the *bon enfant* Jacques Duclos—one of their few leaders with a human face—and campaigned on the need for the 'petits' and the 'moyens' to join forces against the 'gros'. The fact that the PCF proposed a candidate at all showed how firmly the institutions of the Fifth Republic had taken root. The constitutional issue, like the clerical issue, was fast fading as a political divisor. As it happened, Duclos's 21.5 per cent on the first ballot was a fillip for the PCF and a humiliation for Defferre, who, with a mere 5.1 per cent, epitomized the misery of the non-Communist Left, including Mendès France who had supported him. Rocard obtained 3.6 per cent.

Pompidou triumphed on the second ballot with 57.6 per cent against Poher's 42.4, a third of the Communist voters supporting Poher, despite party advice to abstain.

Pompidou's choice of Chaban-Delmas as Prime Minister (20 June 1969) was a skilful one, in that he combined a solid record of loyalty to de Gaulle with a healthy respect for parliament's sensibilities. He kept on half of his predecessor's government, but broadened it in the direction of the Centre by bringing in several former MRP luminaries, notably Maurice Schumann at the Quai d'Orsay. Giscard d'Estaing regained Finance and Pleven was appointed to Justice.

Addressing the Assembly, Chaban-Delmas described France as 'une société bloquée', a concept made familiar by the writings of Michel Crozier and Stanley Hoffmann; the aim must be to replace it with 'la nouvelle société'—a phrase that led critics to accuse him of raiding the American locker for cheap crusading slogans. Following his mentors, Chaban-Delmas spoke of the need to create 'a culture of dialogue', to lessen the rigidity of state control, and to encourage flexible, competitive responses in the economy. In pursuit of these aims, the various information sources of the ORTF were given greater autonomy, as were the railways, Paris public transport, and the EDF–GDF. Their relations with the State became the subject of periodic contracts. Moreover, a Ministry for the Protection of Nature and the Environment was established in January 1971, emblematic of Gaullism with a caring face. Cosmetics and good intent were also evident in the closer meshing of the SMIG with changes in the cost of living (January 1970), though whether its new Micawber-like title, Salaire Minimum Interprofessionnel de Croissance, promised growth or inflation remained to be seen.

Certainly, the return of Giscard to the Ministry of Finance seemed to promise orthodoxy and a commitment to economic growth, even if his devaluation of 12.5 per cent on 8 August 1969 was too gentle to make the price of French exports significantly more attractive abroad. Undaunted, the government replaced weekly wages with monthly wages for a growing number of workers, linking them with improved social security entitlement. (The intention was to extend the scheme to 40 per cent of the labour force by 1974.) In the meantime, Renault and the aeronautical industry introduced modest schemes of profit-sharing.

Nor was regionalism forgotten. Towards the end of 1971, a bill provided for the indirect election in 1973 of twenty-two regional assemblies and the location of matching advisory bodies of socio-professional nominees—all, admittedly, with very limited powers and guaranteed to cause the government little trouble. The regions' financial resources were to be modest, most of them based on taxes on vehicles and property sales.

Although Chaban-Delmas was popular with the public at large, Gaullist fundamentalists accused him of 'closing the great book of Gaullism'; and, in an attempt to mollify them, Pompidou replaced him with Pierre Messmer on 5 July 1972—with Jacques Chirac as Minister of Agriculture. The grim-faced Gaullist image of the new government undoubtedly deepened the unease of such Giscardians as Michel Poniatowski, who wished the Independents to live up to their name. But Giscard, still ensconced at the Ministry of Finance, saw the party's future—and, more particularly, his own—as lying in close co-operation with the Gaullists. And the 1973 elections were approached on this assumption.

Co-operation was likewise the watchword of the Left. Following their humiliating showing in the presidential election of 1969, the Left made efforts to present a more convincing challenge. Meeting in July 1969, the SFIO and a number of the *groupuscules* formed the Parti Socialiste, with Alain Savary as its First Secretary. But when Mitterrand's Convention des Institutions Républicaines formally joined in June 1971, Savary relinquished the post to him. The next move was an agreement with the Communists on a common electoral programme (27 June 1972) and a common strategy on the second ballot. The programme envisaged the abolition of private education and the liberalization of the laws on divorce, contraception, and abortion. The President's term of office would be reduced to five years, his emergency powers under Article 16 would disappear, and his use of referendums under Article 11 would be sharply curtailed. On nationalization, however, the Common Programme halved the Communists' shopping-list.

Within a fortnight, the programme had also attracted the support of Robert Fabre's left-wing Radicals—and optimists once again hailed 'a new Popular Front' in the making. The Radicals, in fact, were undergoing one of their periodic searches for an identity. In 1970, the flamboyant former editor of *L'Express*, Jean-Jacques Servan-Schreiber, had tried briefly to galvanize the party behind a sparkling programme of participation, decentralization, and the 'democratizing of opportunity'. But, after a severe trouncing by Chaban-Delmas in a Bordeaux by-election in September 1970, he changed tack once again, and eventually turned to Jean Lecanuet of the Centre Démocrate with the offer of a new Mouvement Réformateur, retaining the Servan-Schreiber slogans but emptying them of innovatory content (3 November 1971). This move towards the Centre antagonized the left wing of the party, who, under Robert Fabre, formed the Mouvement des Radicaux de Gauche, with aspirations towards co-operation with the Socialists and Communists.

As it turned out, these various manœuvres were but stages in the slow death-throes of the political Centre. Observing the inexorable logic of the situation, Lecanuet accepted Messmer's offer of an electoral strategy which would guarantee the Mouvement Réformateur a basic thirty seats in the new Assembly of 1973—the minimum required to secure formal status as a parliamentary group—in return for which the Mouvement would withdraw its candidates wherever this would facilitate the defeat of the Left. The strategy paid off. The Mouvement Réformateur emerged with thirty seats, and was duly listed as the next item on the menu of the Right. Its predecessor in this role, Jacques Duhamel's Démocratie et Progrès of July 1969, just made the qualifying threshold of thirty seats by taking like minds on board, assuring the government of an overall majority that it would otherwise not have had (Table 17.2).

As for the Gaullists and Giscardians, their margin of victory had been curtailed by the simple fact that 1973 shared with 1967 the unique distinction of being the only electoral year of the Fifth Republic that had not been over-shadowed by some major crisis or constitutional issue; and, as in 1967, this favoured the government's critics.

Pompidou retained Messmer as Premier, with Giscard and Chirac keeping their former portfolios. To steal the thunder of the Left, however, Pompidou proposed to reduce the presidential term of office to five years—making it commensurate with the life of the Assembly. This would ensure that both President and parliament reflected the public opinion expressed in the election year, and would thereby minimize the risk of their representing opposing majorities. Although approved by both the Assembly and the Senate in October 1973, the level of support was not strong enough to guarantee the requisite three-fifths majority in a joint session of parliament under Article 89. Nor was Pompidou sure that the alternative strategy of a referendum would necessarily succeed. Consequently, and regrettably, an eminently sensible proposal came to nothing.

Table 17.2.
National Assembly Elections of 4 and 11 March 1973
(Results for Metropolitan France only)

	% of vote on first ballot	Seats gained after both ballots	
		Total	%
Communists	21.4	73	15.4
PSU, etc.	3.3	2	0.4
Socialists	17.7	89	18.8
Left Radicals	1.4	11	2.3
Mouvement Réformateur	12.5	30	6.3
Centre Démocratie et Progrès	3.7	21 (see text)	4.4
Giscardians	10.3	54	11.4
Gaullists	23.9	175	37.0
Conservatives, etc.	2.8	15	3.2
		473	

Notes: The percentage of electorate casting valid votes was 79.4.
There were additionally 17 overseas deputies.

Unknown to the public, Pompidou was slowly dying of cancer. His customary ability to keep a firm, benign grip on government was clearly waning, giving rise to puzzled unease—especially after the oil crisis of 1973 and the onset of the economic recession. International relations were increasingly left to the Quai d'Orsay, where Michel Jobert lacked his master's former vigour and finesse, preferring to respond to foreign initiatives rather than launch his own. Moreover, Élysée staff became unhealthily influential in decision-making, notably Pierre Juillet and Marie-France Garaud, who had been factors in the sacking of Chaban-Delmas in 1972 and also in the rapid rise of Jacques Chirac, who was now given the Ministry of the Interior in the cabinet reshuffle of 27 February 1974.

Pompidou died on 2 April 1974; and the electoral campaign that followed was in marked contrast to that of 1969. Instead of the Left being in disarray and the Right displaying disciplined solidarity, the situation was reversed. Perched in his eyrie in the Tour Montparnasse, Mitterrand was promised the support of the Communists on the first as well as on the second ballot; and he also had the backing of the PSU, the CGT, and the CFDT. The government majority and the Centre were a different matter. Even before Pompidou was buried, Chaban-Delmas had put himself forward as a candidate without consulting the Giscardians. Accusing Chaban of precipitate haste, Chirac responded by leading a Gaullist rebellion in favour of Giscard, aided by the anti-Chaban elements in the Élysée staff. Others who supported Giscard included the Lecanuet wing of the Mouvement Réformateur.

The first ballot (5 May) gave Mitterrand 43.4 per cent of the metropolitan votes, Giscard 33 per cent, and Chaban-Delmas 14.6 per cent. Among the fringe candidates, the former Poujadist, Jean-Marie Le Pen, obtained 0.7 per cent. But the second ballot inevitably saw the bulk of Chaban-Delmas's voters going to Giscard, giving him 50.7 per cent of the total, compared with Mitterrand's 49.3 per cent. To many observers, the bipolarization of politics now seemed absolute.

But the identity of the victor gave others pause for thought, some claiming that Giscard at the Élysée was a triumph for the Centre—and, as such, a return to the traditions of the Third and Fourth Republics. Indeed, acquaintances were quick to recall a remark Giscard had supposedly made some twenty years earlier: 'France wants to be governed by the centre-right. I will place myself centre-right and one day I will govern France.'

18
Reform and Recession
The Giscardian Era, 1974–1981

THE new President's debonair display of liberal sentiments and his basic concern for financial stability and economic growth were soon to earn his *septennat* the title of the Orleanist presidency. Indeed, his name, his elegantly equine features, and his father's politics in the 1930s—Action Française, Croix-de-Feu, Parti Social Français, and then Pétainism—lent a superficial air of authenticity to the epithet. After a distinguished record at the Polytechnique and the École Nationale d'Administration, Giscard became an *inspecteur des finances*, eventually entering parliament as a CNIP deputy. A protégé of Pinay, he subsequently became Minister of Finance and Economic Affairs in January 1962, at the early age of thirty-five. The 'Swinging Sixties' convinced him of the need to act young as well as be young. The presidential campaign of 1974 displayed photographs of him in a T-shirt, playing the accordion, while his opening speech assured the electorate that: 'I want to look deep into the eyes of France, tell her my message, listen to hers'—music to the ears of a generation imbibing Sacha Distel and other dispensers of easy sentiment. Yet, at the same time, the need to reassure Monsieur Prudhomme was never forgotten—'change without risk' was Giscard's motto for the second ballot.[1]

The early months of the presidency were marked by pointed democratic gestures, all carefully publicized. He invited some road-sweepers in to breakfast, and invited himself out to a number of private homes for dinner—a practice that gave rise to a spate of hoaxes, leaving victims and favoured citizens alike uneasily uncertain as to who, if anyone, would turn up. But the studied informality of the spring of his *septennat* gradually gave way to the cultivation of a more monarchical style. However, he always retained a penchant for personal intervention at the lower levels of government. He saw to it that the Place des Vosges was planted with lime-trees, rather than the patterned flower-beds favoured by the authorities; but after the Pompidolean desecrations to Paris, such concern was no bad thing. Less happy was his emasculation of the 'Marseillaise'. His liberal sensibilities being ruffled by the bellicose bravura of the anthem, he had it shorn of its gusting brass crescendos, and rendered into pastoral, reflective mood—in much the same spirit as the English editor of a children's anthology, who, shocked by the cheerful violence of 'Goosey Goosey Gander', amended it to:

> There I met an old man
> Who couldn't say his prayers.
> So I took him by the left hand
> And helped him down the stairs.

Mitterrand was happily to restore the status quo—blood, brass, and all.

Giscard's credo—later published in *Démocratie française* (1976)—contained few surprises. Believing that class antagonisms were a dying relic of early industrialization, he saw steady, piecemeal progress as the goal of most Frenchmen. Self-fulfilment was to be achieved by giving the individual citizen a greater measure of responsibility within a pluralistic society with enlightened leaders. He came to power with a substantial shopping-list of progressive reforms—many of which he achieved, despite the dispiriting experience of working within the context of an economic recession, and despite the disillusioning experience of working with a largely Gaullist majority that was frequently suspicious of his liberalism.

Given Chirac's role in tripping Giscard's main rival on the Right, Chaban-Delmas, it was not surprising that he should be invited to form the first government (28 May 1974). Giscard had appreciated his qualities as Junior Minister for the budget in 1969, and his eventual assumption of the secretary-generalship of the Gaullist UDR (December 1974–June 1975) was to make him a useful instrument for keeping the President's morose coalition partners in line. Chirac's government was relatively broad-based: four Gaullist ministers, three Giscardians, and four from the Mouvement Réformateur, including Lecanuet (Justice) and the mercurial Servan-Schreiber (Reforms), who was to leave the cabinet ten days later over the issue of French nuclear tests in the South Pacific. Other members included the widely popular Simone Veil at the Ministry of Health, and Françoise Giroud as Secrétaire d'État à la Condition Féminine.

As far as the Élysée staff was concerned, Giscard replaced Pompidou's advisers with his own men, impressing on them the importance of not impinging on the ministers' terrain. But, given Giscard's own desire to increase the power of the President, his advisers became increasingly influential compared with the ministry staff. Like Giscard, over half of them were ENA graduates, but there was a rapid turnover, thereby preventing individuals from acquiring an entrenched personal ascendancy. Giscard disliked criticism and was quick to rid himself of uncongenial figures. Even so, his adviser on African affairs, René Journiac, acquired an almost ministerial authority in this sphere, being credited among other things with the overthrow of Bokassa (p. 351).

LIBERAL REFORM

It was understandable that the bulk of Giscard's liberal reforms should come in the early years of his *septennat*. They were easy, popular gestures that did not involve much in the way of financial outlay, but, once given, they could only be followed by other reforms that ran the risk of alienating vested interests and particular sections of the electorate. Hence the cautious character of Giscard's later years, which was not merely the reflection of disillusion or fatigue. The *septennat* nevertheless presented the curious profile of a regime whose reforming period coincided with the rule of Chirac and his Gaullist colleagues (27 May 1974–25 August 1976), while the later era of Raymond Barre's Giscardian-style governments (27 August 1976–13 May 1981) seemed relatively barren.

The programme started impressively enough with the extension of the franchise to eighteen-year-olds in June 1974. While Giscard was arguably doing no more than follow in the footsteps of other Western democracies, he was braving colleagues' fears that this would strengthen the Left in sub-sequent elections. Indeed, the first ballot of the 1981 presidential election was to see 42.5 per cent of the eighteen- to twenty-year-olds' vote go to the Left, compared with 35 per cent to the Right; but this was a less striking difference than many anticipated. Giscard came to power pledged to reduce the presidential term of office. But, as with Mitterrand after him, the fear of an opposition victory was enough to kill off the proposal—and parliament, sharing the same fear, did not seek to press him. Nevertheless, the government showed deference to parliament by its more sparing use of its contingency weapons, such as the 'blocked vote' beloved of previous administrations (p. 287).

Despite Giscard's talk of enlightened leadership, he displayed a cautious concern to remain in step with public opinion on matters of legal change. Although a majority of parliament was thought to favour the abolition of the death penalty, an opinion poll of June 1979 showed 55 per cent of the public wishing to retain it. Consequently, Giscard and his Justice Minister, Alain Peyrefitte, twice refused to allow a parliamentary vote on the matter, despite their own profound aversion to capital punishment. Similarly, Giscard decided in February 1980 to abandon proposals to modify some of the fiercer aspects of the penal code, notably the specification of minimum sentences and the permanent entry of even trivial offences on job applicants' *casiers judiciaires*. On the other hand, legal aid was improved and public court costs were abolished in civil cases, leaving litigants with only their lawyers' fees to meet. Other welcome developments included the extension of the powers of the ombudsman (24 December 1976)—a creation of the last months of Pompidou's presidency—and the abolition of cinema censorship.

Broadcasting

Censorship of a more extensive kind had traditionally been exercised over the government-controlled radio and television service, and in January 1975 Giscard made a point of specifically repudiating Pompidou's dictum of 1972 that 'television is regarded as the voice of France'. The government accordingly made much of the claim that it no longer sought to control the content of programmes—although the short contracts of its network directors gave it an effective means of indirect pressure. Even so, the new Ministry of Culture and Communication that was set up in 1978 was at pains not to intervene in news programming; and the time allotted to opposition politicians came increasingly close to that accorded to government supporters. But if censorship was theoretically at an end, the fact that the staff remained government employees was still reflected in the ultra-deferential handling of government personnel in interviews. Yet, as in so many things, it was ultimately a matter of political culture and tradition rather than statute. In terms of how its staff were appointed, the BBC was theoretically just as vulnerable to government pressure as the French broadcasting services, yet its independence of attitude was much greater—and the amount of time devoted by the BBC to current affairs programmes was twice that of its French counterparts.

Nor did Giscard make any attempt to relax the government's monopoly of broadcasting, despite the professed commitment of most Giscardians to commercial competition. Admittedly, the law of 7 August 1974 replaced the monolithic ORTF with seven separate organizations; and it was likewise true that TF1 and Antenne 2 had carried commercial advertising since 1968—which meant that they were partially competing for viewers and private finance. But this was scarcely a challenge to state control.

The State and Interest Groups

Indeed, the professed liberalism of the Giscardian government could not disguise the fact that the executive continued to remain more powerful under the Fifth Republic than in any other Western democracy—and Frenchmen of all parties accepted the fact. Vincent Wright has accurately described the British system as one where:

the Government is essentially a broker which, in its search for social consensus, is obliged to modify its policies and to bring about concessions in the light of evolving circumstances. Decision-making is rarely 'heroic' and only rarely results from the imposition of rationally calculated policies, but is rather the negotiation of marginal adjustments to the *status quo*. It necessarily precludes major shifts in policy which would be offensive to certain groups who could jeopardise the harmony of social institutions.[2]

Fewer organizations in France, however, qualified as *interlocuteurs valables* in the eyes of government and the general public; and the government's traditional claim to represent the general rather than the sectional interest of the public was regarded with far less scepticism in France than it would have been in Britain or America, where a plurality of interests was regarded as a sign of health rather than disintegration. Admittedly, the numerical weakness of the unions, together with their political divisions and their revolutionary rhetoric, disposed both government and public to take them less seriously than in Anglo-Saxon countries; as Edgar Faure sardonically observed: 'we have the good fortune to have revolutionary trade unions'. But the influence of other interest groups tended to depend on how far they happened to match the government's own concerns. Farmers were a case in point. Although they had lost the numerical electoral strength that had made them such a force under the Third and Fourth Republics, the more progressive elements among them had acquired new influence as a result of the EEC, making them into the allies of French Eurocrats in the competitive, international field, rather than just another domestic rival for government favour.

Women

Among the various Giscardian reforms that did not involve the government in major expenditure, the President set great store by those improving the lot of women. In the manner of God acceding to Adam's request for a companion, he told *L'Express* in May 1980: 'Je crois avoir été celui qui a inséré la femme française dans la vie de notre société.' Not that there was any lack of prompting from women themselves. Although feminism had been slow to take off in France compared with the United States and several neighbouring European countries, 1969–70 had seen the formation of the Mouvement de Libération des Femmes; and many women who had earlier been prepared to subordinate the cause of women's rights to the general achievement of a more just society, became increasingly impatient with the failure of the Socialist and Communist parties to give their demands sufficient attention. Simone de Beauvoir—'Notre Dame de Sartre', as more militant feminists unkindly called her—was prominent among those who saw the need for a specific women's campaign; and she was also a signatory of the *Manifeste des 343 'salopes'* of 1971, which, bearing names such as Catherine Deneuve and Delphine Seyrig, was a list of women who admitted to having had an abortion and were advocating women's right to them. The outcome was a blossoming of feminist organizations, including such extremist and untypical minority groups as Psychanalyse et Politique, which looked to Jacques Lacan's reworking of Freud, and Front Lesbien, which asserted that heterosexual women were collaborating with the enemy.

Giscard's appointment of Simone Veil to the Ministry of Health resulted in the Veil Act of 17 January 1975, legalizing abortion within the first ten weeks of pregnancy. The initial restriction to cases of personal distress soon became a mere formality; but the patient bore the whole cost of the operation, and doctors had the right to opt out of performing abortions on grounds of conscience. Predictably, the abortion law passed the Assembly only with the help of left-wing votes—less than a third of the Gaullists and Giscardians voting for it, and only half the Centrists. The effect of the law was to reduce the number of back-street abortions, which had accounted for a large proportion of the half-million that still took place each year in the early 1970s—abortions equalling a third of the live births in France. Even so, the increasing efficiency and knowledge of contraception brought the abortion rate down to a fifth of the live birth-rate by the end of the Giscardian presidency—with 150,000 of them performed according to the provisions of the Veil Act, the majority in public institutions where the average fee was about £55. Although the review of the Veil Act in November 1979 found parliamentary opinion divided much as it had been in 1974, an opinion poll of 1981 showed 66 per cent of the public in favour of it and only 24 per cent against.

With the main answer to these problems lying in the field of contraception, a law of December 1974 had appropriately removed most of the remaining obstacles to the sale and prescription of contraceptives. The initial step in this direction had been taken by Lucien Neuwirth's law of 28 December 1967, which had permitted the establishment of family-planning clinics, and the sale of contraceptives in pharmacies; but the Giscardian legislation authorized 'pill' prescriptions for up to a year's duration, while teenagers were no longer required to produce evidence of parental consent before buying contraceptives over the counter. In the same spirit, divorce by mutual consent was legalized by the law of 11 July 1975, which effectively put an end to the harrowing and often ludicrous process of establishing infidelity or misconduct. This, too, found public favour, with 78 per cent expressing approval in a poll in 1981, and only 10 per cent against.

Despite a couple of laws in 1975 stipulating equal pay and employment opportunities, women's average earnings in 1980 were still 20 to 30 per cent lower than those of men, and their position on the promotion ladder was markedly worse. While to some extent this was the consequence of maternal 'time-out' while children were of pre-school age, it also reflected the concentration of women in unskilled and semi-skilled occupations. The tendency for women to seek employment in the same vicinity as their husband's work— rather than vice versa—likewise resulted in many of them accepting jobs that were beneath their capabilities. They made up only 15 per cent of the liberal professions in the early 1970s while only 2 per cent of the candidates in the 1973 elections were women, only nine of whom obtained seats, compared with the thirty or so in the Assembly of 1946. Moreover, until Giscard's presidency, only three women had reached ministerial rank in post-war France; and although by 1976 the government contained five women, most of them had been co-opted as specialists from other walks of life rather than as career politicians. Nor did the outlook for the future necessarily look any brighter. Although the proportion of women students had risen from about 25 per cent in 1930 to 47 per cent in 1977, this was still substantially lower than their numerical position in the population as a whole.

Regionalism and the Environment

Giscard's genuflections to pluralism had created expectation that regionalism would see significant advances. But the modest nature of his intentions was made all too clear in a television interview of 16 June 1976: 'I am . . . hostile to political regionalism. On the other hand, regional economic development has always seemed to me something very reasonable and very efficient. I am also, unlike some others, entirely favourable to France conserving all her cultures.' None of this surprised Anglo-Saxon observers, who were all too aware that anything analogous to the powers of British local authorities would strike Frenchmen of all parties as virtual anarchy—Bakuninism run riot. Much more influential in the development of regional policy was the Paris-based Délégation à l'Aménagement du Territoire et à l'Action Régionale. The development of the Languedoc–Roussillon coast as a holiday region, first launched in the early 1960s, was a typical DATAR project—not a happy advertisement many might feel, but without the plan, the destruction of fine scenery would almost certainly have been worse. Indeed, Giscard's circumspect dealings with regionalist aspirations were continually complicated by his deeply felt environmental concerns. 1975–6 saw a series of laws and regulations limiting building density and preserving green spaces, while a directive of August 1979 restricted building on undeveloped coastlines. Further desecrations that Giscard was anxious to contain were the sprawling ski resorts of Pompidou's *plan neige*, with their shanties and chair-lifts, transforming the mountains into a semi-industrial landscape of rubble and pylons. Despite the annual pilgrimage of four million French skiers to the slopes in the late 1970s, and the bonus of foreigners' spending, development was henceforth dependent on the approval of Paris.

Giscard was also quick to put a brake on the skyscraper building in Paris that had started to mushroom during Pompidou's presidency. Although more carefully controlled than the random desecration of the London skyline in the 1960s, the damage was already causing concern. He likewise stopped the expressway that was planned for the left bank of the Seine, which would have transformed the celebrated riverside walks into a replica of the undulating speedway that had already disfigured the right. As Mayor of Paris, Jacques Chirac was shortly to contest these aesthetic curbs on the thrusting modernity which was all part of the Gaullist legacy—pointing out that without expressways the beauty of the streets would have become increasingly lost in bumper-to-bumper bad temper. Only by aggressive new building could the old be preserved.

It was Giscard's misfortune that the considerable economies and social achievements of his presidency should have taken place against the background of a world recession. The expansion of the 1950s and 1960s had led business circles to assume that economists and governments had learnt to control the cyclical fluctuations that had dogged the economy in the past. Not only was this assumption undermined by the *malaise* of the 1970s, but confidence was shaken by the fact that government economists and advisers were now singing more austere tunes, at odds with recent orthodoxies and reminiscent of an earlier, more pessimistic age. The decision of the Arab oil states to put up the price of oil was a particular blow to France. Oil currently supplied about two-thirds of France's energy requirements; and the quadrupling of oil prices in 1973 was equivalent to a loss of 2 to 3 per cent of GDP in 1974, even if it later tapered off to about 1 per cent in 1978–9.

Yet France managed to ride out the difficulties of the 1970s rather more successfully than many comparable countries. In the first place, the continual, if slackening, growth of her population ensured both an expanding domestic market and a steady supply of labour (p. 382). Secondly, the fact that families were devoting a larger proportion of their increased earnings to saving helped to offset the decline in investment by industry itself. Then, thirdly, the development of the electronically based industries provided France with new sectors of expansion which helped to compensate for the vicissitudes of the old.

Planning

Even so, the French government was unprepared for the economic problems of the mid-1970s—despite warning signs of a recession well before the oil crisis of 1973. The German government, by contrast, had addressed itself to curbing inflation some six months before the crisis, yet France was much slower to do so. Indeed, it was only in the autumn of 1976 that French planning placed it in the forefront of its concerns; and until then the government had mainly relied on a series of stop-go short-term measures. Thus, the First Fourcade Plan preached deflation from late 1974 to late 1975; but Chirac then slammed the engine into forward gear and insisted on a reversion to expansionist policies under the Second Fourcade Plan. The result was that inflation grew at an average annual rate of 10.7 per cent; and the overall balance of payments remained in deficit between 1973 and 1977. The only consolation was that both these developments still remained slightly below the European average.

The Barre government sought salvation in attempting to stimulate rapid economic growth within the limits of the current balance of payments—thereby hoping to kill unemployment and inflation with the same strategy. At the same time, the brave-faced optimism of the Seventh (Barre) Plan of 1976–80 was also a response to the prospect of the 1978 elections, where the left-wing parties were trying to seduce voters with promises of a better deal for the victims of the recession. Remarkably enough, Barre's optimism was largely justified by the results. The Barre Plan sensibly concentrated most of its power for positive stimulus on the development of new technology: electronics, information systems, robots, offshore technology, bio-industries, and energy-saving equipment. And inevitably, nuclear energy and telecommunications loomed large in its priorities—the French telephone systems having long been a byword for chaotic inefficiency.

Economic growth maintained a steady 3 per cent annual expansion, not only converting the trade deficit to a surplus, but transforming the overall balance of payments from a deficit of $3.3 billion in 1977 to a surplus of $4 million in 1978. And all this was achieved without recourse to devaluation or deficit budgeting. With France heavily dependent on oil imports, Barre dismissed devaluation as self-defeating—and argued that the chronic state of the international market offered little compensation in the way of increased exports of French manufactures. Deficit budgeting, on the other hand, would merely have provided investors with an alternative market, thereby reducing the funds available for industrial expansion. Productive investment was encouraged instead, by granting income-tax credits on purchases of French shares, as well as by corporation-tax concessions and subsidized loans; and as a result of these vigorous measures, the fall in private investment was kept below 15 per cent during the presidency.

Exports and Industry

The success of the government's overall economic strategy was reflected in France's position in the league table of major exporting countries—her share of their industrial exports rising from 8.4 per cent in 1968–70 to 10 per cent in 1977–9. Despite the age-old fears of German then of Japanese imports, France exported twice as many vehicles and chemical and pharmaceutical products as she imported, and aircraft and armaments showed an enormous lift in this period. On the other hand, her showing was poor in household consumer durables, where the influx of such items as Italian refrigerators weighed the balance against her—one of the few areas where the EEC had adversely affected her industry.

The Giscard presidency undoubtedly saw a conscious dropping of the 'France can build it' ethos of the 1960s, and a preference for pragmatism and selective stimulation in furthering French economic expansion. The Gaullist principle of there always being a strongly supported French runner in every important individual field was significantly eroded in the 1970s. Symptomatically, the Plan Calcul, designed to further an independent national computer industry, was dropped in 1975 in favour of merging the French CII with the

American Honeywell company. And in the same way, the French graphite-gas process for generating nuclear power was allowed to run down, while the future development of the programme was geared to the American Westinghouse pressurized water reactor. Autonomy was consciously sacrificed to the goal of enabling the nuclear programme to produce a fifth of the nation's total energy needs in 1985—and thereby reduce French dependence on foreign oil to perhaps two-fifths of its overall energy requirements.

As in the case of arms exports, France was frequently blamed for supplying foreign countries with nuclear technology, particularly reprocessing techniques that could be used to produce plutonium with military capability. But, in general, objections to nuclear energy came primarily from environmentalists and from a section of the Socialist party. The Communists, by contrast, were solidly in favour of its domestic development; and organized opposition to power stations came mainly from local interests. Yet, with the exception of the demonstration at Plogoff, it was never on a comparable scale to that in Britain; and the government treated it with thinly disguised contempt.

THE SOCIAL OUTCOME

Yet if the economic climate of the 1970s scarcely favoured bold social initiatives, it was profits and entrepreneurs that bore the brunt of the recession—with profits dropping from 17 per cent to well under 10 per cent of 'value-added' (i.e. output minus basic costs) between 1975 and 1981. The labour force, by contrast, continued to see its standard of living increase, if less markedly than before. Real wages for manual workers rose by more than a quarter, while white-collar wages rose by an average of 11.4 per cent—from what was already a fairly high floor. In fact, the lessening gulf between high and low wages was further reduced by changes in social security contributions—and even the SMIC managed to keep pace with the increases elsewhere. At the same time, a growing number of workers now received a monthly salary, with its attendant advantages of better holidays and job security.

Even so, that other instrument of social justice, taxation, was still a long way from playing the role that it had in several of France's northern neighbours. As always, it was the heavy reliance on indirect taxation that was so prejudicial to the poorer classes. Typically, in the 1980 budget, income tax provided only 45 per cent of the sum that VAT produced, and company tax a mere 20 per cent. It was therefore of small consolation that a married man with two children and an annual income of 80,000 francs paid only 7 per cent of it in income tax—or 10 per cent if he earned 120,000 francs—when virtually every purchase or transaction he made cost him a greater proportion of his income than it did for a company director or a stockbroker. Yet parliament virtually amended out of existence the attempt to institute a capital gains tax in 1976; and even the sad shadow that eventually emerged in July was not implemented until three years later.

At the same time, the rise of unemployment during Giscard's presidency, from 2.8 per cent of the working population to 7.5 per cent, created a growing sense of insecurity—despite improvements in redundancy pay in January 1979 and an increasing number of inducements to early retirement. Indeed, retirement itself was made more bearable by the fact that the basic state pension went up by nearly two-thirds in real terms during the presidency.

Not surprisingly, unemployment radically changed French attitudes to immigration. In contrast to the welcoming policies of the 1960s, immigrants from outside the EEC were prohibited as early as 1974, including the families of existing residents. Moreover, financial incentives (10,000 francs) were given to those willing to be repatriated; and a number of Algerians availed themselves of the offer, causing the net balance of immigration to fall briefly below zero in the course of 1976. But the subsequent lifting of the ban on the immigration of relatives caused numbers to rise once more; and by the late 1970s, immigrants made up a third of the work-force in the building industry and close on a quarter in car assembly.

The much publicized password of the Fifth Republic, 'participation', continued to be as remote from realization on the shop-floor as elsewhere. Although May and October 1980 saw measures encouraging profit-sharing among the workers—with the government offering to finance two-thirds of shares instituted for this purpose—such operations were formally restricted to a mere 3 per cent of company capital. The Sudreau Report of 1975 on workers' rights recommended the involvement of workers in management on a consultative basis, and the extension of profit-sharing to all companies; but it bore little fruit in practice. Nor was the attitude of the workers themselves encouraging. Although nearly 1,200 firms had some form of profit-sharing by the end of 1980, most of the workers had sold their shares as soon as it became legal to do so.

Moving from the factory floor to domestic life, the 1975 Barre Report on housing recommended that market forces should be allowed to play a greater role. Rent controls had traditionally acted as a disincentive to building and maintaining rented property; and in 1977 the government changed its method of providing housing for the low-paid by giving subsidies directly to the family rather than to the builder, as hitherto. By the end of Giscard's presidency, some 130,000 families were benefiting from this; and, given this safety net for the poor, the gradual phasing-out of rent controls could now be set in motion (July 1979). Since four-fifths of French homes were still rented, the potential repercussions were wide-ranging; but time ran out for the presidency before much could come of this in Giscard's term.

Education

Perhaps the most far-reaching of all the Giscardian reforms was René Haby's *loi cadre* of July 1975 on primary and secondary education. Introducing the comprehensive principle into secondary education, it arguably did no more than follow what Britain and other European countries had initiated a decade earlier. Yet it was a courageous step for a right-wing government to take in a country which had traditionally prided itself on the high academic standards of its state schools. It essentially reflected Giscard's concern that social origins should not prejudice educational opportunity; and it represented a serious attempt to tackle the problem at all levels of the school structure.

Starting at the bottom, children of pre-primary age were to be offered optional nursery classes, in an attempt to compensate for any initial cultural deprivation. Then, at six years old, the child went to an unstreamed primary school—after which he entered a *collège*, where, at the age of eleven or twelve, he followed a common syllabus with all children of his age for the first two years (6ème and 5ème). At thirteen or fourteen, children with a technical bent could then take vocational classes alongside the normal curriculum for the next two years (4ème and 3ème); and in practice, about 20 per cent of the intake were to do so. On emerging from 3ème at the age of fifteen or sixteen, the pupil's compulsory education was now formally complete; and about 40 per cent of the original intake left school altogether. But of the remaining 60 per cent, about half went on to a *lycée d'enseignement professionnel*, a technical college which was a springboard for apprenticeships, and where the more able could take the *Certificat d'Aptitude Professionnelle*. As for the other half, most of them went to a *lycée* to prepare for the *baccalauréat* and entry into one or other of the various forms of higher education.

POLITICS

It is conceivable that the 1978 election might have created a left-wing Assembly, if the Communists had not broken with the Socialists the previous year. Mitterrand's rejuvenation of the Socialist party in the 1970s and his determination that it should dominate the Left had caused the PCF to reconsider its strategy. Rather than become the prisoner of the Socialist party, it preferred to pose once more as the party of intransigent purity, the only true apostle of social transformation. Accordingly, the PCF embarrassed Mitterrand by publishing its own understanding of the Left's economic programme—timing it to appear just before Mitterrand's television debate with Barre, thereby enabling the Premier to portray the prospect of a left-wing government as a recipe for national bankruptcy. The ensuing negotiations between the two parties became increasingly bitter, breaking down completely on 22 September 1977.

The result was that the Left went into the legislative elections of 1978 without even a joint strategy for the second ballot.

Yet the Right was likewise a prey to dissension. Giscard's break with Chirac and his appointment of Raymond Barre as Prime Minister in August 1976 had embittered Gaullists, who proceeded to display their confidence in the former Premier by electing him President of their new formation, the Rassemblement pour la République, in the following December. This in fact had long been Chirac's personal goal, with the 1981 French presidential election in mind. He had increasingly found his governmental duties irksome as he piloted Giscard's uncongenial legislation through parliament, risking his popularity with his fellow Gaullists, who were loud in their criticism of the capital gains tax bill of 1976 and other pieces of Giscardian philanthropy. At the same time, tensions between Chirac and Giscard encouraged several ministers to treat the Premier with disrespect. Chirac responded by planning a reshuffle of the government and an early dissolution of the Assembly; but Giscard was not prepared to risk the gamble, and Chirac resigned.

Chirac's decision to go may also have been influenced by the growing difficulty of the economic tasks that were facing the government; and in choosing Barre, Giscard was primarily concerned to find an economic expert who might contrive to put France on course before the electoral campaign of 1978. The Gaullists, however, were not disposed to forgive easily; and Barre soon found himself obliged to reshuffle the government in a Gaullist direction (29 March 1977), dropping Poniatowski and Lecanuet and bringing in the Gaullist hatchet man, Alain Peyrefitte, with a mandate to organize the forthcoming election campaign. None of which inhibited the Gaullists from attacking Barre's economic policies, which they claimed were too *étatist* and insufficiently stimulating to private industry. Indeed, their belligerence received further encouragement when Chirac was elected Mayor of Paris, roundly defeating the Giscardian Michel d'Ornano, after the spring municipal elections of 1977. Thenceforth, the government had to live surrounded by irritatingly assertive evidence of the new mayor's dynamism—even to the ubiquitous exhortations to dog owners to teach their pets to use the gutter. It was small consolation that in 1981 critics of the new mayor were able to embellish these notices with Chirac election stickers, causing them to read: 'Chirac, apprenez-lui le caniveau'. (It was generally recognized, however, that Chirac's concern on this point was justified. A third of French homes possessed a dog—compared with a quarter in Britain—and pharmacies were full of dog-tonics, labelled: 'Protégez votre chien contre la fatigue'.)

Despite these divisions, the governmental alliance entered the legislative elections of March 1978 with a greater sense of discipline than their opponents. The government's precarious position was reflected in the unprecedentedly large vote of the Opposition on the first ballot—45.3 per cent against the government's 46.5 per cent. Yet the Giscardians could take heart from the fact that their recent absorption of the Réformateurs had helped them to capture 21.4 per cent of the vote, bringing their share close to the Gaullists' 22.5 per cent and disposing the RPR to treat them with more respect. Indeed, the Right's closing of ranks on the second ballot brought its reward in the distribution of seats, the Gaullists and their associates obtaining 150 and the Giscardians 137 (including overseas deputies and subsequent adjustments)— 90 more than their opponents (Table 18.1). The Left, by contrast, had been unable to follow up its first-ballot advantage. Communists and Socialists widely rallied to each other's support on the second round. The first ballot had been a sobering experience for the Communists, in that for the first time in post-war France their share of the vote (20.6 per cent) was less than that of the Socialists (22.6 per cent). Moreover, in those constituencies where the left-wing candidate in the second ballot was a Communist, a quarter of the Socialist voters chose to transfer their allegiance to the government candidate rather than to vote for a Communist. Such switches posed fewer agonies of conscience when the government candidate was a Giscardian rather than a Gaullist—and the RPR had reluctantly to admit that the cause of the Right had been strengthened by the growing electoral power of the UDF. Even so, the Communists obtained more seats than many had anticipated—86, compared with the Socialist–MRG's 114 (including overseas deputies). But this was to be the last time that they were to enjoy representation of this order.

Table 18.1.
National Assembly Elections of 12 and 19 March 1978
(Results for Metropolitan France only)

	% of vote on first ballot	Seats gained after both ballots	
		Total	%
Communists	20.6	86	18.1
Socialists	22.6	102	21.5
Other socialist groups	3.3	0	0.0
Left Radicals	2.1	10	2.1
Giscardians, Centre, etc.	21.4	125	26.4
Gaullists	22.5	141	29.7
Conservatives, etc.	3.2	9	1.9
Others	5.1	1	0.2
		474	

Notes: The percentage of electorate casting valid votes was 81.7.
There were additionally 17 overseas deputies.

None of this, however, deflected Georges Marchais from his current ghetto strategy. He continued to praise the Soviet Union, and consistently defended Russian policy in Afghanistan. At the same time, he pushed his patronage of working-class militancy to the point of backing popular resentment against immigrant labour. All of which helped to speed the desertion of the remaining intellectuals in the party.

FOREIGN POLICY

Afro-Asian Affairs

It is often asserted that Giscard was less interested in foreign policy than de Gaulle or Pompidou. As a self-styled man of the Centre Right, his declared aim was to achieve consensus in domestic policies, and it was arguable that he did not need to seek this in national ambitions on the world stage. Certainly, his policies in the EEC were primarily economic in their concerns, while his activities in Africa were largely dictated by the perennial French desire to maintain strong links with the former French empire and with the francophone world in general. Admittedly, only 3 per cent of France's former African subjects actually spoke French, but at least they did not speak English—as yet. And there was the additional factor of French investment in African mines and other economic interests—all of which was enough to sharpen a lively French sensitivity to any changes in the balance of power in Africa. France therefore chose to keep 14,000 troops ready for instant use there, some stationed in Senegal and Djibouti, but most of them in her remaining Indian Ocean dependencies. France actively intervened—with uncertain intent and mixed results—in the interminable internal wrangles of Mauritania and Chad—and more spectacularly in the former colony, Oubangi Chari, where the notorious self-appointed Emperor Bokassa first received, and then forfeited, French support. Rumours of cannibalism and other unedifying activities on the part of Giscard's protégé eventually led to a French expedition (20 September 1979) and his replacement by the former President Dacko—notwithstanding Bokassa's earlier gifts of diamonds to his patrons. Giscard was shortly to repeat Marie-Antoinette's discovery that diamonds are not always a girl's best friend.

Equally chequered in its logic and outcome was France's intervention in the former Belgian territory of Zaïre. Claiming to be the protector of francophone interests, the government twice acceded to General Mobotu's appeal for help to suppress rebels in 1977–8. Although militarily successful and a good advertisement for French efficiency, it created deep resentment in Belgium, whose troops were upstaged in the bid to protect European residents during the second intervention. Nor was the French departure from her own last mainland colony, Afars and the Issas, an occasion of unmixed joy. A referendum on 8 May 1977 produced a 98 per cent majority in favour of independence—

which was granted—but the tribal tensions continued between the Ethiopian Afars and the Somali Issas, just as they had done under French rule.

Giscard made much publicized visits to distant countries—including China, India, Pakistan, and Russia—and there was much use of the word 'mondialiste' in the rhetoric of the Élysée and the Quai d'Orsay; but it cannot be pretended that his activities outside Europe and Africa had the impact or significance of de Gaulle's initiatives in the 1960s. Relations with the oil-producing countries of the Middle East were strengthened by extensive arms deals, notably with Iraq and Saudi Arabia; but the effective role of France in the affairs of the Middle East and the eastern Mediterranean was small compared with the gestures and declarations that accompanied them.

Giscard's rhetoric on France's mission to humanity went much further than de Gaulle's. Even when talking for foreign consumption, de Gaulle had always been conscious of what he saw as the limitations of the French. Like Moses, he had been continually exasperated by the people he was called upon to lead, and it would be hard to imagine him making Giscard's bland statement of May 1975: 'my fundamental idea is that the superiority of France is a superiority of the spirit. It is not a superiority of force, it cannot be an economic superiority . . . It is a superiority of . . . a country which understands best the problems of its times and which brings to them the most imaginative, the most open, the most generous solutions.'

East–West and French Defence

It is arguable that it was as much the times that had changed as the man. The freezing of *détente* between East and West in the late 1970s left little room for the bold manœuvres that de Gaulle had attempted a decade earlier. Giscard's flamboyant assumption of the role of mediator at the time of Russia's invasion of Afghanistan (1979–80) was an empty exercise—nor did France pick up any major medals at the Moscow Olympics to add lustre to her decision to ignore America's plea for a boycott. On the other hand, French co-operation with NATO grew markedly closer under Giscard, partly because of the complexities of NATO's recently developed concept of a flexible response to a possible Russian attack. To combine flexibility with effectiveness required a close understanding and co-ordination between the members of the Western alliance. And the matter was further complicated by the development of tactical nuclear weapons, which added an extra dimension of fine tuning in the escalation from conventional to nuclear war—all of which necessitated even closer understanding between partners if disaster was to be avoided. Indeed, in June 1976 France announced its commitment to 'sanctuarisation élargie', extending the boundaries of its defensive concern to embrace not merely France but NATO's eastern frontier.

France's tactical nuclear weapons consisted of the Pluton, which was brought into service in May 1974 and was initially carried by the AMX 30 tank—with the intention of extending it to the Super-Étendard carrier-based planes when these became operational in the early 1980s. Beyond these, however, lay the strategic nuclear force, the main strength of which was the flotilla of five nuclear submarines, each of which was intended eventually to carry sixteen nuclear warheads. There were also eighteen S2 ballistic missiles on the Plateau d'Albion and patently vulnerable to pre-emptive attack—while more mobile but outdated were the thirty-six Mirage IV nuclear bombers. The striking power of the French nuclear force had quadrupled during Giscard's presidency, with the intention of reaching ninety megatons by 1985—or 4,500 times the strength of the Hiroshima explosion. The overall defence budget had reached 3.8 per cent of GNP by 1980, compared with 3.4 per cent in 1974—conventional weapons also being among the beneficiaries.

Giscard was admittedly more prepared than his predecessors had been to talk the language of disarmament, even if it did not add up to anything tangible. His proposal for a European disarmament conference was perhaps no more than a verbal attempt to upstage the Vienna-based talks on Multilateral Balanced Force Reductions, which France was not attending. And, true to the policy of his predecessors, Giscard would have nothing to do with negotiations on strategic-arms limitation. Indeed, critics were quick to accuse him of cynicism, given France's rapidly growing exports of armaments to the Third World, which by 1978 made France the third largest arms exporter in the world, ranking next to the two superpowers. Although sales to South Africa nominally ceased in August 1975, French arms continued to be made there under licence—and sanction-bound Rhodesia was another grateful client. With 4.5 per cent of the French working population employed by the armaments industry, France was in no hurry to convert it all to ploughshares—which no one would buy at a quarter of the price.

Europe

The story of Giscard in the EEC was also dominated by economics. He played a major part in the establishment of the European Monetary System in 1979, and in persuading the Irish and the Italians to enter. But, characteristically, the French government delayed the implementation of the system until the Germans gave way on concessions to French farmers. And it also had no compunction in imposing illegal tariffs on Italian wine in September 1975, and an equally illegal embargo on British lamb in 1979. At the same time, it was the most adamant of the member states in opposing any easing of the Community Budget burden on Britain.

Giscard established a good working relationship with Chancellor Helmut Schmidt of West Germany, who shared many of his assumptions on financial and economic matters. But their close terms were a source of unease to other member states—reminiscent of Franco-German relations in the early 1960s. Particular misgivings were felt in February 1977, when Giscard stated that if the EEC were enlarged any further, it might require an inner consultative circle of leading members—which some of the smaller states saw as a hint of future Franco-German domination.

The French government, like most other EEC governments, showed an almost Gaullist contempt for the European Parliament. It assumed that the bulk of the general public cared little about it—with most people not even knowing who their Euro-MP was. The public tended to see its main champion as the national government minister who was dealing with the EEC—even if he belonged to a different political persuasion. As in 1914, national interests tended to override ideals of international class-brotherhood. Indeed, there was general approval for the EEC ministers when they delivered a withering attack on the European Parliament for contesting the budget in 1979. The language that the ministers used was almost identical to that of the Prussian chief minister in 1848, when he denounced the Frankfurt Parliament as an 'in-disciplined rabble'.

The European parliamentary elections in France were conducted according to a national system of proportional representation, with one ballot. Since there was no government to form, alliances were unnecessary; and such interest as the elections aroused was principally as a barometer of party ratings in the country. Even so, only 60 per cent of the electorate bothered to vote in 1979.

DEFEAT

It was unfortunate for Giscard that the presidential election of 1981 followed France's worst economic year since 1975, with overall growth down to the European average of 1 per cent, living standards down by 0.8 per cent, and the balance of payments in heavy deficit. But Giscard had lost so much of his reformist image by 1981, while on the other hand a number of his conservative supporters in business and agriculture were swinging towards Chirac. Indeed, Chirac was to obtain 18 per cent of the vote on the first ballot (26 April 1981), whereas his true-blue Gaullist colleague, Michel Debré, received only 1.7 per cent. Even so, Giscard still retained the largest share of the right-wing vote (28.3 per cent) and was to go forward as its champion into the second round on 10 May 1981.

The Left, for their part, principally looked to Mitterrand. Michel Rocard had briefly offered himself as a presidential candidate in the autumn of 1980, following his flattering showing in opinion polls; but when Mitterrand made it clear that he intended to stand, Rocard withdrew. The Communists were unsparing in their attacks on Mitterrand prior to the election, yet a quarter of their customary clientele voted for him on the first ballot, fearing that to vote for Marchais might leave Giscard and Chirac as the two top candidates going forward to the second ballot. The consequent poor showing of Marchais on the first ballot obliged the leadership to follow their troops and back Mitterrand on the second. The final result on 10 May gave Mitterrand a somewhat clearer lead over Giscard (51.75 to 48.24 per cent) than Giscard had had over Mitterrand in 1974, but it was nevertheless a close result by the standards of the 1960s. On the whole, the various occupational groups voted as expected, although Mitterrand obtained a larger slice of the lower-management and white-collar vote than he had done in 1974.

Left-wing euphoria was tumultuous—with young enthusiasts proclaiming a new Popular Front, without perhaps realizing what a doom-laden slogan that was. In many ways, the Mitterrand experiment was to do rather better than the Popular Front; but basically it was fated to become entangled in analogous problems. And in so far as it did do better, this was largely a reflection of the more advantageous hand of cards that it picked up.

19
Keynesianism in One Country
The Socialist Experiment

'TAKE a good look at this; you'll never see anything like it again', was Mitterrand's ambivalent comment on the celebrations of May 1981. The Socialist manifesto had spoken of 'making our country the crucible of the liberation of man', and Pierre Mauroy forecast the creation of 'a new style of citizenship'. But Mitterrand was well aware that the Socialists' electoral victory primarily represented public disappointment with the Giscard achievement—and that this itself was the product of international economic circumstance rather than fundamental failures on Giscard's part. He could see that there was no mounting ground swell of popular demand for major changes in the structure of society or the economy, and that the left-wing vote essentially reflected a desire for higher wages, longer holidays, secure employment, and better social security benefits: in other words, the sort of improvements that Giscard would have been happy enough to give, had he felt confident that the economy could stand the strain and remain competitive. Mitterrand was also well aware that France's membership of the EEC and her trading links with other countries in which monetarism and similar purgative policies were in fashion, made a whole-hog strategy of socialism in one country a recipe for economic disaster.[1]

Critics to his Left, however, pointed out that Mitterrand's embracing of socialism had been a slow business, not unconnected, they suggested, with personal ambition; and it was not part of his credo to attempt a break with capitalism. He had been a minister eleven times during the Fourth Republic, when democratic liberties rather than redistributive policies had been his prime concern. No one questioned his brave Resistance record, or his courageous refusal to be a party to government complicity in colonial injustice and chicanery in the years that followed. It was rather a matter of whether his ultimate socialism reflected a personal evolution or merely an appreciation of where the political future lay. Apologists pointed out that his mentor, Mendès France, followed a similar path, and that the move from liberalism to socialism was a commonplace of political biography. If opportunism had been the prime motive, there were much swifter roads to power in the 1970s than socialism. To which critics replied that Mitterrand was a master of the waiting game. Certainly, his sallow, chiselled face and dispassionate, Pétainesque voice frequently invited the overworked epithet, 'the sphinx without a riddle', and it was mainly at election time that he affected a proletarian growl when uttering phrases such as: 'moi, je le trouve injuste', as he strode around in hat and scarf like Aristide Bruant at the Ambassadeurs. Yet, if he had the patience and stamina to pursue long-term aims, he also had a quasi-Bismarckian penchant for sudden gestures or ruses to intimidate or side-step his opponents in the short-term—and not only his opponents. Obvious examples were his threat to institutionalize referendums on a wide variety of issues in 1984, largely to counter the Right's opposition to the private schools bill; or his introduction of *scrutin de liste* simply to weaken the Right's expected gains in the 1986 election. It was manœuvres of this sort, pregnant with hazardous implications for the future, that particularly aroused accusations of opportunism. His readiness to abandon election pledges that proved inconvenient was common currency among politicians, and aroused less comment. No one seriously expected him to fulfil his promise to reduce the presidential term to five years, or to abandon his predecessors' recourse to *ordonnances* to bypass parliament.

The man he appointed as Prime Minister, Pierre Mauroy, was generally regarded as a moderate pragmatist. A former teacher and active trade-unionist, he was best known as the linchpin of the party in the northern provinces. His government was largely composed of Socialists, with the concessions to allies confined to three portfolios for the MRG and one for the Mouvement Démo-cratique. Socialist stalwarts with ministerial experience under the Fourth Republic occupied the Interior (Gaston Defferre) and Education (Alain Savary), while Defence and Telecommunications were given to old friends of Mitterrand, Charles Hernu and the appropriately named Georges Fillioud. The 1970s generation of active Socialists was represented by the CERES militant, Nicole Questiaux (Social Security—all too predictably redesignated as National Solidarity), Édith Cresson (Agriculture), and Laurent Fabius (Budget). Sig-nificantly, Finance and Planning were given to pragmatic 'moderates', Jacques Delors and Michel Rocard—Delors belonging to the technocratic section of the government, which also included Claude Cheysson (Foreign Affairs), and a month later Pierre Dreyfus (Industry) and Robert Badinter (Justice). By contrast, the far Left of the party was represented by the CERES leader, Jean-Pierre Chevènement, assigned to the important, but politically uncontentious, Ministry of Research and Technology. The government was also remarkable for the presence of six women—one more than the previous record under Giscard.

Faced with a hostile parliamentary majority, Mitterrand had no desire as yet to find out whether government could work in such circumstances. Calling for elections forthwith, his decision was overwhelmingly justified by the first-ballot results of 14 June, when the PS/MRG obtained 37.5 per cent of the vote, compared with 24.7 per cent in 1978 (Table 19.1). Their gains were as much at the expense of the Communists—down from 20.6 per cent in 1978 to 16.1 per cent in 1981—as at that of the Gaullists and Giscardians, both of whose overt losses were less marked—the first-ballot strategy of the Gaullist–Giscardian

Table 19.1.
National Assembly Elections of 14 and 21 June 1981

	% of vote on first ballot	Seats gained after both ballots	
		Total	%
Communists	16.1	44	9.0
Socialists	37.5	270	55.0
Left Radicals		15	3.0
Other Left	3.1	5	1.0
Giscardians	19.2	62	12.6
Gaullists	20.8	88	17.9
Other Right	3.2	7	1.4
		491	

Notes: The percentage of electorate casting valid votes was 69.8.
Unlike the tables in previous chapters, these figures include the 17 overseas deputies.

Union pour une Nouvelle Majorité saving them from worse disaster. The explanation for this, however, was also to be found in the low poll (70.4 per cent instead of 82.8 per cent), a frequent phenomenon in France when one election followed closely on another. Whatever the explanation, the abstainers included a large number of Right-of-Centre voters who were disillusioned with Giscard. The swing of advantage was compounded on the second ballot (21 June), when the PS/MRG obtained an absolute majority of seats—285 (of which 270 were Socialist), compared with 114 in 1978 (including overseas deputies), thereby coming near the Gaullist landslide of 1968, but without the aid of a tail wind such as the 'grande peur des bien pensants' in 1968. As on the first ballot, the Communists were as much the losers as the Gaullists and Giscardians—all three parties emerging with only half their former representation. The experience underlined the increasingly presidential nature of the regime. The Socialist party appeared to all as the party of the new President, and its capture of votes from both sides reflected the fact.

Mitterrand could now afford to be magnanimous and take up the Communists' offer to participate in government—an offer that was all the more interesting, in that it ran counter to Moscow's advice that they should support the new ministry but not join it. Accordingly, four Communists were appointed; but, as at the Liberation, none was given a key post that carried responsibility for internal order or national security. Otherwise, the new ministry was much the same as before the election.

Mauroy himself was not a Mitterrand man. Indeed, he and Rocard had unsuccessfully attempted to oust him from the Socialist leadership in December 1980, yet they worked well enough in harness. Mauroy initially favoured a loose, supervisory role in his dealings with ministerial colleagues, and looked to the President to enforce discipline when difficulties arose. But after the reshuffle of March 1983, he exercised a tighter personal control.

Superficially, it might now seem that France was 'la République des professeurs' in a much more absolute sense than it had ever been under the Third Republic. Over a third of the ministers were teachers—three times the average for 1970–80, and four times the average under Giscard. Over a third of the Assembly were also teachers—compared with less than a fifth before the election—all of which reflected the influx of new Socialist deputies, of whom more than half were teachers, mostly from the secondary sector. On the other hand, the Socialist victory saw the proportion of working-class deputies rise to a mere 4.3 per cent—the hopes of nineteenth-century socialists for a workers' parliament remaining almost as remote in 1981 as when they had first been conceived.

By contrast, the number of senior civil servants in Mauroy's ministerial team was down from two-fifths under Giscard to a quarter, and the proportion of *grands corps* figures in the President's personal cabinet was cut from 60 per cent to 19. But, while commitment to the Socialist party was the prime passport to influence in the early days of the New Jerusalem, the dawn of economic realism in the following years saw the resurgence of the indispensable technocrats, notably in Mauroy's third government of March 1983. And although the government conceded that they were indispensable, decrees of September and October 1982 attempted to democratize entrance into the École Nationale d'Administration by opening a parallel stream of entry for those promising civil servants in employment who had not been through the Institut d'Études Politiques or other prestigious antechambers of membership. This was followed by the law of 19 January 1983 which opened up similar opportunities for established trade-union officials and other people with comparable experience.

Yet the Socialist purge of the administration was rather less severe than most civil servants had anticipated. As in the past, it was the Ministry of the Interior that saw the most upheaval, with two-thirds of the prefects being shifted, most of them sideways to other appointments, but a dozen of them out of the prefectoral corps altogether. Indeed, by the end of its first year in office, the government had made 117 prefectoral moves—compared with 190 during the whole Giscard *septennat*. Yet the situations were very different, in that Giscard had inherited a broadly sympathetic administration, while Mitterrand was the first left-wing President of the Fifth Republic. By the same token, the higher echelons of the police also saw many changes. Alice Saunier-Seïté's filling of academic rectorships with political friends during the Giscard era was swiftly countered by the Socialists' removal of two-thirds of the entire body of rectors. As for the traditional slaughter of ministerial *directeurs*—well over half the total number—this was uneven in its incidence, the convictions and personality of the minister being a deciding factor in most cases. Thus, the Communist ministers made a clean sweep of all their *directeurs*, while other ministers anxious to assert a new-broom image behaved similarly. Of the replacements, about a quarter were active left-wing supporters. That the proportion was not higher reflected the difficulty of finding enough party militants with the necessary administrative expertise. On the other hand, as under previous regimes, the more technical ministries and the Ministry of Defence saw relatively few changes.

A minister's *cabinet* was traditionally a close advisory body of personal cronies, and 1981 saw the customary clean-out of the five hundred or so hangers-on who made up these variegated groups. The newcomers were markedly more party men than their Giscardian predecessors, most of whom had been known primarily for their technical expertise; and the proportion of *énarques* fell from a half to well under a third.

The Socialists, like Giscard, made much of their virtuous intention to distance the media from government control, and the Fillioud Act of 1982 sought to achieve this by setting up a new high authority for broadcasting. Yet, on the pretext that pure beginnings required the removal of old dirt, the government put indirect pressure on the seven television and radio heads to resign, despite the fact that their contracts still had a couple of years to run. Only one held out; and, similarly, all but one of the four directors of news were bullied into resignation.

It is tempting to see this as an American-style spoils system, which up to a point it was. But it also represented a state of affairs that the higher civil servants had to some extent brought upon themselves by seeking diversified careers with political dimensions. The appointment of technocrats to ministerial office in the early years of the Fifth Republic had created ambitions in the hearts of *fonctionnaires*. No longer content to continue the 'impartial' routine of a Whitehall-style administrator, they had sought political contacts, and moved in and out of private business. The presidential system and the bipolarization of politics increasingly made such contacts something of an indelible hallmark, which were either an asset or a liability, depending on which party was in power.

ECONOMIC AND SOCIAL POLICY

The government's initial policy was what has been called 'redistributive Keynesianism'. Economic expansion would be stimulated by government investment, which would provide employment for the jobless and raise the wages of the low-paid, thereby increasing the spending power of the mass of the population and the domestic market for French products. Mitterrand confessed in July 1983: 'I was carried away by our victory; we were intoxicated. Everyone ... predicted the return of growth by 1983. Honestly, I lack the necessary knowledge to say they were wrong.' It is perfectly true to say that the OECD predicted growth, and that opponents such as Chirac were just as committed as Mitterrand to achieving it as rapidly as possible. But France's misfortune was to loosen her belt at a time when her neighbours were tightening theirs, with deeply damaging consequences for the balance of payments, inflation, and the rate of exchange. If it is easy to draw comparisons with the mistakes of the Popular Front, at least Mitterrand's team was more conscious than Blum's of the need to improve per capita productivity.

Barre's Eighth Plan was scrapped, and Rocard presented an Interim Plan (1982–3) in December 1981, to be followed eventually by the Ninth Plan (1984–8). Yet in practice the plans were to be largely ignored by the President and his other ministers. Unlike the days of short-lived governments under the Fourth Republic, ministers now had both the time and the inclination to concern themselves with long-term economic strategies. They also had stable parliamentary majorities to implement them; and it was arguably no longer so necessary for these matters to be left to a commission that was insulated from the pressures of political life. At the same time, the difficulty of predicting long-term economic trends in the 1980s made planning of the traditional kind a hazardous undertaking in itself; and as early as 1982, Rocard was wrily referring to the plan as 'a theatrical exercise in collective psychodrama'.

The first nine months of the Mitterrand presidency were dominated by the nationalization programme, culminating in the law passed on 13 February 1982.[2] The State took over all the shares in six major industrial parent companies: the Compagnie Générale d'Électricité, the Compagnie Générale de Constructions Téléphoniques, Thomson-Brandt (electronics and telecommunications), Péchiney-Ugine-Kuhlmann (aluminium and chemicals), Saint-Gobain-Pont à Mousson (glass, paper, and textiles), and Rhône-Poulenc (textiles and chemicals). The only concession in the operation was that foreign shareholders in subsidiary firms were given the option of retaining part of their holdings.

Running parallel to the programme of outright nationalization was the acquisition of majority holdings in a number of other major companies. As in all established industrial countries, the shrinking demand for steel was a prime economic and social problem. Third World countries increasingly produced their own steel, while modern manufactures used a much wider variety of basic materials than had been the case when steel was the backbone of industry. The Giscardian State had already been helping out the ailing giants, Usinor and Sacilor—the Fafner and Fasolt of the old heavy sector of the French economy; and the Mauroy government, in effect, was largely formalizing the situation by taking a majority holding of their shares. Then, following the railways precedent of 1937, the State took a 51 per cent holding in the arms and aeronautical firms, Dassault-Breguet and Matra. Part foreign-owned firms were a more difficult matter, but the government negotiated substantial control over Roussel-Uclaf (pharmaceuticals), ITT-France (telecommunications), and CII-Honeywell Bull (computers). Thus, with thirteen of France's twenty largest companies now in government hands—and with a dominating share in many others—the State's control of industrial turnover (including energy) had risen in a matter of months from 16 per cent to 30 per cent. It was especially marked in key growth industries. But all this cost the nation a hefty sum in compensation to former owners: 39 billion francs in capital, and 47 billion in interest over fifteen years.

Apart from the cost of compensation, the State was assuming a heavy burden of responsibility for future investment in the newly acquired industries. As at the Liberation, it was strongly argued that the State needed greater control of the money supply—and the *grève du milliard* that had supposedly defeated Blum in the 1930s was invoked once more. The government accordingly took over the remaining private shares in those large banks that had been nationalized at the Liberation, and proceeded to nationalize thirty-six smaller banks, together with the investment giants, Paribas and Suez. This, in effect, left only the French subsidiaries of foreign banks in the private sector. As before, the nationalized banks were initially promised autonomy in the running of their affairs. But, given their intended role in the government's industrial strategy, this autonomy fast became a fiction, reflected in the fact that their industrial lending rate was restricted by the government. In the short run, this enabled the government to finance industrial growth by affordable loans rather than by major increases in taxation that would alienate the electorate. But this could not continue indefinitely. Nationalized banks were as much at the mercy of market forces as other banks, and could not survive if interest rates were kept artificially low. So, sooner or later, the government would be faced with cutting its aid to industry or presenting the taxpayer with a much larger bill.

There was considerable disagreement within the government as to how, and to what purposes, the State was to use its extended control of industry. Some favoured making the nationalized industries exemplars of forward thinking and efficiency, whose main function would be to act as leaders and pace-makers for the rest of the economy. Success and self-sufficiency were therefore to be prime objectives; and it was assumed by many proponents of this school of thought that this would entail the conferring of a considerable degree of autonomy on the nationalized firms. The example of the nationalized Renault firm—as yet unsullied by the enormous losses that it was shortly to suffer—was high in the minds of those who favoured this concept; and it was significant that Renault's former director, Pierre Dreyfus, was brought in as Minister for Industry in Mauroy's post-election government.

Others, however, saw the State's control of industry as serving a much wider purpose, giving it the power to ameliorate the problems of society both in the long term and on a day-to-day basis. The public sector could provide employment and money to tide the nation over its present difficulties, until the widely predicted international recovery materialized in 1983. This view predicated frequent government intervention, as well as facing managers with the disconcerting threat of suddenly being called on to help solve wider problems that were not part of the firm's intrinsic concerns or allowed for in its budget. The main proponent of this concept was the CERES left-winger, Jean-Pierre Chevènement, who replaced Dreyfus as Minister for Industry in 1982. So unpopular was his interventionist treatment of the public sector, that Mitterrand replaced him with Laurent Fabius in March 1983, with a mandate to restore decision-making to the firms themselves.

As far as industry as a whole was concerned, the Socialists made it clear that they were renouncing Giscard's *politique de crénaux* of 1978, which had concentrated aid on strong runners and what were hoped to be the growth industries of the future. Henceforth, all viable industry was to be encouraged to modernize, with industrial self-sufficiency as the ultimate target. This so-called *politique de filières* was aimed at creating an unbroken series of made-in-France labels, from the simplest component to the most sophisticated finished product. It was back to the ethos of the Gaullist and Pompidolean periods, with government aid lavished on national champions in each sector, as in the days of the Sixth Plan (1971–5). Large size, high technology, and mass production would supposedly enable leading French firms to withstand foreign competition at home and abroad, with all that that promised for national prosperity and for preserving employment. The level of direct government loans and grants to private industry was accordingly doubled in the first two years of Socialist government. Critics, however, questioned the appropriateness of encouraging large-scale mass production in a period of restricted international markets, and asked whether the immediate future for French exports did not lie in specialized quality goods, where small firms of skilled craftsmen making short runs of models, and responding rapidly to changes in taste and requirements, could reach markets that were less accessible to the giant factories of Japan and elsewhere, whose heavy investment in producing a restricted range of mass-produced models made them less able to respond to specialist needs. Government apologists responded by pointing out that exports were only part of the government's concern in promoting industry. There were also the parallel problems of employment and expensive foreign imports.

Shortage of money, however, brought the *politique de filières* to an end; and 1984 found the government increasingly obliged to restrict its resources to anticipated growth areas. The depressed nature of the steel, coal, and motor industries was progressively accepted as a fact of life, as was the mounting level of unemployment; and it was no longer thought appropriate to continue flinging money at it.

If the Socialists of the 1980s had a more sophisticated understanding than Léon Blum had had of how economic growth could be stimulated, they shared his uneasy awareness that social benefits could not be postponed until such time as the economy started to boom. It was, in any case, part of their economic strategy that the internal market should be strengthened by increased employment and improved wages for the lower paid. Yet they were determined not to repeat Blum's mistakes of June 1936, and hinder productivity by too sudden a reduction in hours. Although the eventual aim was a thirty-five-hour week, the initial cut proclaimed in July 1981 and realized in January 1982 was a nominal drop from forty to thirty-nine—and there it stuck until the Socialists fell from power in 1986. It compared badly with the three and a half hours that Giscard had unobtrusively given the workers during the course of his *septennat*; and it only created 28,000 extra jobs. But, more attractively, January 1982 also saw holidays with pay extended from four to five weeks; and, given the August shutdown in France, this was a less damaging concession economically than it would have been elsewhere.

As for wages, the government was likewise anxious not to duplicate Blum's error and to reduce the attractiveness of French goods to foreign buyers. Accordingly, the SMIC was given a succession of modest upward turns, which, when offset by inflation, gave the lowest paid a real increase of about 15 per cent in their living standards in 1981–2—with knock-on improvements for the 1.7 million workers whose wages were geared to it. The government also made some effort to shift the burden of direct taxes away from the lower income groups, as well as to limit the overall rise in taxation—never an easy task for a reforming government. In fact, tax as a proportion of GDP rose only 2 per cent in the first three years of Socialist rule, whereas it had risen 6 per cent during Giscard's *septennat*.

When it came to other methods of redistributing wealth, family allowances and low-income housing allowances were increased by 25 and 50 per cent respectively; and among various medical benefits, the State also undertook to reimburse 70 per cent of the cost of an abortion (December 1982). Yet the average annual increase in overall 'social transfers' during the reforming period of the Mauroy ministry was still slightly less than the 6.6 per cent annual increase for the Giscard period as a whole. And pensioners' real income grew by only a quarter under the Socialists, compared with nearly two-thirds under Giscard. Indeed, the frustrated Nicole Questiaux had long since been removed from the Ministry of Social Affairs and National Solidarity (June 1982).

At the end of the day, however, job creation and job security were the main worries of the bulk of the population. The government itself took 200,000 extra people on to its pay-roll in its first year in office. But obviously expansion of this kind could not continue without entailing crippling increases in taxation and government borrowing. As in other countries, early retirement was a further means of job creation that the government was anxious to encourage. Voluntary retirement at sixty was rewarded with a pension ranging upwards from 80 per cent of the SMIC to 50 per cent of a middle-management salary (with effect from 1 April 1983), in the hope that some 90,000 workers with the requisite amount of contributions would take advantage of it. And a firm 100,000 new jobs were created through government aid for voluntary retirement at fifty-five (January 1982)—the agreements between State and management being characteristically entitled 'solidarity contracts'—'solidarity' still being the password of the Mitterrand regime. As a result of these and other measures, especially the injection of capital into industry, unemployment rose by only 4 per cent in 1982, compared with 29 per cent in West Germany and 22 per cent in the United States.

But all this cost money. The government could legitimately point out that it had inherited from Giscard a 60 billion franc deficit in the balance of payments, and an inflation rate of 13 per cent. But the increase in wages and social security contributions saddled industry with 34 billion francs of extra expenditure in the first year of Socialist rule—a sum equivalent to 8 per cent of total French exports of goods and services. Moreover, in real terms government expenditure went up by 11.4 per cent in 1981 and 1982—deepening the budget deficit from 0.4 per cent to 3 per cent of GDP. Yet critics forgot that this was still a relatively smaller proportion than in most Western countries—only half of America's (6 per cent) and a sixth of Italy's (17 per cent). That the proportion was not worse was itself a tribute to the Socialists' success in stimulating GDP, which increased by over 2 per cent in 1981–3, whereas elsewhere in Western Europe it was largely static.

Yet most economists would still argue that too high a price was paid for this expansion. The prime problem was that the higher incomes of the population were being spent on foreign imports rather than on French products. This, of course, was the in-built hazard of Keynesianism in one country—unless there were high tariffs to protect it. But such tariffs were precluded by French membership of the EEC and her increasing dependence on international trade, which now represented 23 per cent of GDP instead of 13 per cent as in the early 1950s. The result was that French industrial production, instead of being stimulated by the rapid growth in the public's purchasing power, was expanding at only half the rate of purchasing power. 1982 saw car imports rise by an alarming 40 per cent, electrical appliances by 27 per cent, and consumer goods by 20 per cent—causing the overall trade deficit to grow by two-thirds. And although inflation fell from 12.5 per cent to 9.3 per cent between 1981 and 1983, this deceleration was much less marked than in other comparable countries.

Moreover, the downward slide of the franc against the dollar—a staggering 60 per cent in three years—gave France all the disadvantages of a devaluation with very few of the compensating advantages. Since over a third of France's total imports were valued in US dollars, the result was disastrous, halving the potential growth of the French economy, and keeping the rate of inflation 3 to 5 per cent higher than it would have been, as well as increasing the cost of borrowing abroad. Long before the full impact of these losses began to be felt, Finance Minister Jacques Delors had been obliged to announce that 'it will be necessary to pause before announcing further reforms' (29 November 1981)— language all too reminiscent of Blum's in February 1937. And, bowing to the logic of economic pressure, a major austerity package was introduced on 13 June 1982, with wage freezes as its dominant characteristic.

At the same time, there was a marked change in the rhetoric of government. In its first year of office, ministers, including the Premier, had often used the old Socialist slogans of the 1930s, many of them borrowed from the Communists—'the wall of money', 'make the rich pay', 'the men of the châteaux', etc. The only tribal chant that still retained a strong ring of truth was 'la grève du milliard'. Not only had inflation upstaged the 'milliard', but it was an alarming fact that panic-stricken investors had transferred large amounts of capital abroad to escape whatever socialism might have had it in mind to do. The reality, if not the extent, of the problem was later recognized by the right-wing parties, when they offered a financial amnesty for repatriated wealth when they came to power in 1986. But the change of economic strategy in 1982–3 largely put an end to the vilification of the propertied classes. Indeed, entrepreneurs and profit-makers were increasingly spoken of with respect, as embodying an attitude of mind that the government was anxious to encourage. Part of the cost of family allowances was transferred from employers to workers, and Mauroy openly declared in January 1983: 'We want to have wages rise more slowly than prices in order to curb consumer purchasing power and increase profitability'— language of which any dyed-in-the-wool conservative could have been proud.

25 March 1983 brought a second austerity plan, raising taxes and social security contributions and lowering the corresponding benefits. Taken together, the two packages represented the greatest planned reduction in the people's purchasing power since the Liberation. The same week had also seen the franc devalued against the Deutschmark for the third time in eighteen months—albeit by a very modest 2.5 per cent, paltry in the context of the franc's overall decline of 27 per cent against the mark. As Mauroy objectively confessed a fortnight later: 'Quite simply, a real left-wing policy can be applied in France, only if the other European countries also follow policies of the left'— something that both critics and well-wishers had tirelessly been telling the government for two years. Nor were the rewards of austerity slow in coming: the trade deficit was halved in 1983 and the inflation rate subsided more rapidly.

The relative success of what was sardonically called the New Economic Policy, and the likely necessity of continuing it for some time, suggested the advisability of appointing a Premier who was publicly considered a professional exponent of its principles, rather than a belated convert to it through bitter experience like Mauroy. The current Minister of Industry, Laurent Fabius, fitted the bill. A proverbial 'brillant énarque', who might have come out of Giscard's stable, he represented the glossy, technocratic, social democratic wing of the party, far removed from the avuncular, heart-in-the-right-place militancy of Mauroy. Son of a rich Jewish art-dealer,thirty-seven, elegantly dressed, and prematurely bald, Fabius appeared to old-style picket-men as some globe-trotting emissary of a discreetly ruthless merchant bank. And the circumstances of his coming to power endeared him to neither the Left of the party nor its semi-forgotten coalition partners, the Communists. Apart from the uncongenial nature of Fabius's mandate, the Mauroy government had been prodded into resignation (17 July 1984) by a series of presidential decisions and announcements, made with little or no reference to the ministers concerned (p. 373). The Communists, to Mitterrand's relief, made it clear on 19 July that they would have nothing to do with the new government, and so, at long last, disappeared of their own volition.

In the same month, Fabius told the National Assembly: 'The state has reached its limits: it ought not to try to go beyond them'. And, indeed, the next twenty months saw the State in gentle retreat, following policies that went considerably further than Giscard's had done in respecting market forces and the discretion of management. By the elections of March 1986, inflation was down to 4 per cent, with only Belgium (2.5 per cent), Switzerland (2.2 per cent), and West Germany (0.7 per cent) doing better among her near neighbours. On the other hand, compared with the other major currencies, the franc was down 13 per cent on its position when the Socialists came to power in 1981; and France's indebtedness to other nations had tripled in the same period. Moreover, the budgetary deficit remained an obstinate fact of life, despite the attempts of the new Finance Minister, Pierre Bérégovoy, to match his tax cuts with reductions in government spending. Yet both the trade balance and the overall balance of payments were now in marginal surplus. Economic growth in 1985 was of the order of 1.3 per cent, which, if poor compared with the OECD's average of 4.75 per cent in the previous year, was at least positive, despite the deflationary policies currently pursued.

Remarkably, the six major firms that the Socialists had nationalized in 1982 made a collective profit of 5.3 billion francs in 1985, with only one of them in deficit. But the older nationalized industries had a less happy financial record—notably steel and coal—despite the shedding of tens of thousands of jobs. Renault continued to be a sad story of decline, hampered by debts and an ageing range of models, while Peugeot, its private-sector pace-maker, was likewise in difficulties. And even if the doldrums of the motor industry were common to most Western economies, France expected to have to get rid of a third of its car-workers over the next two years.

Indeed, unemployment was the principal worry of government on the eve of the 1986 election. Running at 10.7 per cent, it was high by French standards, even though it was lower than any of her EEC partners except West Germany (9.2 per cent). The proportion of young people without jobs (a quarter of those under twenty-five) was especially worrying—particularly in a country which had no equivalent of supplementary benefit for the long-term unemployed. Nevertheless, for those in work, the SMIC and related wages had remained 12 per cent higher in real terms than they had been in 1981, despite the general attempt from 1983 to keep wage increases below the level of inflation. Yet Giscard could coolly point out that he had produced twice as high a rise during his *septennat*, even though he, too, had been governing in a period of recession; and, in proportion to the net output of French business, overall wages under Mitterrand fell from 58.5 per cent in 1981 to 55.6 per cent by 1985—lower than Giscard's 57.4 per cent in 1980.

While finance was the traditional deathbed of left-wing governments in France, the Socialists made much of the structural changes they had brought about in their five-year span. They tirelessly reminded the electorate of the Auroux laws of 1982, which had made works committees compulsory for virtually all firms, and had strengthened workers' methods of registering complaints and obtaining redress. Firms with fifty or more employees were now obliged to negotiate annually on wages and working conditions—although the actual award of pay increases would still largely depend on higher authority and the state of the economy, as the austerity policies of 1983–5 demonstrated all too forcibly. Moreover, it was not regarded as part of the function of these committees to discuss redundancies and long-term planning.

The CGT and the CFDT both behaved fairly co-operatively during the government's austerity years, much to the Communist party's annoyance. Overall union membership had probably sunk from about 20 per cent to 15 per cent of the labour force between 1974 and 1984, despite union claims to less discouraging figures. The labour movement, as always, was further weakened by its divisions, the CFDT welcoming government proposals for a flexible working week, while the CGT and FO were firmly opposed to it. On the other hand, there was general union approval for the re-establishment in December 1982 of the workers' right to elect administrators to the social security funds—a tradition which de Gaulle had discontinued in 1967.

REGIONALISM

It was de Gaulle, however, who had given wide currency to the concept of regionalism. And while his immediate successors had done relatively little to give it any substance, it was essentially the Socialists who made decentralization a reality, even if some of their measures were based on their predecessors' initial sketches. Both the Socialists and the Communists had formally espoused decentralization after the events of May 1968 had made it an apparent vote-catcher; and their enthusiasm was particularly marked in those areas where they were strongly entrenched in local government. Indeed, they ostentatiously gave the Ministry of the Interior the suffix 'and Decentralization' on coming to power, a designation that would have pleased Proudhon if not Marx. The right-wing parties likewise became interested, once the local elections of 1982–3 showed the Right in the ascendant.

The Defferre Act, which became law on 2 March 1982 removed the prefect's traditional right to vet and veto the decisions of local elected bodies; he could henceforth only challenge them on legal grounds in an administrative court. Moreover, the prefect became in title, and largely in fact, the *Commissaire de la République*, retaining responsibility for the Paris-based services in the locality, but handing over all other local matters to the elected president of the *départemental* council, together with the administrative staff who handled them. By contrast, as far as the Paris-based services were concerned, the new *commissaire*'s role was much greater than the prefect's had been, in that he became their director as well as co-ordinator, with full executive responsibility. A similar division of functions was laid down for the embryonic regional government, with the stipulation that the twenty-two regional councils would henceforth be directly elected by universal suffrage.

As for the responsibilities of the different levels of local government, these were redefined in two subsequent laws of 7 January and 7 July 1983. The region was formally given the task of drawing up a four-year economic plan for the locality, interlocking with the overall national plan. Alongside this, the municipalities acquired extensive powers of urban planning and control over building responsibility that had hitherto been exercised by the Ministry of Urbanism. The government had also contemplated devolving the housing programme to local level, but, given its recognized urgency, decided otherwise. Sadly, the economic problems of central government prevented it realizing anything more than a faint shadow of its housing schedule.

Nowhere was decentralization more of an issue than in Corsica, which continued to be the scene of occasional bomb-blasts and minor outrages. Breaking with the intransigence of their predecessors, the Socialists granted Corsica a specific statute, giving it a greater measure of autonomy (5 February 1982). In addition to its two existing *départemental* councils, it now proceeded to elect its first regional assembly for the island as a whole (8 August 1982). Based on proportional representation, its political opposition was so fragmented that effective government was increasingly difficult, finally obliging the prefect in 1984 to administer the island on a week-to-week basis. The electoral system was marginally changed for the next election on 12 August 1984, but the Radicals and conservatives were split between party bosses, making the conservatives uncomfortably reliant on the support of the handful of National Front members. Paris could nevertheless continue to take comfort from the fact that all these groups were committed to the

French connection and that the separatists received only 5.2 per cent of the public vote.

Decentralization was also a feature of the Socialists' prescriptions for the arts, under their enterprising Minister of Culture, Jack Lang, whose impish features suggested an amalgam of Harpo Marx and Tom Baker of *Dr Who* fame. He had made a name for himself in the 1960s as the founder of Nancy's international drama festival; and his appointment to the ministry was seen as evidence of the government's desire to help and encourage local initiative, without seeking to control it. In the spirit of his earlier assertion that 'culture is a battle for the right to live freely', Lang wished to transform André Malraux's provincial *Maisons de la Culture* into centres of innovation and self-expression rather than mission stations of French national culture, as Malraux had envisaged them (p. 303). State aid to the arts was tripled in real terms during the Socialists' tenure of power, some of it taking the shape of financial help to these private companies whose aims and achievements accorded with the government's ambitions for the provinces. At the same time, the Fifth Republic's tradition of joint funding of the arts by State and locality spurred local government into matching the Socialists' generosity. The fruits of this included several new regional orchestras, and the expansion of the Opéra de Lyon, which, under John Eliot Gardiner, quickly established an international reputation for fresh, elegant performances of French opera. Lang made a number of imaginative appointments to regional arts centres; but, predictably enough, some of the recipients preferred to return to Paris, where reputations could be made more quickly, even if it meant fore-going the lifebelt of a public salary.

Paris could not complain of neglect under Lang's ministry. Substantial plans for reorganizing the capital's opera and ballet were put under way, the Cinéma-thèque Française was expanded, the Picasso museum was established, and several major new museums were built, including the Gare d'Orsay conversion. Nor did Lang forget the more private world of the arts—books and the individual reader. He regarded it as a matter of shame that only 14 per cent of the population belonged to a library, compared with over 50 per cent in Britain. Spending on public libraries was quadrupled and the number of library loans grew by a third during his ministry. Moreover, parliament gave virtually unanimous support to Lang's bill removing the obstacles to cheaper book-retailing (10 August 1981).

In the realm of individual liberties, the Socialists were arguably more cour-ageous than Giscard, despite his professed concern for this field. They notably abolished the death penalty (30 September 1981), despite the opposition of 56 per cent of the general public; and they also limited the traditional powers of the *juge d'instruction* in the preparation of criminal cases. Legal aid was extended, and discrimination against homosexuals was made more difficult. But these various reforms predictably caused rumblings on the Right, including police demonstrations in June 1983 against what were claimed to be the over-liberal attitudes of the Minister of Justice, Robert Badinter. The government, however, asserted its determination not to be intimidated by these disturbing echoes of April 1958, and the Minister of the Interior, Gaston Defferre, sacked the organizers.

EDUCATION

Yet these demonstrations were to pale into insignificance compared to those aroused by the government's modest proposals for educational change. Like many good men before him, the Education Minister, Alain Savary, addressed himself to the problem of wastage and drop-outs in universities, proposing that there should be selection at the end of the second year—by which time the aptitude of students should be clearly assessable. But his intention that this intermediate hurdle should also be geared to the number of employment oppor-tunities available in each field met with much vociferous dismay—as did several other of his related proposals. After various modifications, however, the Savary law emerged from parliament bloody but unbowed in December 1983.

Savary was also concerned by the paucity of students from working-class backgrounds in the *grandes écoles*. While the proportion in universities had risen from under 5 per cent in the 1960s to 15 per cent in 1981, it was only 4 per cent in the *grandes écoles*, entrance to which largely depended on a strong performance in the mathematics *baccalauréat* ('Bac C'), which was taken by three-fifths of children from professional families, but only by a fifth of those from working-class backgrounds. Savary's solution was to replace 'Bac C' with 'Première S', a general science programme, in December 1981; but sceptics pointed out that he was merely replacing one privileged path with another, and that the key to equality of opportunity lay in changing expectations and attitudes in the family itself.

Savary's greatest cross, and the one that broke his back, was the private schools issue. In retrospect, there was a case for leaving the Debré law of 1959 and subsequent legislation much as it was. (The Guermeur law of 25 November 1977 had notably given private schools under *contrat d'association* much greater control over the appointment of their staff.) An opinion poll of the early 1980s revealed that 67 per cent of the general public were happy enough with the principle behind these laws; and the ultimate outcome of the wearisome confrontation that subsequently took place saw little change to the status quo.

But the Socialist party was a party of schoolteachers, mostly from the public sector, and they could not forbear from reliving their youth by resurrecting the threats they had made twenty years earlier and which they were now at long last in a position to realize—or so they thought. Savary thought otherwise, and had no intention of either nationalizing the private schools or withdrawing state subsidies from them. But he recognized the existence of real anomalies, and believed that the State as paymaster had a right to make the beneficiaries more accountable. Under the Third and Fourth Republics, Catholic schools had tirelessly pointed to the example of Britain in their campaign to get state help; and yet British Catholic schools were, in effect, part of the state system, with all the obligations that this entailed. With some exceptions, those British schools that remained outside the state system received no money—and this was generally recognized as logical and just. It was Savary's intention to come closer to the British situation, while still leaving the private schools with a measure of independence. The state of affairs he inherited was one in which the private sector held 14 per cent of the primary-school and nursery-school population and 21 per cent of the secondary; and 95 per cent of the private sector was specifically Catholic. Of these Catholic schools, 99 per cent of the secondary schools and 29 per cent of the primary schools had entered into the full *contrat d'association* with the State (pp. 296–7), while 71 per cent of the primary schools were under the less comprehensive *contrat simple*. The anomalies came from the fact that the State periodically found itself financing courses which it would have ruled out as luxuries or inappropriate in the public sector.

Savary's initial proposals of December 1982 envisaged that the governing bodies of state-supported private schools should contain representatives of the State and local government, and that their teachers should formally become state employees on a par with the state teachers—except for the 7 per cent who were in holy orders, who would continue to be paid under contract, more or less as before. Savary also proposed that these schools should be taken into account in the overall logistics of the French educational system, and their location borne in mind when adjusting state resources to take account of shifts in population. Nevertheless, the independence of the schools would continue to be recognized, in that ownership of their buildings would remain in private hands and the schools' religious character would be specifically guaranteed. In so far as there was any legitimate cause for disquiet on the part of the owners, it was perhaps in the long-term implications of being included in the State's geographical strategy. On the other hand, they were not required to accept overspill from the state schools; and it was arguably in their own and everyone's interest that state funding should take account of changes in local requirements. At the same time, private teachers had personally more to gain than lose in becoming state employees—and the schools themselves would still have a strong voice in the matter of appointments.

The storm of anger that these proposals aroused revealed concerns in the demonstrators that had little to do with religion. The irony was that only a fifth of the parents with children at Catholic private schools were themselves practising Catholics. Of the other 80 per cent, a fair number genuinely wished their children to be brought up as Catholics, with freedom to continue or discard their faith on leaving school. And there was a proportion of those in Brittany, the Massif Central, and other areas of high religious observance who were merely following the tradition of the locality in choosing a Catholic school. But elsewhere, many who chose the private sector did so for much the same reasons as British parents: they might be in the catchment area of a state school with social problems or where there was a large immigrant community with language difficulties. Similarly, parents with a child at a state school might disagree with the school's assessment of their child's streaming on leaving *cinquième* or *troisième* (p. 348)—and therefore see the private sector as a means of obtaining the level of education they sought. Significantly, the aspects of the Savary package that caused the most furore were those that appeared to draw the private schools and their teachers under the umbrella of the state structure, with the unspoken fear that they might become socially indistinguishable from state schools—or that there might eventually be less freedom of movement from one school to another. It was true that state aid had already rendered private-school fees relatively negligible; but they were still sufficient of a deterrent to keep the lowest-paid sections of society in the state sector, unless they were religiously very highly committed.

But the onslaught on Savary took on the guise of a crusade to protect religious freedom—much to the embarrassment of the bishops, many of whom were privately disposed to accept the package with certain modifications that Savary was prepared to make. The RPR and the UDF leapt on to the bandwagon, Chirac playing host to a monster rally in Paris on 24 June 1984 which drew over a million participants—the largest demonstration in French history. Large numbers of bewildered clergy and Catholic associations loyally followed in the wake of what was, in effect, a gigantic political manœuvre, playing on the fears of parents, whose prime concerns were social and academic rather than religious. Mitterrand decided that enough was enough, and on his own initiative he withdrew the bill on 12 July, provoking Savary's resignation and ultimately that of the government as a whole a few days later (p. 367). An anodine bill, passed on 20 December 1984, largely consecrated the status quo, arousing little comment either in or outside parliament. Its main feature was to give the State a stronger voice in the appointment of staff to private schools under *contrat d'association*, thereby restoring the situation that had existed between 1959 and 1977.

Nevertheless, the episode was seized on by journalists as a demonstration that Catholicism was still a major force in French politics and that, on balance, it was a phenomenon that favoured the Right rather than the Left. Opinion polls were cited, indicating that in the 1981 presidential and legislative elections, there were at least three practising Catholics supporting Giscard's coalition for every two supporting Mitterrand's. But these figures left unanswered the question of whether the Catholics who voted for Giscard saw him first and foremost as a Catholic, prepared to defend Catholic schools—even though he was the President who had liberalized the abortion laws. Alternatively, it might be asked whether their vote sprang from the fact that they belonged to a social milieu that was broadly sympathetic to Giscard's financial and economic policies, and shared his attitude to America and the EEC. And what about the minority of practising Catholics who voted for Mitterrand? Were they some-how less Catholic than those who voted for Giscard? Did their possible prefer-ence for Mitterrand's social policies take greater precedence over their Catholicism than did the preference of the Catholic majority for Giscard's financial and economic policies? Observers were no longer in a situation in which they could point to specific votes in parliament or in the electorate at large and say '*that* is the Catholic vote'. Even on such issues as abortion, which came nearest to providing an identifiable pattern of voting behaviour, other factors played a part and confused head-counting.

Paradoxically, as the number of practising Catholics continued to decline, the beliefs and attitudes of those who survived tended to become more varied as the Church in France became increasingly reluctant to give a strong lead on public issues. A few desperate die-hards chose to put it down to the fact that the new Archbishop of Paris, Aaron Lustiger, was a converted Jew. Had Charles Maurras still been alive, he would doubtless have exclaimed: 'Les israélites sont partout! C'est la revanche de Dreyfus!'

His astonishment would have been even greater if he had seen the French stamp commemorating the revocation of the Edict of Nantes in 1685, on which 'Liberté, Égalité, Fraternité' was replaced with 'Tolérance, Pluralisme, Frater-nité'. Even the Socialists had forgotten their principles—as demonstrated among other things by their authorization of independent television channels in July 1985. The only reassuring feature was that the franchise for Channel 5 was eventually given to personal friends of the President, thereby demonstrat-ing that there were at least some traditions of political behaviour that had with-stood the erosion of time.

FOREIGN AND DEFENCE POLICIES

Otherwise, tradition was principally kept alive in Mitterrand's foreign and defence policies, where the elements of continuity with past governments were much stronger than the differences. The Socialists had been keen promoters of NATO in 1949, and had censured de Gaulle for withdrawing in 1966. Admit-tedly, it could be argued that the censure was on balance more of a reflection of hostility to the independent nuclear deterrent than a warm feeling for NATO as such. Mitterrand himself, in the 1965 presidential campaign, urged that the independent nuclear deterrent should be scrapped. Yet, by 1974, he was asserting that one could not scrap the nuclear deterrent until France obtained a cast-iron international guarantee of her security—a view advocated earlier by Charles Hernu. Indeed, Mitterrand's think-tank claimed that it was in the interests of socialism for France to retain a certain 'minimum deterrent force', if only to defend a French socialist society from outside aggression by reactionary or other foreign forces. And this was a view that was generally accepted by the party's *comité directeur* by November 1976. The Communists, for their part, came to realize that an independent French nuclear deterrent might be a method of detaching France from the Atlantic alliance, since a France with a deterrent would be less reliant on American help. Accordingly, Jean Kanapa's report on security in May 1977 stated that until her conventional weapons were stronger, France ought to retain her independent nuclear force.

In power, Mitterrand's defence thinking showed itself more akin to de Gaulle's than Giscard's. Whereas Giscard had a personal leaning to close cooperation with NATO—which only his dependence on Gaullist votes prevented from being more explicit—Mitterrand was more committed to traditional, if largely theoretical, 'independence'. Indeed, he said of de Gaulle: 'il a vu juste'. At the same time, the balance of France's tightly strapped defence budget was increasingly directed towards strategic nuclear weapons and away from conventional forces. In fact, it was conventional weapons that suffered the brunt of the 30 per cent cut-back in the military equipment grant for 1982—coupled with a reduction in manpower. The government's seemingly magnanimous gesture towards environmentalism, in dropping the scheme to extend the Larzac tank-training ground, was made easier by its doubts concerning the project's necessity. And if the Defence Minister, Charles Hernu, decided to leave National Service at twelve months, he was as much motivated by keeping down the dole queues as by keeping up military preparedness.

France, however, could not afford to be on anything other than friendly terms with America. The economic consequences of her social engineering required the safety-net of American goodwill, while her own independent deterrent was heavily dependent on American technology—despite the much-publicized role of French research and development. Mitterrand was therefore disposed to back Reagan in the deployment of Pershing 2 and Cruise missiles in Europe; and in January 1983 he personally urged the West German parliament to give the Pershings house-room. At the same time, he initially refused to meet with Soviet leaders as long as Russian troops were on Afghan soil, and when his resolution eventually flagged, he used his visit to Russia in June 1984 to criticize the Soviet record on human rights. Yet he was at pains to impress on Reagan that *détente* with the Eastern bloc should be actively and sincerely pursued; and there was reproof in his voice when he spoke of the affairs of

Central America. He was not disposed to regard left-wing regimes there as *ipso facto* a threat to the United States.

He nevertheless emphasized to both sides that the French nuclear deterrent could neither be bartered, nor even considered in the calculations of East–West strengths in Europe occasioned by the current Soviet–American bouts of bid-calling on disarmament. And Mitterrand's concept of 'global deterrence' was markedly different from NATO's 'flexible response'. In the event of a threat to France's 'vital interests'—deliberately left undefined—there was a hierarchy of warning responses, in which the use of tactical nuclear weapons would indicate an immediate readiness to resort to strategic weapons—a much more rapid escalation than that envisaged by NATO.

Moreover, as in Giscard's presidency, France's remnants of empire were still regarded as part of her 'vital interests', despite the fact that they totalled only one and a half million people and were an economic liability. Most of these distant islands had horrendously adverse trade balances—with imports running from five to twenty times greater than exports, and with foreign countries getting most of the benefit. Independence would mean economic extinction, for the government had no intention of propping up their economies if they threw off the French connection. Only New Caledonia, with its nickel-mining, had an independence movement of any significance. But even New Caledonia had a trade deficit of six to one; and it was essentially its large minority of Melanesian Kanaks that wanted independence. Even so, the tension between them and the European majority was a matter of continuous concern throughout the presidency, involving sporadic loss of life in spite of various French attempts to resolve the matter through administrative and economic changes.

Elsewhere in the Pacific, there was widespread resentment against French use of the Mururoa atoll for nuclear explosions. When the environmental organization, Greenpeace, announced its intention of penetrating the danger area during the tests, French agents blew up its ship, *Rainbow Warrior*, while it was in harbour in New Zealand, killing a Dutch photographer (10 July 1985). Faced with an international outcry, the government protested its innocence. But as the foreign journalists untangled the tissue of complicity, Charles Hernu was made to resign his portfolio at the Ministry of Defence, before the trail led higher (20 September 1985). A close colleague of Mitterrand, he had nevertheless embarrassed his master in the past by ill-considered assertions; and now he had no option but to lay down his political life for his fellows. The bulk of French opinion was intrigued rather than outraged by what had happened; and even the right-wing opposition, for whom superficially it seemed an electoral windfall, sensed that they would lose rather than gain support by attempting to trace responsibility to the Matignon or the Élysée. Besides, the election to all intents and purposes had already been won—as the opinion polls clearly testified.

AGRICULTURE

Within the EEC, Mitterrand took on Giscard's mantle of protecting French agriculture and keeping financial concessions to Britain to a minimum—even if outwardly there was rather less *froideur* with the redoubtable *dame de fer*. Not that the 8 per cent of the population involved in farming showed him much gratitude—demonstrators chased his Minister of Agriculture, Madame Édith Cresson, across the fields when she sought to justify the government's record to them. In fact, the farmers were doing well out of the Socialist experiment—just as they did under the Popular Front, despite the fact that many of them expected to be its victims. As under Blum, much of the increase in consumer spending was on food, most of it home-grown, although the ratio was the other way round as far as purchased manufactures were concerned. Indeed, 1982 was the farmers' best year since the early 1970s. Moreover, the government did what it could to shield them from the more direct effects of the austerity programme of March 1983. And, once more as under Blum, government intervention continued to guarantee them artificially high prices for grain and sugar-beet.

Indeed, the spirit of the Third Republic seemed to be much abroad, in that it was now government policy to try to prevent the number of farms falling below a million. With growing unemployment in the towns, there seemed little point in continuing the post-war policy of amalgamation and economies of scale. The small, inefficient, labour-intensive holdings that had been the despair of the planners in the 1950s and 1960s now had more than nostalgic attraction; they seemed the answer to many problems—at least in the short run. Yet in fact, the number of full-time farms was already down to 800,000, and the only chance of holding the million line was to encourage the part-time smallholders who ran their plots as income supplements or as hobbies. But that had little relevance to the ultimate aim of the policy—to keep down the level of unemployment. And as the 1986 elections approached, unemployment remained the government's greatest liability.

REARGUARD TACTICS

Mitterrand's prime concern in 1986 was to keep the expected victory of the Right as marginal as possible; and to this end he resurrected an old election pledge to restore proportional representation. Dating back to the days when many assumed that the Left could not win an overall majority under the existing system of single-member constituencies, the proposal had been invalidated by the Socialist victory of 1981. It could now serve no other purpose than to frustrate the returning swing of the pendulum—which was precisely Mitterrand's undisguised aim.

Although the President had been seriously talking of reintroducing some sort of proportional representation since 1984, it came as a shock when the scheme he eventually proposed in April 1985 was that of 1946, which many thought had been a major element in the French Republic's constitutional problems (p. 144). Admittedly, its worst effects were curbed by a proviso that only those parties that obtained 5 per cent or more of the *départemental* vote could obtain a seat in parliament. And, more cogently, it could be argued that the nature of politics was very different in the 1980s from what it had been under the Fourth Republic. The increased powers of the President and the fact that he was elected by universal suffrage had exerted a bipolarizing effect, which would help to prevent the re-emergence of the multiplicity of parties characteristic of *scrutin de liste*. Moreover, the waning of three of the four Cs in politics since the 1950s would likewise militate against a mushrooming of parties. To which critics replied that the monster rally of 24 June 1984 on the private schools issue had clearly demonstrated how naïve it was to imagine that these issues were dead.

A number of the President's colleagues conceded that times had changed and that clericalism was no longer the main issue in the private schools question. But they were nevertheless uneasy about the effect of *scrutin de liste* on the political fortunes of extremist parties, such as the Communists on the Left and the National Front on the Right. The Communists were now no longer allies; and if there was only one ballot in the election, there would be no opportunity for the Communists to rally to the support of the Socialists on the second ballot to keep out the less congenial parties of the Right. As for the National Front, they had obtained 11 per cent of the vote in the elections to the European Parliament on 17 June 1984, and likewise obtained a similar proportion in the cantonal elections of 10 and 17 March 1985—admittedly both on low polls. Under the existing parliamentary system of single-member constituencies they would probably obtain no seats in the National Assembly, but under *scrutin de liste* they might obtain upward of thirty. Given the Front's intransigent policies of repatriating immigrants and its echoes of inter-war fascism, it was strongly argued that Mitterrand was assuming a grave responsibility in giving Le Pen's followers the opportunity of entering parliament in significant numbers. Alarmists pointed out that Mussolini had only had thirty-five deputies in parliament when he took power in 1922.

The Socialists' embarrassment was all the greater in that it was precisely part of Mitterrand's strategy to allow the Front into parliament, where they would prove an even greater embarrassment to the Gaullists and Giscardians. If, as Mitterrand calculated, the loss of respectable right-wing seats to the National Front would deprive the Gaullists and Giscardians of an absolute majority, they would be faced with an awful choice. They could either come to terms with the Front—and run the serious risk of losing the respect of liberals of all persuasions—or they could simply reconcile themselves to not having an overall majority and restrict their policies to those issues on which they could do deals with the Radicals and Socialists. This would leave Mitterrand with a strong role as arbiter, as long as he chose to remain at the Élysée.

The main drama of the situation lay in the fact that whatever electoral system was used, the Right would probably gain sufficient seats to oblige the President to choose the Prime Minister from among their number. And, since Mitterrand's term of office did not expire until 1988, this would create the unprecedented situation that critics of the seven-year term had continually warned would happen one day. The Premier and the President would be from opposing parties, and this would raise a host of difficulties as to who was responsible for what. It immediately posed the question as to whether the President, who was constitutionally responsible for Defence and Foreign Affairs, would be able to pursue policies in those areas that were not to the liking of the cabinet. Pessimists predicted that he would rapidly be reduced to the role of a constitutional monarch, wearily assuming nominal responsibility for measures that were not of his making. But the smaller the majority of the Right, the more likely it was that Mitterrand could retain a strong hand in government; and it was on this assumption that he imposed *scrutin de liste* on an uneasy electorate.

Making the same assumption, Raymond Barre let it be known that he would not accept office as Premier in such a situation. In his opinion, it would be the duty of the President to resign if faced with a hostile majority. Chirac and Giscard d'Estaing, on the other hand, felt that such a display of pure principle was all very well in someone who had less claim than they did to head a right-wing government—even if Barre's current showing in the popularity polls was better. Faced with the probability of exercising power, they indicated their readiness to do so, even under Mitterrand.

The frustrating nature of the situation was that *scrutin de liste* did not easily lend itself to a simple policy of letting the electorate decide which of the two main right-wing parties should provide the Prime Minister. In theory, it was the President who made the choice; but it was generally assumed that he would have little option but to choose the man with the largest parliamentary following. Under the traditional *scrutin uninominal*, it would have been possible for the Gaullists and Giscardians to compete separately for the electorate's favour on the first ballot, and then unite behind mutually agreed right-wing candidates on the second. But to compete against each other on a one-ballot *scrutin de liste* ran the obvious risk of splitting the right-wing vote and giving victory to the Left. Faced with this daunting possibility, the Gaullists and Giscardians chose to offer joint lists of candidates in most *départements*—but this prudent strategy none the less obliged them to try to assess in advance which party was likely to do best in a particular locality, since this would determine whether Gaullists or Giscardians came at the head of the alliance's list. Mitterrand calculated that this invidious exercise would help to sow further dissension between the opposition parties, to say nothing of subsequent bitter recrimination in those *départements* where the prediction proved inaccurate. All of which would strengthen his role and freedom of manœuvre.

The results of the election on 16 March 1986 testified to Mitterrand's astuteness as a tactician, even if they arguably displayed his readiness to sacrifice the long-term interests of national political stability to the short-term advantage of his party and, more particularly, of himself (Table 19.2). The Gaullists (147), the Giscardians (130), and their allies managed between them to scrape an absolute majority of two—leaving the Socialist–MRG alliance and its satellites as the largest single group in parliament (216). The Communists and the National Front each had 35 deputies, infuriating the opponents of *scrutin de liste*, who pointed out that under the former system the National Front would have had no candidates and the Communists only about 25. Moreover, the left-wing parties as a block would probably have received much the same number of seats as under *scrutin de liste*, even though the Gaullists and Giscardians would probably have won a further 50 seats between them. But although Mitterrand's change of the electoral system had not succeeded in depriving the Right of an overall majority, it was so slender that his presidential room for manœuvre remained considerable.

Table 19.2.
National Assembly Election of 16 March 1986

	% of vote on first ballot	Seats gained after both ballots	
		Total	%
Communists	9.8	35	6.1
Socialists	31.2	210	36.4
Left Radicals	0.4		
Other Left	4.0	6	1.0
UDF	41.0	130	22.5
Gaullists		147	25.5
Other Right	4.0	14	2.4
National Front	9.7	35	6.1
		577	

Notes: The percentage of electorate casting valid votes was 74.9.

Unlike the tables in previous chapters, these figures include the 22 overseas deputies. Note also the larger number of deputies under the new electoral system.

The logic of the results left him with little option but to call Chirac to the Matignon; but Chirac was all too well aware that the chalice handed to him could well be poisoned. With his eyes set firmly on the Élysée for 1988, it was vital for Chirac that the forthcoming experiment in cohabitation should redound to his credit. If it broke down, it was important that the blame should be seen to lie fairly and squarely with Mitterrand. If, alternatively, it worked well, it must appear to be largely the result of Chirac's skill and integrity and Mitterrand's reluctant recognition of his own moral defeat. While other countries, notably the United States, had prolonged experience of Presidents governing with hostile parliamentary majorities, there remained significant constitutional differences between theirs and the French situation. And, much more importantly, neither the experience nor the political culture of France made such an experiment easy.

It is at this point that most books on contemporary subjects subside, like an expiring firework, into an intermittent splutter of 'it remains to be seen if ...', and 'only time will tell whether ...'. And this particular damp squib can only echo them.

APPENDIX I

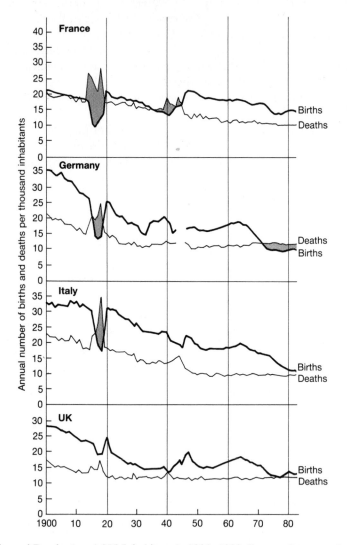

1. Births and Deaths (per 1,000 Inhabitants), 1900–1982: France, Germany, Italy, UK
Based on: Jean-Marcel Jeanneney and Elizabeth Barbier-Jeanneney, *Les Économies, occidentales du xix siècle à nos jours*, i (Paris, 1985), 20–1.

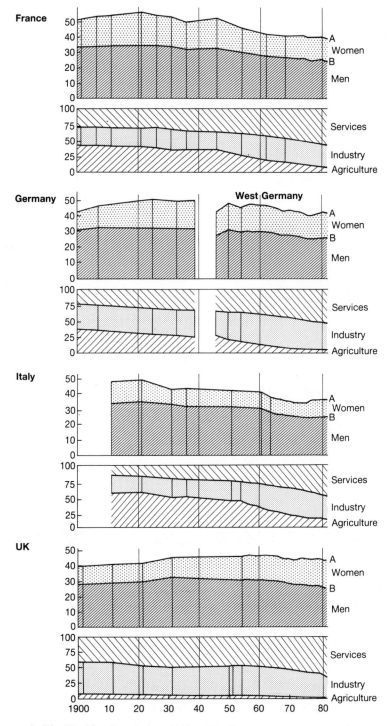

2. The Working Population, 1900–1982: France, Germany, Italy, UK

Based on Jean-Marcel Jeanneney and Elizabeth Barbier-Jeanneney, *Les Économies occidentales du xix siècle à nos jours*, i (Paris, 1985), 132–3.

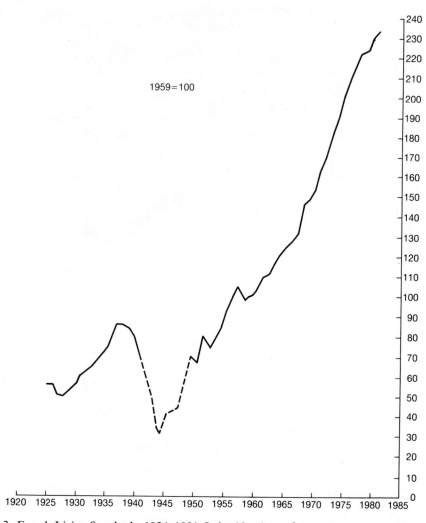

1959=100

3. French Living Standards, 1924–1981: Index Numbers of Average Industrial Take-
home Pay (Hourly Rates) in Real Terms

Based on: Jean-Marcel Jeanneney and Elizabeth Barbier-Jeanneney, *Les Économies occidentales du
xix siècle à nos jours*, i (Paris, 1985), 145.

APPENDIX II

Table 1
Age Distribution in Selected Years: France, Germany, UK

	Total population (millions)	Under 15 (%)	15–64 (%)	65 and over (%)
France				
1901	38.49	25.7	65.9	8.4
1936	41.18	24.4	65.6	10.0
1980	53.71	22.3	63.7	14.0
Germany				
1900	56.05	32.8	62.3	4.9
1939	69.31	21.6	70.6	7.8
1980 (West)	61.57	18.2	66.3	15.5
United Kingdom				
1900	41.16	32.5	62.7	4.8
1938	47.49	21.9	69.5	8.6
1980	55.95	21.1	64.0	14.9

Table 2

Size Distribution of French Agricultural Holdings, 1892–1970

Area of Holdings (hectares)	Number of Holdings (thousands)						Proportion of Arable Land (%)				
	1892	1929	1955	1963	1967	1970	1955	1963	1967	1970
0–4	4,064	2,160	792	547	447	420	5.6	4.0	3.4	2.9
5–19	1,217	1,310	1,004	849	724	506	34.3	30.0	26.2	22.2
20–49	335	380	375	393	399	394	34.7	36.8	38.1	38.2
50–99	52	81	74	84	92	101	15.4	17.5	19.2	21.3
100 and over	33	32	20	23	26	30	10.0	11.7	13.1	15.2
TOTAL	5,702	3,966	2,267	1,899	1,689	1,552				
Cultivated area in thousands of hectares:			32,426	32,193	32,007	31,683				

Table 3
Industrial Growth, Selected Items, 1910–1969: France, Germany, UK, Italy

Output of Coal and Lignite: Annual Averages (in million metric tons)

	France	*Germany*		*UK*	*Italy*
1910–14	37.4	247.1		274.3	0.7
1915–19	23.7	244.1		247.0	1.5
1920–4	33.9	237.8		240.9	1.2
1925–9	52.2	316.8		226.8	1.1
1930–4	50.0	264.7		223.5	0.7
1935–9	47.3	334.0		233.6	2.0
1940–4	43.5	421.0		208.7	3.6
		East	*West*		
1945–9	45.9	113.8	134.6	201.6	2.3
1950–4	55.1	163.1	192.4	225.5	1.9
1955–9	58.7	212.5	243.8	221.3	1.6
1960–4	54.2	246.7	246.3	197.3	2.0
1965–9	46.9	250.5	221.4	172.5	2.1

Output of Electric Energy: Annual Averages (in milliard kilowatt hours)

	France	*Germany*		*UK*	*Italy*
1910–14	[1.5]	7.10		[1.4]	1.76
1915–19	[1.8]	9.20		[2.6]	3.45
1920–4	7.54	16.30		9.55	4.71
1925–9	13.20	25.04		14.39	8.09
1930–4	16.12	26.93		20.00	10.75
1935–9	19.78	48.98		31.26	14.20
1940–4	19.62	55.68		42.50	19.42
		East	*West*		
1945–9	25.36	13.07	24.90	51.70	17.38
1950–4	39.76	22.88	56.37	72.23	27.65
1955–9	57.40	32.95	93.16	101.40	40.50
1960–4	82.75	45.29	139.97	151.00	60.47
1965–9	113.54	59.77	192.85	199.56	90.11

Output of Pig-iron: Annual Averages (in million metric tons)

	France	*Germany*		*UK*	*Italy*
1910–14	4.3	14.4		9.6	0.4
1915–19	1.4	10.0		8.8	0.4
1920–4	5.0	7.4		6.2	0.2
1925–9	9.5	11.8		6.1	0.5
1930–4	7.2	6.7		4.8	0.5
1935–9	6.7	15.9		7.6	0.8
1940–4	3.5	12.8		7.6	0.8
		East	*West*		
1945–9	4.9	0.2	4.0	8.4	0.3
1950–4	8.8	0.7	13.7	10.8	1.0
1955–9	11.8	1.7	20.6	13.3	1.9
1960–4	14.6	2.1	25.1	15.5	3.3
1965–9	16.3	2.3	28.8	16.5	6.9

Output of Steel: Annual Averages (in million metric tons)

	France	*Germany*		*UK*	*Italy*
1910–14	3.8	15.0		7.3	0.8
1915–19	1.8	12.9		9.1	1.1
1920–4	4.4	10.4		7.2	1.0
1925–9	8.7	16.2		7.8	1.8
1930–4	7.1	10.9		6.8	1.6
1935–9	7.0	20.4		11.8	2.2
1940–4	4.3	20.4		12.9	1.8
		East	*West*		
1945–9	5.6	—	4.9	13.8	1.5
1950–4	10.0	2.1	17.4	17.2	3.3
1955–9	14.0	3.2	26.9	20.7	6.2
1960–4	17.9	4.0	33.8	23.5	9.3
1965–9	20.4	4.6	39.1	25.9	15.1

Output of Motor Vehicles: Annual Averages (in thousands)

	France	Germany		UK	Italy
1925–9	207	90		206	58
1930–4	193	101		265	38
1935–9	200	304		450	64
1940–4	—	—		144	30
		East	*West*		
1945–9	150	4	—	417	45
1950–4	480	26	451	816	162
1955–9	998	55	1,275	1,265	368
1960–4	1,301	88	2,416	1,868	925
1965–9	2,037	114	3,042	2,120	1,469

Cotton Spindles: At Intervals (in thousands)

	France	Germany		UK	Italy
1913	7,400	11,186		55,653	4,600
1923	9,600	9,605		56,583	4,570
1929	9,880	11,250		55,917	5,210
1938	9,783	10,323		37,340	5,395
		East	*West*		
1950	8,148	750	5,785	29,580	5,566
1959	6,071	1,150	5,948	14,104	4,854
1969	3,621	1,625	4,349	3,635	4,125

Source: Official government statistics, cited in Carlo M. Cipolla (ed.), *The Fontana Economic History of Europe*, vi, pt. 2 (Glasgow, 1976), 696–715.

Table 4

Long-term Trends in Economic Production, 1870–1979: France and Selected European Countries (Average Annual % Changes in Gross Domestic Product)

	1870–1913	1922–1937	1953–1973	1973–1979
France	1.6	1.8	5.3	3.0
Germany	2.8	3.2	5.5	2.4
Italy	1.5	2.3	5.3	2.6
UK	1.9	2.4	3.0	1.3
Spain	—	1.7	6.1	2.8
Austria	3.2	0.8	5.7	3.1
Belgium	2.0	1.4	4.3	2.3
Denmark	3.2	2.9	4.3	2.1
Finland	2.8	4.4	5.0	2.3
Ireland	—	1.5	3.3	3.6
Netherlands	1.9	1.9	4.9	2.5
Norway	2.1	3.4	3.9	4.4
Sweden	2.8	3.5	3.9	1.8
Switzerland	2.1	2.1	4.6	—0.4
TOTAL	2.0	2.5	4.8	2.4

Source: Andrea Boltho (ed.), *The European Economy: Growth and Crisis* (Oxford, 1982), 10.

Table 5

The Growth of French Labour Productivity, 1896–1969 (% per year)

Sector	1896–1913	1913–1929	1929–1938	1938–1949	1949–1963	1963–1969
Agriculture and forestry						
Production per worker	1.5	0.9	2.1	1.3	6.0	6.0
Production per man-hour	1.9	1.4	2.5	1.7	6.4	6.4
All industry						
Production per worker	1.7	2.1	1.0	0.2	5.2	5.3
Production per man-hour	2.0	3.4	2.9	—0.9	5.1	5.3
All productive sectors						
Production per worker	1.7	1.5	0.8	0.9	5.1	5.0
Production per man-hour	2.0	2.5	2.2	0.3	5.2	5.2
Gross national product						
Production per worker	1.6	1.4	0.7	0.8	4.5	4.4
Production per man-hour	1.9	2.3	1.9	0.3	4.5	4.6

Source: J.-J. Carré, P. Dubois, and E. Malinvaud, *French Economic Growth* (London, 1976), 79.

Table 6
Investment Ratios,[a] 1928–1938 and 1950–1970:
France and Selected European Countries

	1928–1938 (%)	1950–1970 (%)
France	11.8	15.6
Germany	9.7	17.7
Italy	13.6	14.8
UK	5.7	12.9
Austria	6.1[b]	18.3
Denmark	8.9	15.5
Norway	12.4	23.7
Sweden	10.5	16.4
Western Europe	9.6	16.8

[a] Non-residential gross fixed investment in % of GNP at current prices.
[b] 1924–37.

Source: Andrea Boltho (ed.), *The European Economy: Growth and Crisis* (Oxford, 1982), 11.

Table 7
Distribution of the French Industrial Labour Force by Size of Establishment,
1906–1966 (%)

No. of People Employed	% of Labour Force				
	1906	1931	1954	1962	1966
1–10	32.2	19.7	16.0	14.0	13.1
11–100	27.6	30.1	30.8	30.8	31.9
101–500	21.7	23.6	25.9	27.7	29.0
Over 500	18.5	26.6	27.2	27.5	26.0
Number of industrial employees (in thousands)	3,679	5,442	5,918	6,475	6,831

Source: François Caron, *An Economic History of Modern France* (London, 1979), 280.

Table 8
Evolution of Employment in France, 1954–1982 (% of Active Population)

	1954	1962	1968	1975	1982	Proportion women (%) 1962	1982
Farmers	20.7	15.8	12.0	7.6	6.1	39.2	37.5
Farm-workers	6.0	4.3	2.9	1.7	1.3	11.5	15.8
Industrial and commercial employers	12.0	10.6	9.6	7.8	7.9	36.7	34.6
Senior management and liberal professions	2.9	4.0	4.9	6.7	7.7	15.9	26.5
Middle management	5.8	7.8	9.8	12.7	13.8	39.6	48.8
White-collar workers	10.8	12.4	14.7	17.6	19.9	58.8	65.5
Blue-collar workers	33.8	36.7	37.8	37.7	35.1	21.6	24.0
Service workers	5.3	5.4	5.7	5.7	6.5	80.9	79.0
Other categories	2.7	2.9	2.6	2.4	2.1	23.4	13.0
(Clergy)	(0.9)	(0.9)	(0.7)	(0.5)	(0.3)	(64.4)	(44.5)
TOTAL (millions)	19.185	19.251	20.398	21.775	23.525	34.6	40.7

Principal source: INSEE.

Table 9
Take-home Pay plus Social Benefits in Cash, 1979–1983: France and Selected Countries
(expressed as a percentage of gross earnings) [a]

	Single people			Two-child families		
	1979	1981	1983	1979	1981	1983
France	79.7	79.5	77.4	93.2	93.4	92.9
W. Germany	68.4	67.4	66.1	80.1	80.2	77.8
UK	70.6	69.0	68.6	83.1	81.2	81.8
Italy	81.5	79.2	77.8	83.0	81.3	80.1
Belgium	74.7	71.2	68.7	95.0	90.9	89.2
Netherlands	64.5	64.2	59.7	75.2	74.9	70.2
Luxemburg	70.7	71.2	70.5	91.9	92.7	94.1
Denmark	59.9	57.8	55.2	69.6	66.9	64.5
Sweden	63.5	63.9	64.5	74.1	74.1	74.1
Ireland	71.9	71.9	66.9	86.6	85.1	81.0
Spain	83.9	82.9	81.6	89.5	88.8	87.2
USA	73.3	69.8	70.3	82.3	79.0	78.1
Japan	87.5	86.8	86.3	93.3	92.4	92.0

[a] Take-home pay refers to gross earnings minus income tax and employees' social security contributions.

Source: OECD, *The Tax/Benefit Position of Production Workers, 1979–1983* (Paris, 1984), table 4, p. 51.

Table 10
Gross Domestic Product per Head, 1973 and 1983: France and
Selected Countries (Volume at 1980 Purchasing Power Standard)

	1973	1983
France	7,400	9,000
Belgium	7,300	8,600
Denmark	8,000	9,200
Greece	3,800	4,400
Ireland	4,300	5,400
Italy	6,000	6,900
Luxemburg	9,000	9,100
Holland	7,200	8,200
UK	7,100	7,800
EEC average (10 countries)	7,000	8,100
Spain	5,200	5,800
Portugal	3,300	3,800
USA	10,300	11,200
Japan	6,700	8,700

Source: Eurostat, *Basic Statistics of the Community* (Luxemburg, 1985), 18.

Table 11
Average Gross Hourly Earnings of Manual Workers, 1978–1983:
France and Selected European Countries
(Current Purchasing Power Standards—PPS)

	1978	1981	1983
France	3.0	4.4	5.4
Belgium	4.2	6.4	7.4
Denmark	4.8	6.9	7.8
W. Germany	4.0	5.8	6.8
Greece	2.0	3.3	4.2
Ireland	3.6	5.2	6.0
Italy	3.7	5.4	6.3
Luxemburg	5.0	6.4	7.7
Holland	4.0	5.7	6.8
UK	3.6	4.9	6.1

Source: Eurostat, *Basic Statistics of the Community* (Luxemburg, 1985), 22.

Table 12
Strikes, 1975–1978: France and Selected Countries
(Number of Working-Days Lost per 1,000 Workers)

	1975	1976	1977	1978
France	390	420	260	200
W. Germany	10	40	—	370
UK	540	300	840	840
Italy	1,730	2,315	1,560	890
Belgium	340	560	420	650
Netherlands	—	10	140	—
Denmark	110	220	240	90
Sweden	20	10	23	10
Spain	370	2,540	3,350	1,820
USA	990	1,190	1,070	1,080
Japan	390	150	70	60

Source: Documentation Française, *Français, qui êtes-vous?* (Paris, 1981), 281.

French Governments, 1936–1986
Dates of Formation and Resignation

NB Caretaker responsibilities often continued beyond the date of resignation, hence the variation between some of these dates and those in the main text.

Third Republic

Léon Blum: 4 June 1936–21 June 1937
Camille Chautemps: 22 June 1937–14 January 1938
Camille Chautemps: 18 January 1938–10 March 1938
Léon Blum: 13 March 1938–8 April 1938
Édouard Daladier: 10 April 1938–20 March 1940
Paul Reynaud: 21 March 1940–16 June 1940
Philippe Pétain: 16 June 1940–12 July 1940

État Français (Vichy)

Philippe Pétain, Chef de l'État
Pierre Laval, Vice-président du Conseil: 12 July 1940–13 December 1940
Pierre-Étienne Flandin, François Darlan, Général Huntziger, members of the Comité Directeur du Gouvernement: 14 December 1940–9 February 1941
François Darlan, Vice-président du Conseil: 10 February 1941–17 April 1942
Pierre Laval, Chef du Gouvernement: 18 April 1942–September 1944

Gouvernement Provisoire de la République Française

Charles de Gaulle: 3 June 1944 (Algiers), re-formed 9 September 1944 (Paris)–9 November 1945
Charles de Gaulle: 21 November 1945–20 January 1946
Félix Gouin: 26 January 1946–12 June 1946
Georges Bidault: 23 June 1946–28 November 1946
Léon Blum: 16 December 1946–16 January 1947

Fourth Republic

Paul Ramadier: 22 January 1947–19 November 1947
Robert Schuman: 24 November 1947–19 July 1948
André Marie: 26 July 1948–28 August 1948
Robert Schuman: 5 September 1948–7 September 1948

Henri Queuille: 11 September 1948–6 October 1949
Georges Bidault: 28 October 1949–24 June 1950
Henri Queuille: 2 July 1950–4 July 1950
René Pleven: 12 July 1950–28 February 1951
Henri Queuille: 10 March 1951–10 July 1951
René Pleven: 10 August 1951–7 January 1952
Edgar Faure: 20 January 1952–29 February 1952
Antoine Pinay: 8 March 1952–23 December 1952
René Mayer: 8 January 1953–21 May 1953
Joseph Laniel: 27 June 1953–12 June 1954
Pierre Mendès France: 19 June 1954–5 February 1955
Edgar Faure: 23 February 1955–24 January 1956
Guy Mollet: 1 February 1956–21 May 1957
Maurice Bourgès-Maunoury: 12 June 1957–30 September 1957
Félix Gaillard: 5 November 1957–15 April 1958
Pierre Pflimlin: 14 May 1958–28 May 1958
Charles de Gaulle: 1 June 1958–8 January 1959

Fifth Republic

Michel Debré: 8 January 1959–14 April 1962
Georges Pompidou: 14 April 1962–5 October 1962
Georges Pompidou: 28 November 1962–8 January 1966
Georges Pompidou: 8 January 1966–6 April 1967
Georges Pompidou: 6 April 1967–11 July 1968
Maurice Couve de Murville: 11 July 1968–20 June 1969
Jacques Chaban-Delmas: 20 June 1969–5 July 1972
Pierre Messmer: 5 July 1972–28 March 1973
Pierre Messmer: 2 April 1973–27 February 1974
Pierre Messmer: 27 February 1974–27 May 1974
Jacques Chirac: 27 May 1974–25 August 1976
Raymond Barre: 27 August 1976–28 March 1977
Raymond Barre: 30 March 1977–31 March 1978
Raymond Barre: 5 April 1978–13 May 1981
Pierre Mauroy: 22 May 1981–22 June 1981
Pierre Mauroy: 23 June 1981–22 March 1983
Pierre Mauroy: 23 March 1983–17 July 1984
Laurent Fabius: 19 July 1984–17 March 1986

APPENDIX IV

The Arts and the Public in the 1930s

The broad currents of French cultural life in the 1930s have been briefly outlined in Chapter 1 (pp. 22–7). This appendix merely seeks to indicate some of the writers, painters, and composers who were influential in these developments, and how they were viewed by the general public.

The large sales of Georges Bernanos's *Journal d'un curé de campagne* (1936)—over half a million in its first thirty years—reflected what might seem to be the paradoxical appeal of 'Christian' novelists, such as Bernanos (1888–1948), François Mauriac (1885–1970), and Julien Green (1900–), to a largely secular-minded middle-class readership. In John Cruickshank's words:

> they respond to the element of mystery in human experience, setting hints of transcendence against the drab rationalism and suburban scientism of the age. A powerful writer like Bernanos may not command the 'philosophic belief' of his readers, but he often compels their 'poetic consent'. He may not fill that 'God-shaped blank' which Huxley finds even in the secular mind, but he often arouses a response in it.[1]

This was part of the appeal of Mauriac's *Le Nœud de vipères* (1932), arguably the finest French novel of the 1930s.

Yet the desire to fill the vacuum left by 'the death of God' was paralleled by the anxieties created by 'the death of Man'. Earlier writers had indirectly questioned traditional concepts of man as a coherent entity by portraying personality as a succession of disparate thoughts and feelings, with no discernible co-ordination or goal. These intimations of what came to be the familiar 'absurdity of the human condition' generated a variety of responses. Action or commitment was the prime prescription of a major strand of French literature throughout the 1930s and 1940s. The individual could escape his loneliness and pointlessness by devotion to ideals or causes that transcended his personal limitations. For some writers, the risk and adventure involved in such commitment were enough; indeed, the very uncertainty of the future and the intrinsic value of the cause itself might enhance its liberating potential. Saint-Exupéry (1900–44) exemplified this attitude, even if the million readers of his *Vol de nuit* (1931) were predominantly attracted by its simple, direct style and strong story-line (p. 24). André Malraux (1901–76) explored similar themes in *La Condition humaine* (1933) and *L'Espoir* (1937), but with no promise of ultimate fulfilment for his heroes—only an intermittent desperate pride in the stoical pursuit of their elusive goals. A century earlier, Georg Büchner and Gustave Flaubert had described their own reaction to the meaninglessness of life as 'nausea'; and Jean-Paul Sartre's brief novel of that name (1938) reaffirmed the verdict. But it was not until five years later, in *L'Être et le néant* (1943), that Sartre (1905–80) proposed a solution—partly built on ideas developed by Martin Heidegger and Karl Jaspers in the 1920s—the fortunes of which are looked at in Chapter 10 (pp. 216–17).

Running parallel with these novelists, who consciously saw themselves as the products and prophets of their time, were those who merely sought to describe it, albeit with a perception and humour that some of their more ambitious contemporaries might justifiably have envied. Present-day readers, looking for a fresco of French life in these years, would find much to their profit in the later volumes of Jules Romains's *Les Hommes de bonne volonté* (1932–46). Not that the more philosophically inclined writers were incapable of casting a mordant eye on society in the 1930s. Sartre's story, *L'Enfance d'un chef* (1939), included among much else a devastating sketch of certain aspects of the *patronat* mentality.

Much of the notable poetry of the 1930s was incomprehensible to the bulk of the middle-brow public. If with Mallarmé, half a century earlier, poetry had started to slip beyond the horizon of the ordinary educated man, it had at least been recognizable as potentially rewarding; and the baffled reader regretfully felt that the loss was his. With the Surrealists, however, he was not so sure. It was disheartening to be told that to look for meaning was to miss the point—and disturbing when a sizeable segment of respectable critics spoke of the emperor's new clothes. Even so, growing public interest in Freudian psychoanalysis gave Surrealism a wider currency than it might otherwise have had. André Breton (1896–1966) and Philippe Soupault (1897–) had experimented with 'automatic writing' in *Les Champs magnétiques* (1921), in which their attempt to yield to the unconscious had excited contemporary writers by the wealth of startling images it produced. Echoing familiar elements in psychoanalysis, Surrealism saw the unconscious as the link between man's body and his conscious thought and experience, and believed that its exploration might uncover the working relationship between the material realm of physiology and the subjective world of ideas and aesthetic experience. The immediate result was a brilliant fireworks display of imagery; but disillusion with its limitations as an investigative process caused a number of Surrealists, including Louis Aragon (1897–1982) and Paul Éluard (1895–1952), to join the Communist party, asserting that the writer's duty was to change the world, not merely his way of perceiving and interpreting it. Yet if Surrealism was the most significant movement in French poetry between the wars, there remained outstanding talents outside it, notably Alexis Saint-John Perse (1887–1975) and Francis Ponge (1899–), content to cultivate conscious perceptions and associations, and enjoying a following commensurate with their greater accessibility to the ordinary reader.

Surrealist painting was more widely known to the general public than Surrealist poetry—if not necessarily regarded with any greater affection. Walking past a row of pictures entailed less investment in time and concentration than reading a poem or listening to music—and it could be done in congenial company, encouraging conversation rather than silencing it. At the same time, the capacity of the visual arts to make an impression within seconds made painting a more accessible vehicle for Surrealist intentions—especially when the affinity between their images and the viewer's own experience of dreams was emphasized by the sharp-edged realism of their draughtsmanship. But few of the best-known Surrealist painters who exhibited in France between the wars were French—even René Magritte was Belgian. Similarly, the bulk of the great artists of other persuasions who were resident in France were foreign, notably Picasso, Soutine, and Chagall. Among indigenous genius, Georges Braque (1882–1963), Henri Matisse (1869–1954), and Fernand Léger (1881–1955) enjoyed the most sustained reputation. Exemplifying different tendencies during their creative evolution, their work tended to greater abstraction in their later years. Foreign Francophiles, however, were to feel a particular gratitude to Maurice de Vlaminck (1876–1958), André Derain (1880–1954), Albert Marquet (1875–1947), Maurice Utrillo (1883–1955), and Raoul Dufy (1877–1953), whose familiar landscapes evoked so much of France between the wars.

Dufy's frothy world of regattas and house-parties was the backcloth to much of French theatre in the 1930s. The decline and death of Sergei Diaghilev (1872–1929) deprived theatrical production of an organizing creative impulse that had galvanized not only ballet but had indirectly stimulated theatrical directors of all genres. His gift for bringing together widely differing musical, artistic, and entrepreneurial talent, with startling results, created a taste for surprise and paradox which in less discriminating hands was in danger of becoming an end in itself. The inventive, mercurial Jean Cocteau (1889–1963) was one such exponent, whose plays and films have continued to intrigue and entertain audiences, while leaving them short of substance to ponder. The same might be said of Jean Anouilh (1910–), whose brittle, amusing dialogue could always sustain interest in the theatre, but left one feeling somewhat hungry afterwards. Even Jean Giraudoux (1882–1944), who engaged his listeners in themes of serious substance, felt obliged to lace them copiously with easy charm, the net effect suggesting G. B. Shaw with generous helpings of sugar and cream. All these playwrights shared the prevalent predilection for taking historical or classical figures and making them the vehicle for what they had to say on issues of current concern.

If this had also been a favourite device of Shaw, it likewise ran parallel with the neo-classicism of much French music of these years, where clarity of texture and lightness of touch were also in danger of becoming ends in themselves. There was no innovator of the stature of Claude Debussy (1862–1918). Maurice Ravel died in 1937; and the inter-war generation of composers, such as Francis Poulenc (1899–1963), Jacques Ibert (1890–1962), and Darius Milhaud (1892–1974), seemed happy enough to charm and amuse—which they did with enviable skill—whereas their Russian contemporary, Stravinsky, was using neo-classicism to give a sharp profile to his exploration of new rhythms and sonorities. Olivier Messiaen (1908–) was yet to make a major impact on French audiences—although it was already clear in the 1930s that his experiments in tonality and rhythm, partly influenced by his study of bird-song, were to open new perspectives for later composers.

NOTES

A book of this kind inevitably draws on a wide range of secondary works, many of which duplicate each other on particular points. Books recommended for further reading are listed in the bibliography; and the following references are confined to works that have been cited in the text, or which have been the prime source for material or opinions expressed there.

1. *France in the 1930s*

   1. Most persuasively by Stanley Hoffmann (ed.), *In Search of France* (New York, 1963). Michel Crozier's *La Société bloquée* (Paris, 1970) uses the phrase to signify more recent issues than those originally envisaged by Hoffmann and his French translators.
   2. *The State of France: A Study of Contemporary France* (London, 1955), 74–5.
   3. *Promenades: A Historian's Appreciation of Modern French Literature*, pbk. edn. (Oxford, 1986), 45–6.
   4. 'The Postwar Resurgence of the French Economy', in Hoffmann (ed.), *Search of France*, p. 129.
   5. Ibid., p. 141.
   6. These various issues are well discussed by Laurence Wylie, 'Social Change at the Grass Roots', in Hoffmann (ed.), *Search of France*, pp. 159–234.
   7. Laurence Wylie is likewise stimulating on these questions (ibid.).
   8. Philippe Bernard and Henri Dubief, *The Decline of the Third Republic, 1914–1938* (Cambridge, 1985), 251–2.
   9. *Pas d'histoire, les femmes* (Paris, 1977), 18.
   10. Conditions in the empire in these years are well summarized in ch. 10 of Christopher Andrew and A. S. Kanya-Forstner, *France Overseas: The Great War and the Climax of French Imperial Expansion* (London, 1981), 237–51.
   11. *Listener*, 24 Aug. 1978, p. 234, and 7 Sept. 1978, p. 300. The order of the citations has been altered here in the interests of compression.

2. *'La République des Députés'*

   1. *La Démocratie française* (Paris, 1976), 155, cited in J. R. Frears, *Political Parties and Elections in the French Fifth Republic* (London, 1977), 263.
   2. Elliot Paul, *A Narrow Street*, 2nd edn. (London, 1947), 206.
   3. *Crisis and Compromise: Politics in the Fourth Republic*, 4th edn. (London, 1972), 4.
   4. Ibid., pp. 4–5.
   5. During the 1930s, elections to the Chamber of Deputies were conducted by *scrutin d'arrondissement*—a two-ballot procedure in single-member constituencies. On the first ballot, the leading candidate obtained the seat only if he had received more than half of the votes cast, representing at least a quarter of the registered electorate. In

the majority of constituencies, a second ballot was usually necessary (a week later), in which the seat would be given to whichever candidate had the most votes, even if they fell short of half the poll. The political consequences of this and other French electoral systems are discussed on pp. 142–4 and 269–70. Senators were elected for nine years by a complicated electoral college in each *département*, made up principally of delegates from various layers of local government plus the deputies of the *département* in the lower house of parliament. Although the population size of each *département* broadly determined the number of senators it could elect, small rural *départements* remained disproportionately well represented. Every three years, a third of the membership of the Senate vacated their seats, necessitating elections.

6. Cited by Laurence Wylie in Hoffmann (ed.), *Search of France*, p. 222.
7. *State of France*, pp. 39–40.

3. *The Popular Front*

1. *State of France*, pp. 30–2.
2. Quoted in Alexander Werth, *The Destiny of France* (London, 1937), 296–7.
3. Cited in Georgette Elgey, *La République des illusions (1945–1951)* (Paris, 1965), 123.

4. *The Road to Compiègne*

1. Cited in Robert J. Young, *In Command of France: French Foreign Policy and Military Planning, 1933–1940* (Cambridge, Mass., 1978), 4.
2. At one extreme, R. H. S. Stolfi lists 2,574 armoured vehicles under German command against 3,254 French, while at the other extreme, Vincent J. Esposito calculates the confrontation as 2,439 German against 2,689 Allied: Stolfi, 'Equipment for Victory in France in 1940', *History*, 55 (1970), 10–12; Esposito, *A Concise History of World War II* (New York, 1964), 57.
3. The exodus is well described in Colin Dyer, *Population and Society in Twentieth-century France* (London, 1978), 108–13, and in H. R. Kedward, 'Patriots and Patriotism in Vichy France', *Trans. Roy. Hist. Soc.*, 32 (1982), 175–92.

5. *The Occupation*

1. The most useful general books on the Occupation are Robert Paxton, *Vichy France: Old Guard and New Order, 1940–1944* (London, 1972); Jean-Pierre Azéma, *From Munich to the Liberation, 1938–1944* (Cambridge, 1984); H. R. Kedward, *Occupied France: Collaboration and Resistance, 1940–1944* (Oxford, 1985).
2. Cited by Richard Cobb, *Promenades*, p. 56.
3. *Listener*, 20 Apr. 1978, p. 514.
4. On this and other smells, see Richard Cobb, *French and Germans, Germans and French* (Hanover, NH, 1983).
5. The whole issue of youth under Vichy is closely examined in W. D. Halls, *The Youth of Vichy France* (Oxford, 1981).
6. Jacques Duquesne, *Les Catholiques français sous l'Occupation* (France, 1966), is the most comprehensive account of the Church's behaviour in these years.

7. Cobb, *French and Germans*, is a rich anthology of Vichy vocabulary.
8. *Rural Revolution in France: The Peasantry in the Twentieth Century* (Stanford, 1964), 91.
9. France and the German economy are extensively treated in Alan S. Milward, *The New Order and the French Economy* (Oxford, 1970).
10. Vichy and economic planning are carefully examined in Richard Kuisel, *Capitalism and the State in Modern France* (Cambridge, 1981), ch. 5.
11. Living standards under the Occupation are excellently treated in his *La Vie économique des français de 1939 à 1945* (Paris, 1978).
12. See especially, Dominique Rossignol, *Vichy et les francs-maçons: La liquidation des sociétés secrètes, 1940–1944* (Paris, 1981).
13. The standard work is M. Marrus and R. Paxton, *Vichy France and the Jews* (New York, 1981).
14. *Decline or Renewal? France since the 1930s* (New York, 1974), 33–4.
15. Ibid., p. 40.
16. *French and Germans*, p. 168.
17. Ibid., pp. 95–7.
18. The issue is discussed in Hoffmann, *Decline or Renewal*, pp. 3–25.

6. *Resistance and Liberation*

1. For the Resistance in the Unoccupied Zone, see H. R. Kedward, *Resistance in Vichy France: A Study of Ideas and Motivation in the Southern Zone, 1940–1942* (Oxford, 1978).

7. *Retribution and the New Jerusalem*

1. The genesis and posthumous fortunes of the CNR charter are examined in Claire Andrieu, *Le Programme commun de la Résistance* (Paris, 1984).

8. *The Constitution of the Fourth Republic*

1. The working of the constitution under the Fourth Republic is closely analysed in Philip Williams, *Crisis and Compromise: Politics in the Fourth Republic*, 4th edn. (London, 1972).

9. *The Pattern of Politics, 1946–1954*

1. The most useful analytical book on this period remains Philip Williams, *Crisis and Compromise*—and the most readable account Alexander Werth, *France, 1940–1955* (London, 1956). Recent research is drawn upon in Jean-Pierre Rioux, *La France de la Quatrième République*, 2 vols. (Paris, 1980–3). Georgette Elgey, *République des illusions*, and *La République des contradictions (1951–1954)* (Paris, 1968), provide a fund of lively detail and anecdote.

10. *Economic Growth and Social Change, 1947–1973*

1. In Hoffmann (ed.), *Search of France*, p. 155.
2. Ibid., p. 162.
3. Laurence Wylie is particularly informative: *Village in the Vaucluse*, 2nd edn. (New York, 1964), and 'Social Change at the Grass Roots', in Hoffmann (ed.), *Search of France*, pp. 159–234.
4. *La Grande Peur des bien-pensants* (Paris, 1931), 454, cited by Paxton, *Vichy France*, p. 146.
5. *Promenades*, pp. 3–4.
6. *State of France*, p. 70.
7. Ibid., pp. 439–42.
8. *The New France: A Society in Transition, 1945–1977* (London, 1977), 549–50.

11. *France Overseas*

1. *State of France*, pp. 205 and 277.

12. *Janus and Cassandra*

1. *State of France*, p. 462.

14. *'La République des Dupes'*

1. *State of France*, p. 171.
2. *A Savage War of Peace: Algeria, 1954–1962* (London, 1977), 537.

15. *'La République des Citoyens'*

1. André Malraux, *Les Chênes qu'on abat . . .* (Paris, 1971), 43. This and the following issues are admirably treated in Philip G. Cerny, *The Politics of Grandeur: Ideological Aspects of de Gaulle's Foreign Policy* (Cambridge, 1980), and in Douglas Johnson, 'The Political Principles of General de Gaulle', *International Affairs*, 41 (1965), 650–62—to which the first part of this chapter is heavily indebted.
2. A major theme of de Gaulle's *Le Fil de l'épée* (Paris, 1932).
3. 8 July 1942, cited in Cerny, *Politics of Grandeur*, p. 51.
4. *Fil de l'épée* (1973 edn.), p. 28.
5. William C. Andrews and Stanley Hoffmann (eds.), *The Fifth Republic at Twenty* (Albany, 1981), 38–9.
6. If a bill was shuttling to and fro between the houses without agreement being reached, it was open to the government to refer it to a *commission mixte*, consisting of 7 senators and 7 deputies, for the differences to be ironed out. If these were settled, the bill was then reread in both houses before becoming law. If an agreed version failed to emerge from this procedure, the government could choose either to let the bill drop or to submit it to the Assembly for a final verdict, which only required a simple majority to become law. The government was thereby able to give a helping hand to bills that had its approval, or let die those that did not. Thus the

deputies lost the power they had obtained in 1954 (p. 244) to make their will prevail against the upper house—except in cases where it suited the government for this to happen. The government's powers were further enhanced by the fact that it could now refer a bill to a *commission mixte* after just two readings in each house—or merely one, if the government deemed the bill 'urgent'. Moreover, when the bill emerged from the *commission mixte*, it was up to the government to decide in what form it should be resubmitted to parliament; the minister, if so minded, could add further amendments of his own. In the first decade of the regime, 90.6 per cent of bills that became law did so without recourse to a *commission mixte*. A further 5.4 per cent did so as a result of a *commission mixte*, while the remaining 4 per cent were rescued by the government, which then referred them to the Assembly for a final verdict. It was only when faced with a Socialist-dominated Assembly in the 1980s that the Senate became obstreperous—opposing no less than a quarter of the legislation passed on to it, albeit without affecting the eventual outcome.

In the same fashion, the Constitutional Council (the Fifth Republic's equivalent of the American Supreme Court) gave governments very little trouble until the 1980s. During the de Gaulle and Pompidou presidencies, the Council handled on average less than 1 case a year; under Giscard, the average rose to 7, but with the Council giving its approval to three-quarters of the measures referred to it. Under the Socialists, by contrast, the average number of cases considered by the Council doubled, with over half of them ending in unfavourable verdicts for the government. The 9 members of the Council were appointed for nine-year terms by the President of the Republic and the Presidents of the Assembly and the Senate—with each President choosing 3. The Socialist governments were thus faced with the conservative appointments of the previous decade.

7. *State of France*, p. 172.

16. *De Gaulle's Foreign Policy*

1. Jack Hayward, *The One and Indivisible French Republic* (London, 1973), 227.
2. Hoffmann, *Decline or Renewal*, p. 389.
3. 8 January 1958, cited by Cerny, *Politics of Grandeur*, p. 80.
4. Hayward, *One and Indivisible*, p. 228.
5. Hoffmann, *Decline or Renewal*, p. 434.
6. Ibid.

17. *Challenge and Response, 1968–1974*

1. *The Long March of the French Left* (London, 1981), 64.
2. *Decline or Renewal*, pp. 145–84, presents an excellent brief analysis of the events of 1968.
3. Daniel Cohn-Bendit, *The French Student Revolt*, trans. B. Brewster (New York, 1968), 58.

18. *Reform and Recession*

1. This chapter is heavily indebted to J. R. Frears, *France in the Giscard Presidency* (London, 1981), and to Vincent Wright (ed.), *Continuity and Change in France* (London, 1984).
2. *The Government and Politics of France* (London, 1978), 184.

19. *Keynesianism in One Country*

1. This chapter owes much to Philip Cerny and Martin Schain, *Socialism, the State and Public Policy in France* (London, 1985); John S. Ambler (ed.), *The French Socialist Experiment* (Philadelphia, 1985); Howard Machin and Vincent Wright, *Economic Policy and Policy-making under the Mitterrand Presidency, 1981–1984* (London, 1985); and to Wright (ed.), *Continuity and Change*.
2. The initial nationalization legislation of Dec. 1981 was delayed for a couple of months by the Constitutional Council which considered the compensation terms to be inadequate. On the Council's role in the 1980s, see p. 405 n. 6.

Appendix IV. The Arts and the Public in the 1930s

1. *French Literature and its Background*, vi: *The Twentieth Century* (London, 1970), 188.

SELECT BIBLIOGRAPHY

The best British source of up-to-date information on current books and articles on twentieth-century France is the quarterly review of the Association for the Study of Modern and Contemporary France, *Modern and Contemporary France*.

The four decades before the starting-point of this book are succinctly surveyed by James F. McMillan, *Dreyfus to de Gaulle: Politics and Society in France, 1898–1969* (London, 1985), which also has an excellent annotated bibliography.

The Period as a Whole

Stanley Hoffmann (ed.), *In Search of France* (New York, 1963)—also available as *France: Change and Tradition*—and his *Decline or Renewal? France since the 1930s* (New York, 1974), are a fund of perceptions and stimulating ideas. Equally invigorating, but more frequently in need of qualification, are the trenchant assertions of Herbert Lüthy, *The State of France: A Study of Contemporary France* (London, 1955). David Thomson, *Democracy in France since 1870*, 5th edn. (London, 1969), still remains a reliable guide to the major issues, while Maurice Duverger's *La République des citoyens* (Paris, 1982) is a cogent demonstration of the impact of constitutional change on the nature of politics, even if it tends to underestimate other factors.

J.-J. Carré, P. Dubois, and E. Malinvaud, *French Economic Growth* (London, 1976), is arguably the best of the economic surveys, and easier to use than the fragmented sections of the encyclopaedic Fernand Braudel and Ernest Labrousse (eds.), *Histoire économique et sociale de la France*, iv, pts. 2–3 (Paris, 1980–2). Much shorter accounts may be found in Carlo M. Cipolla (ed.), *The Fontana Economic History of Europe*, v and vi (Glasgow, 1976), and the thoughtful if eclectic François Caron, *An Economic History of Modern France* (London, 1979). See also, R. F. Kuisel, *Capitalism and the State in Modern France: Renovation and Economic Management in the Twentieth Century* (Cambridge, 1981). French economic performance is usefully put in a comparative international context in Jean-Marcel Jeanneney and Élisabeth Barbier-Jeanneney, *Les Économies occidentales du xix siècle à nos jours*, 2 vols. (Paris, 1985), and in Thelma Liesner, *Economic statistics, 1900–1983* (London, 1985).

Colin Dyer, *Population and Society in Twentieth-century France* (London, 1978), and Pierre Sorlin, *La Société française*, ii (Paris, 1971), provide brief readable surveys of social change, while the disabilities of women are persuasively discussed in Huguette Bouchardeau, *Pas d'histoire, les femmes* (Paris, 1977).

For religion, see A. Coutrot and F.-G. Dreyfus, *Les Forces religieuses dans la société française* (Paris, 1965), and François Lebrun (ed.), *Histoire des catholiques en France du xv siècle à nos jours* (Paris, 1980).

The standard work on education is Antoine Prost, *Histoire générale de l'enseignement et de l'éducation en France*, iv (Paris, 1981), and on the Press, C. Bellanger *et al.*, *Histoire générale de la presse française*, iii–v (Paris, 1972–6).

On electoral history, see Claude Leleu, *Géographie des élections françaises depuis 1936*

(Paris, 1971), and for their international context, see Thomas T. Mackie and Richard Rose, *The International Almanac of Electoral History*, 2nd edn. (London, 1982). Among the long-perspective political surveys, much can be learned from René Rémond, *The Right Wing in France from 1815 to de Gaulle*, 2nd edn. (Philadelphia, 1969), while his *Atlas historique de la France contemporaine, 1800–1965* (Paris, 1966), compiled with a team of colleagues, is commendably wide-ranging in its thematic coverage.

Richard Cobb, *Promenades: A Historian's Appreciation of Modern French Literature* (Oxford, 1980), provides a pleasurable and rewarding pursuit of the author's enthusiasms.

The 1930s

General surveys of the period include Philippe Bernard and Henri Dubief, *The Decline of the Third Republic, 1914–1938* (Cambridge, 1985), and the thought-provoking Theodore Zeldin, *France, 1848–1945*, ii: *Intellect, Taste, and Anxiety* (Oxford, 1977), while the standard economic survey remains Alfred Sauvy, *Histoire économique de la France entre les deux guerres*, 3 vols. (Paris, 1984). Life in inter-war Paris is affectionately evoked in Elliot Paul, *A Narrow Street*, 2nd edn. (London, 1947), even if his political comments are crude and wide of the mark.

On the position of women, see Bouchardeau, *Pas d'histoire*, and James F. McMillan, *Housewife or Harlot: The Place of Women in French Society, 1870–1940* (Brighton, 1981); and for religious issues see René Rémond, *Les Catholiques dans la France des années 30*, 2nd edn. (Paris, 1979), P. Christophe, *1936: Les Catholiques et le Front Populaire* (Paris, 1979), and John Hellman, *Emmanuel Mounier and the New Catholic Left, 1930–1950* (Toronto, 1981).

Books on cinema are notoriously uneven, but the basic facts on French cinema in the 1930s can be obtained from Roy Armes, *French Cinema* (New York, 1985), René Prédal, *La Société française (1914–1945) à travers le cinéma* (Paris, 1972), and François Garçon, *De Blum à Pétain: Cinéma et société française (1936–1944)* (Paris, 1984).

Colonial questions are well surveyed in Christopher Andrew and A. S. Kanya-Forstner, *France Overseas: The Great War and the Climax of French Imperial Expansion* (London, 1981).

Even after half a century, the savour of French politics in the 1930s is best conveyed in Denis W. Brogan's vintage masterpiece, *The Development of Modern France (1870–1939)* (London, 1940), and by the journalist Alexander Werth's accounts in *France in Ferment* (London, 1934), and *The Destiny of France* (London, 1937). Among more recent works, see notably Julian Jackson, *The Politics of Depression in France, 1932–1936* (Cambridge, 1985), Edward Mortimer, *The Rise of the French Communist Party, 1920–1947* (London, 1984), Daniel Brower, *The New Jacobins: The French Communist Party and the Popular Front* (Ithaca, 1968), David Caute, *Communism and the French Intellectuals, 1914–1960* (London, 1964), Nathaniel Greene, *Crisis and Decline: The French Socialist Party in the Popular Front Era* (Ithaca, 1969), John Marcus, *French Socialism in the Crisis Years, 1933–1936: Fascism and the French Left* (New York, 1958), Peter J. Larmour, *The French Radical Party in the 1930s* (Stanford, 1964), Serge Berstein, *Histoire du Parti Radical*, ii: *Crise du radicalisme, 1926–1939* (Paris, 1982), William D. Irvine, *French Conservatism in Crisis: The Republican Federation of France in the 1930s* (Bâton Rouge, 1979), P. Machefer (ed.), *Ligues et fascismes en France, 1919–1939* (Paris,

1974), Eugen Weber, *Action Française: Royalism and Reaction in Twentieth-century France* (Stanford, 1962), and James Joll (ed.), *The Decline of the Third Republic* (London, 1959).

For the Popular Front, see Georges Lefranc, *Histoire du Front Populaire (1934–1938)* (Paris, 1965), Paul Warwick, *The French Popular Front: A legislative analysis* (Chicago, 1977), René Rémond and Pierre Renouvin (eds.), *Léon Blum, chef de gouvernement, 1936–1937*, 2nd edn. (Paris, 1981), and Guy Bourdé, *La Défaite du Front Populaire* (Paris, 1977). Julian Jackson, *The Popular Front in France* (Cambridge), is expected to appear in 1988.

The subsequent years are ably covered in Jean-Pierre Azéma, *From Munich to the Liberation, 1938–1944* (Cambridge, 1984), while useful discussion of specific issues is to be found in René Rémond and J. Bourdin (eds.), *Edouard Daladier, chef de gouvernement, avril 1938–septembre 1939* (Paris, 1977), and *La France et les français en 1938–1939* (Paris, 1978).

For the vicissitudes of French foreign and defence policies, see Robert Young, *In Command of France: French Foreign Policy and Military Planning, 1933–1940* (Cambridge, Mass., 1978), Jean-Baptiste Duroselle, *La Décadence, 1932–1939*, and *L'Abîme, 1939–1945* (Paris, 1979–82), J. E. Dreifort, *Yvon Delbos at the Quai d'Orsay* (London, 1973), R. Frankenstein, *Le Prix du réarmement français, 1935–1939* (Paris, 1982), Azéma, *Munich to the Liberation*, Vincent J. Esposito, *A Concise History of World War II* (New York, 1964), Alistair Horne, *To Lose a Battle: France 1940* (London, 1979), and R. H. S. Stolfi, 'Equipment for Victory in France in 1940', *History*, 55 (1970), 1–20.

Occupation and Liberation

The outstanding book on the period remains Robert O. Paxton, *Vichy France: Old Guard and New Order, 1940–1944* (London, 1972). Among more recent work, see in particular the following publications by H. R. Kedward, *Occupied France: Collaboration and Resistance, 1940–1944* (Oxford, 1985), *Resistance in Vichy France: A Study of Ideas and Motivation in the Southern Zone, 1940–1942* (Oxford, 1978), 'Patriots and Patriotism in Vichy France', *Transactions of the Royal Historical Society*, 32 (1982), 175–92, and jointly edited with Richard Austin, *Vichy France and the Resistance: Culture and Ideology* (London, 1985). Azéma, *Munich to the Liberation*, contains a useful, lively account, as does Alexander Werth's perennial *France, 1940–55* (London, 1956), which, despite its age and all-pervasive *parti pris*, remains unsurpassed as a panorama of the fifteen years it covers. Richard Cobb, *French and Germans, Germans and French: A Personal Interpretation of France under Two Occupations* (Hanover, NH, 1983), engagingly and imaginatively re-creates the ambience of the period, as does Hoffmann, *Decline or Renewal?* See also John Sweets, *Choices in Vichy France: The French under Nazi Occupation* (New York, 1986), René Rémond (ed.), *Le Gouvernement de Vichy, 1940–42* (Paris, 1972), and Alfred Sauvy, *La Vie économique des français de 1939 à 1945* (Paris, 1978).

On particular issues, see Alan Milward, *The New Order and the French Economy* (Oxford, 1970), Kuisel, *Capitalism and the State*, William D. Halls, *The Youth of Vichy France* (Oxford, 1981), Jacques Duquesne, *Les Catholiques français sous l'Occupation* (Paris, 1966), Xavier de Montclos (ed.), *Églises et chrétiens dans la deuxième guerre mondiale: La France* (Lyon, 1982), Dominique Rossignol, *Vichy et les francs-maçons: La liquidation des sociétés secrètes, 1940–1944* (Paris, 1981), M. R. Marrus and R. O. Paxton,

Vichy France and the Jews (New York, 1981), S. Courtois, *Le PCF dans la guerre* (Paris, 1980), Henri Michel, *Les Courants de pensée de la Résistance* (Paris, 1962), Claire Andrieu, *Le Programme commun de la Résistance* (Paris, 1984), which is especially valuable on the genesis of the social reforms of the Liberation, Jacques Siclier, *La France de Pétain et son cinéma* (Paris, 1981), and R. T. Thomas, *Britain and Vichy: The Dilemma of Anglo-French Relations 1940–42* (London, 1979).

For the Liberation period, see Comité d'Histoire de la Deuxième Guerre Mondiale, *La Libération de la France* (Paris, 1976), Peter Novick, *The Resistance versus Vichy: The Purge of Collaborators in Liberated France* (London, 1968), and the early chapters of Jean-Pierre Rioux's able survey, *The Fourth Republic, 1944–1958* (Cambridge, 1987).

Useful biographies include Richard Griffiths, *Marshal Pétain* (London, 1970), Geoffrey Warner, *Pierre Laval and the Eclipse of France* (London, 1968), Alexander Werth, *De Gaulle* (London, 1965), and Jean Lacouture, *Charles de Gaulle*, 2 vols. (Paris, 1984–5).

Society since the Second World War

D. L. Hanley, A. P. Kerr, and N. H. Waites, *Contemporary France: Politics and Society since 1945*, 2nd edn. (London, 1984), provides a clear and well-organized survey of the period on a thematic basis, with a remarkably rich annotated bibliography. The diversity of daily life and attitudes in France is perceptively surveyed in John Ardagh, *The New France: A Society in Transition, 1945–1977* (London, 1977)—subsequently reissued as *France in the 1980s* (London, 1982)—and, for more recent years, in Theodore Zeldin, *The French* (London, 1983). There are many compilations of social statistics and opinion polls on current social issues, notably Jean-Daniel Reynaud and Yves Grafmeyer (eds.), *Français, qui êtes-vous?* (Paris, 1981), Jean-Yves Potel, *L'État de la France et de ses habitants* (Paris, 1985), and Gérard Vincent, *D'Ambition à Zizanie* (Paris, 1983), who has also written two useful social chronologies, *Les Français, 1945–1975* and *Les Français, 1976–1979* (Paris, 1977–80), giving year-by-year information on a host of subjects, including literary prize-winners and best sellers. See also the following French annuals: SOFRES, *Opinion publique: Enquêtes et commentaires*, INSEE, *Données sociales* and *Annuaire statistique de la France*, not forgetting Dominique and Michèle Frémy's invaluable one-volume annual encyclopaedia, *Quid*.

Living conditions and social benefits are examined in La Documentation Française, *Les Institutions sociales de la France* (Paris, successive updated edns.), Georges Dorion and André Guionnet, *La Sécurité sociale* (Paris, 1983), Wallace C. Peterson, *The Welfare State in France* (Lincoln, Nebr., 1960), Walter Friedlander, *Individualism and Social Welfare: An Analysis of the System of Social Security and Social Welfare in France* (East Lansing, 1962), and Jean-Pierre Dumont, *La Sécurité sociale toujours en chantier: Histoire, bilan, perspectives* (Paris, 1981).

The inequalities of French life are specifically examined in Jane Marceau, *Class and Status in France: Economic Change and Social Immobility, 1945–1975* (Oxford, 1977), and Peter Morris (ed.), *Equality and Inequalities in France: Proceedings of the Fourth Annual Conference of the ASMCF* (Nottingham, 1984). See also Ezra Suleiman, *Élites in French Society: The Politics of Survival* (Princeton, 1978).

The unions are extensively treated in Jean-Daniel Reynaud, *Les Syndicats en France*, 2 vols. (Paris, 1975), René Mouriaux, *Les Syndicats dans la société française* (Paris, 1983), and *La CGT* (Paris, 1982), George Ross, *Workers and Communists in France: From Popular Front to Eurocommunism* (Berkeley, 1982), Hervé Hamon and Patrick Rotman, *La Deuxième Gauche: Histoire intellectuelle et politique de la CFDT* (Paris, 1982), and A. Bergounioux, *Force Ouvrière* (Paris, 1975).

There have been many descriptions of post-war village life in France, but Lawrence Wylie, *Village in the Vaucluse*, 2nd edn. (New York, 1964), remains in a class of its own.

On education, the Press, and culture, see the following books, in addition to those listed on p. 407, William D. Halls, *Education, Culture, and Politics in Modern France* (Oxford, 1976), Hanley *et al.*, *Contemporary France*, Jean-Noel Jeanneney and Jacques Julliard, *Le Monde de Beuve-Méry, ou le métier d'Alceste* (Paris, 1979), Jacques Brenner, *Histoire de la littérature française de 1940 à nos jours* (Paris, 1978), John Cruickshank (ed.), *French Literature and its Background*, vi: *The Twentieth Century* (London, 1970), John E. Flower, *Writers and Politics in Modern France* (London, 1977), and *Literature and the Left in France* (London, 1983), Vincent Descombes, *Modern French Philosophy* (Cambridge, 1980), the compilation *Les Dieux dans la cuisine: Vingt ans de philosophie en France* (Paris, 1978), and Charles Lemert, *French Sociology: Rupture and Renewal since 1968* (New York, 1981).

On religious issues, see William Bosworth, *Catholicism and Crisis in Modern France: French Catholic Groups at the Threshold of the Fifth Republic* (Princeton, 1962), Émile Poulat, *Une Église ébranlée: Changement, conflit et continuité de Pie xii à Jean-Paul ii* (Paris, 1980), Peter Hebblethwaite, *The Runaway Church*, 2nd edn. (Glasgow, 1978), Guy Michelat and Michel Simon, *Classe, religion, et comportement politique* (Paris, 1977), Danièle Hervieu-Léger, *Vers un nouveau christianisme?* (Paris, 1986), Yves Lambert, *Dieu change en Bretagne* (Paris, 1985), and R. Mehl, *Le Protestantisme français dans la société actuelle, 1945–1980* (Geneva, 1982).

The Post-war Economy

The European context of the French achievement is well described in Andrea Boltho (ed.), *The European Economy: Growth and Crisis* (Oxford, 1982), Sima Lieberman, *The Growth of European Mixed Economies, 1945–1970* (New York, 1977), *The Economist*, *Europe's Economies* (London, 1978), and Eurostat, *Basic Statistics of the Community* (Luxembourg, successive edns.). Among the many books specifically on the French economy, see Hugh D. Clout, *The Geography of Post-war France: A Social and Economic Approach* (Oxford, 1972), Maurice Parodi, *L'Économie et la société française depuis 1945* (Paris, 1981), Jean-Pierre Pagé, *Profil économique de la France au seuil des années 80* (Paris, 1981), Henry Ehrmann, *Organized Business in France* (Princeton, 1957), Robert Gilpin, *France in the Age of the Scientific State* (Princeton, 1968), Gordon Wright, *Rural Revolution in France* (Stanford, 1964), and Pierre Barral, *Les Agrariens français de Méline à Pisani* (Paris, 1968).

Politics since the War

Chronological surveys are provided by Rioux, *The Fourth Republic*, the ever-rewarding veteran, Werth, *France, 1940–1955*, the old and opinionated, but equally lively, Ronald Matthews, *The Death of the Fourth Republic* (London, 1954), and two schematized textbooks, J. Chapsal and A. Lancelot, *La Vie politique en France depuis 1940*, 4th edn.

(Paris, 1975), and Serge Sur, *La Vie politique en France sous la ve République* (Paris, 1977). Georgette Elgey, *La République des illusions (1945–1951)* (Paris, 1965), and *La République des contradictions (1951–1954)* (Paris, 1968), are a mine of illuminating anecdote. Electoral history is analysed in François Goguel, *Chroniques électorales: Les Scrutins politiques en France de 1945 à nos jours*, 3 vols. (Paris, 1981–83).

The political life and institutions of the Fourth Republic are closely investigated in Philip Williams's unsurpassed *Crisis and Compromise: Politics in the Fourth Republic*, 4th edn. (London, 1972), Duncan MacRae, *Parliament, Politics, and Society in France, 1946–1958* (New York, 1967), and L'Université de Nice, *La Quatrième République: Bilan trente ans après la promulgation de la Constitution du 27 octobre 1946* (Paris, 1978).

Those of the Fifth Republic are examined in William G. Andrews and Stanley Hoffmann (eds.), *The Fifth Republic at Twenty* (Albany, 1981)—also available in an abbreviated edition, *The Impact of the Fifth Republic on France* (Albany, 1981)—Philip Williams and Martin Harrison, *Politics and Society in de Gaulle's Republic* (London, 1971), Dorothy Pickles, *The Government and Politics of France*, 2 vols. (London, 1972–3), John R. Frears, *Political Parties and Elections in the French Fifth Republic* (London, 1977), Jack Hayward, *The One and Indivisible French Republic* (London, 1973), Vincent Wright, *The Government and Politics of France* (London, 1978), and Henry W. Ehrmann, *Politics in France*, 4th edn. (Boston, 1983).

On parties, politicians, and movements, see David S. Bell (ed.), *Contemporary French Political Parties* (London, 1982), Stanley Hoffmann, *Le Mouvement Poujade* (Paris, 1956), Malcolm Anderson, *Conservative Politics in France* (London, 1974), S. Guillaume, *Antoine Pinay ou la confiance en politique* (Paris, 1984), Douglas Johnson, 'The Political Principles of General de Gaulle', *International Affairs*, 41 (1965), 650–62, Philip Cerny, *The Politics of Grandeur: Ideological Aspects of de Gaulle's Foreign Policy* (Cambridge, 1980), Jean Touchard, *Le Gaullisme, 1940–1969* (Paris, 1978), Jean Charlot, *Le Gaullisme d'Opposition, 1946–1958* (Paris, 1983), *The Gaullist Phenomenon: The Gaullist Movement in the Fifth Republic* (London, 1971), and *L'UNR: Étude du pouvoir au sein d'un parti politique* (Paris, 1967), Jean-Claude Colliard, *Les Républicains Indépendants: Valéry Giscard d'Estaing* (Paris, 1972), Ronald E. M. Irving's *Christian Democracy in France* (London, 1973), and *The Christian Democratic Parties of Western Europe* (London, 1979), Francis de Tarr, *The French Radical Party from Herriot to Mendès France* (London, 1961), Jean Lacouture, *Pierre Mendès France* (Paris, 1981), François Bedarida and Jean-Pierre Rioux, *Pierre Mendès France et le mendésisme* (Paris, 1985), R. W. Johnson, *The Long March of the French Left* (London, 1981), Olivier Duhamel, *La Gauche et la ve République* (Paris, 1980), Roger Quilliot, *La SFIO et l'exercice du pouvoir, 1944–1958* (Paris, 1972), Harvey G. Simmons, *French Socialists in Search of a Role, 1956–1967* (Ithaca, 1970), B. D. Graham, *French Socialists and Tripartism, 1944–47* (London, 1965), David S. Bell and Byron Criddle, *The French Socialist Party: Resurgence and Victory* (Oxford, 1984), Philip G. Cerny (ed.), *Social Movements and Protest in France* (London, 1982), Arthur Hirsch, *The French New Left: An Intellectual History from Sartre to Gorz* (Boston, Mass., 1981), Jacques Fauvert, *Histoire du parti communiste*, 2nd edn. (Paris, 1977), Caute, *Communism and the French Intellectuals*, Irwin M. Wall, *French Communism in the Era of Stalin: The Quest for Unity and Integration, 1945–1962* (Westport, Conn., 1983), Donald Blackmer and Sidney Tarrow (eds.), *Communism in Italy and France* (Princeton, 1975), Annie Kriegel, *The French Communists: Profile of a People* (Chicago, 1972), Tony Judt, *Marxism and the French Left* (Oxford,

1985), and François Hincker, *Le Parti communiste au carrefour: Essai sur quinze ans de son histoire, 1965–1981* (Paris, 1981).

On other issues in politics and administration, see Janine Mossuz-Lavau et Mariette Sineau, *Enquête sur les femmes et la politique en France* (Paris, 1983), Ezra Suleiman, *Politics, Power, and Bureaucracy in France: The Administrative Élite* (Princeton, 1974), F. Ridley and J. Blondel, *Public Administration in France* (London, 1964), Gérard Belorgey, *Le Gouvernement et l'administration de la France*, 2nd edn. (Paris, 1970), and P. Bernard, *L'État et la décentralisation: Du préfet au commissaire de la République* (Paris, 1983).

The vicissitudes of decolonization and the developments that led to the demise of the Fourth Republic are treated in R. F. Holland, *European Decolonization, 1918–1981* (London, 1985), Ronald E. M. Irving, *The First Indo-China War* (London, 1975), Philip Williams, *Wars, Plots and Scandals in Post-war France* (Cambridge, 1970), which has an excellent chapter on the events of 1958, Alastair Horne, *A Savage War of Peace: Algeria, 1954–62* (London, 1977), John S. Ambler, *The French Army in Politics, 1945–1962* (Ohio, 1966), G. Kelly, *Lost Soldiers: The French Army and Empire in Crisis, 1947–62* (Cambridge, Mass., 1965), André Nozière, *Algérie: Les chrétiens dans la guerre* (Paris, 1979), and René Rémond, *Le Retour de de Gaulle* (Brussels, 1983).

The events of May 1968 and what led to them are analysed in Michel Crozier, *La Société bloquée* (Paris, 1970), and Hoffmann, *Decline or Renewal?*—and in more detail in Roger Absalom, *France: The May Events 1968* (London, 1971), B. E. Brown, *Protest in Paris: Anatomy of a Revolt* (Morristown, NJ, 1974), and Richard Johnson, *The French Communist Party versus the Students: Revolutionary Politics in May–June 1968* (New Haven, 1972).

Political developments since de Gaulle's resignation and their socio-economic repercussions are examined in Fondation Nationale des Sciences Politiques, *La Présidence de la République de Georges Pompidou: Exercice du pouvoir et pratiques des institutions* (Paris, 1983), John R. Frears, *France in the Giscard Presidency* (London, 1981), Vincent Wright (ed.), *Continuity and Change in France* (London, 1984), Philip Cerny and Martin Schain, *Socialism, the State and Public Policy in France* (London, 1985), Howard Machin and Vincent Wright, *Economic Policy and Policy-making under the Mitterrand Presidency, 1981–1984* (London, 1985), John S. Ambler, *The French Socialist Experiment* (Philadelphia, 1985), Sonia Mazey and Michael Newman (eds.), *Mitterrand's France* (London, 1987), George Ross, Stanley Hoffmann, and Sylvia Malzacher, *The Mitterrand Experiment* (Oxford, 1987).

Foreign and Defence Policies since the War

In addition to Cerny's *Politics of Grandeur* and useful chapters in Hoffmann, *Decline or Renewal?*, Hanley *et al.*, *Contemporary France*, and Hayward, *One and Indivisible*, see Alfred Grosser's *La Quatrième République et sa politique extérieure* (Paris, 1961), *French Foreign Policy under de Gaulle* (Boston, Mass., 1967), and *Affaires extérieures: La politique de la France, 1944–1984* (Paris, 1984), Guy de Carmoy, *Les Politiques étrangères de la France, 1944–1966* (Paris, 1967), S. Cohen and M.-C. Smouts, *La Politique extérieure de Valéry Giscard d'Estaing* (Paris, 1985), and Jolyon Howorth, *France: The Politics of Peace* (London, 1984).

INDEX

Entries refer to conditions in France, unless otherwise stated. Entries concerning other countries are grouped under national headings. (Thus French real wages are listed as 'wages, real', but Swedish real wages are listed under 'Sweden: wages, real'.) International agreements are grouped alphabetically under 'treaties, international agreements, etc.'.